The Geology and Tectonics of the Jabal Akhdar and Saih Hatat Domes, Oman Mountains

Geological Society books refereeing procedures

The scientific and production quality of the Geological Society's books matches that of its journals. Since 1997, all book proposals are reviewed by two individual experts and the Society's Books Editorial Committee. Proposals are only accepted once any identified weaknesses are addressed.

The Geological Society of London is signed up to the Committee on Publication Ethics (COPE) and follows the highest standards of publication ethics. Once a book has been accepted, the volume editors agree to follow the Society's Code of Publication Ethics and facilitate a peer review process involving two independent reviewers. This is overseen by the Society Book Editors who ensure these standards are adhered to.

Geological Society books are timely volumes in topics of current interest. Proposals are often devised by editors around a specific theme or they may arise from meetings. Irrespective of origin, editors seek additional contributions throughout the editing process to ensure that the volume is balanced and representative of the current state of the field.

Submitting a book proposal
More information about submitting a proposal and producing a book for the Society can be found at https://www.geolsoc.org.uk/proposals

It is recommended that reference to all or part of this book should be made in one of the following ways:

Scharf, A., Mattern, F., Al-Wardi, M., Frijia, G., Moraetis, D., Pracejus, B., Bauer, W. and Cellegari, I. 2021. *The Geology and Tectonics of the Jabal Akhdar and Saih Hatat Domes, Oman Mountains.* Geological Society, London, Memoirs, **54**, https://doi.org/10.1144/M54

Scharf, A., Mattern, F., Al-Wardi, M., Frijia, G., Moraetis, D., Pracejus, B., Bauer, W. and Cellegari, I. 2021. Tectonostratigraphy of the eastern part of the Oman mountains. *Geological Society, London, Memoirs*, **54**, 11–44, https://doi.org/10.1144/M54.2

Geological Society Memoir No. 54

The Geology and Tectonics of the Jabal Akhdar and Saih Hatat Domes, Oman Mountains

A. Scharf

Department of Earth Sciences, College of Science, Sultan Qaboos University, PO BOX: 36, PC 123, Al-Khod, Muscat, Sultanate of Oman

F. Mattern

Department of Earth Sciences, College of Science, Sultan Qaboos University, PO BOX: 36, PC 123, Al-Khod, Muscat, Sultanate of Oman

M. Al-Wardi

Department of Earth Sciences, College of Science, Sultan Qaboos University, PO BOX: 36, PC 123, Al-Khod, Muscat, Sultanate of Oman

G. Frijia

Department of Physics and Earth Sciences, University of Ferrara, Via Saragat 1, 44122, Ferrara, Italy

D. Moraetis

Department of Applied Physics and Astronomy, College of Sciences, University of Sharjah, PO BOX: 27272, Sharjah, United Arab Emirates

B. Pracejus

Department of Earth Sciences, College of Science, Sultan Qaboos University, PO BOX: 36, PC 123, Al-Khod, Muscat, Sultanate of Oman

W. Bauer

Department of Applied Geosciences, German University of Technology GUtech, PO BOX: 1816, PC 130, Halban, Sultanate of Oman

and

I. Callegari

Department of Applied Geosciences, German University of Technology GUtech, PO BOX: 1816, PC 130, Halban, Sultanate of Oman

2021
Published by
The Geological Society
London

The Geological Society of London

The Geological Society of London is a not-for-profit organisation, and a registered charity (no. 210161). Our aims are to improve knowledge and understanding of the Earth, to promote Earth science education and awareness, and to promote professional excellence and ethical standards in the work of Earth scientists, for the public good. Founded in 1807, we are the oldest geological society in the world. Today, we are a world-leading communicator of Earth science – through scholarly publishing, library and information services, cutting-edge scientific conferences, education activities and outreach to the general public. We also provide impartial scientific information and evidence to support policy-making and public debate about the challenges facing humanity. For more about the Society, please go to https://www.geolsoc.org.uk/

The Geological Society Publishing House (Bath, UK) produces the Society's international journals and books, and acts as European distributor for selected publications of the American Association of Petroleum Geologists (AAPG), the Geological Society of America (GSA), the Society for Sedimentary Geology (SEPM) and the Geologists' Association (GA). GSL Fellows may purchase these societies' publications at a discount. The Society's online bookshop is at https://www.geolsoc.org.uk/bookshop

To find out about joining the Society and benefiting from substantial discounts on publications of GSL and other Societies go to https://www.geolsoc.org.uk/membership or contact the Fellowship Department at: The Geological Society, Burlington House, Piccadilly, London W1J 0BG: Tel. +44 (0)20 7434 9944; Fax +44 (0)20 7439 8975; E-mail: enquiries@geolsoc.org.uk

For information about the Society's meetings, go to https://www.geolsoc.org.uk/events. To find out more about the Society's Corporate Patrons Scheme visit https://www.geolsoc.org.uk/patrons

Proposing a book
If you are interested in proposing a book then please visit: https://www.geolsoc.org.uk/proposals

Published by The Geological Society from:
The Geological Society Publishing House, Unit 7, Brassmill Enterprise Centre, Brassmill Lane, Bath BA1 3JN, UK

The Lyell Collection: www.lyellcollection.org
Online bookshop: www.geolsoc.org.uk/bookshop
Orders: Tel. +44 (0)1225 445046, Fax +44 (0)1225 442836

British Library Cataloguing in Publication Data

A catalogue record for this book is available from the British Library.
ISBN 978-1-78620-493-6
ISSN 0435-4052

Distributors

For details of international agents and distributors see:
www.geolsoc.org.uk/agentsdistributors

Typeset by Nova Techset Private Limited, Bengaluru & Chennai, India
Printed and bound by CPI Group (UK) Ltd, Croydon CR0 4YY, UK

Contents

Preface

The Oman Mountains represent some of the most spectacular exposures of sedimentary, igneous and metamorphic rocks on the Arabian Peninsula with a geological history that spans over 800 million years, including spectacular outcrops that led to a better understanding of the composition of the continental and oceanic lithosphere, as well as sedimentary rocks that form reservoirs and seals in the subsurface to the south. Therefore, the Oman Mountains serve as a prime textbook example for the geological development of the southern margin of the Tethys and geological features related to obduction processes.

This volume reviews the previous geological work carried out in the Jabal Akhdar and Saih Hatat domes of the eastern Oman Mountains (or Hajar Mountains) to synthesize the scientific knowledge of this part of the country which covers an area of approximately 250 km × 125 km. A geological map has been compiled as a comprehensive tectonometamorphic synthesis. The geological map is largely based on the maps of Béchennec *et al.* (1992), Le Métour *et al.* (1992) and Wyns *et al.* (1992). Small inconsistencies found in the previously published maps have been corrected (for details see Appendix D (Scharf *et al.* 2021, Chapter 7, this Memoir)).

The first substantial study of the geology of the Oman Mountains was carried out by George Martin Lees in 1924–25 (Lees 1928) , but it was Ken Glennie and his co-workers from Shell in the 1960s and early 1970s who made the first detailed geological map of the mountains and accounted for the geology in the light of plate tectonics. At that time, the fundamental understanding of the overall architecture of the Oman Mountains and its southern foreland had been established. Marine sediments of the Hawasina Basin and the largest and best-preserved ophiolite sequence of the world form a major part of the Oman Mountains. These rocks have since been a research focus for the international geoscientific community.

The two large tectonic windows and domes (Jabal Akhdar and Saih Hatat) provide insights into the pre-Cretaceous geological development of the southeastern corner of the Arabian Platform. These rocks have received much less attention, despite the fact that their geological record from the Cryogenian to the Cretaceous covers one of the most interesting phases of Earth's history. Here, one can study the diamictites of the 'Snowball Earth' glaciation, the most long-lasting negative carbon isotope excursion of *c.* 17‰ during the Ediacaran ('Shuram Excursion'), the collision between East- and West-Gondwana, and post-tectonic, fossiliferous sedimentary rocks. The Permian breakup of Pangaea also affected the area, and later, during the Late Cretaceous, this eastern margin of the Arabian Platform was dragged down into a subduction zone and then became overridden by the Hawasina nappes and the famous Semail Ophiolite.

Comprehensive geological mapping of the surface outcrops of Oman by teams of the French Geological Survey, BRGM, published many works during the second half of the 1980s and first half of the 1990s, leading to a better understanding of the geology in two tectonic windows (domes) in the mountains and adjoining areas. The remarkable socio-economic transformation of the Sultanate with a rapidly improving infrastructure made once extremely isolated and remote wadis accessible, resulting in an ever-increasing number of publications covering the geology of this region.

The authors of this book are university teachers and researchers in structural geology/tectonics, regional geology, geodynamics, mineralogy/petrology, geochemistry, basin analysis, sedimentology and stratigraphy at two institutions with geoscience programs: the Sultan Qaboos University (SQU) and the German University of Technology in Oman (GUtech). Two authors recently moved to the University of Ferrara (Italy) and University of Sharjah (UAE). A multitude of geoscientific information is spread over an enormously wide range of journals and books, not all of them are readily available. In a first attempt to summarize the knowledge of the Jabal Akhdar and Saih Hatat tectonic windows, a review paper was prepared, but the scope soon exceeded the limitations of many journals, especially regarding the amount of text, illustrations and the inclusion of oversized maps/charts which were required for a better understanding.

This book provides a comprehensive overview and review of the geology and tectonic development of the eastern Oman Mountains and is written for ongoing geologists and professionals who want to work in this area. However, not all aspects can be discussed at length. Thus, we provide the original sources for further details. The compiled geological map depicts an overview of the different formations/units and is coordinated with the text. The original BRGM maps at a scale of 1: 250 000 and 1: 100 000 contain more details and information. Thus, they are indispensable for proper research and field investigations. Our compiled map is not meant to substitute the original maps; rather, it provides a comprehensive overview, ready to hand. We hope that this work will stimulate the continuation of research in the Jabal Akhdar and Saih Hatat domes as well as the surrounding areas and provide a new generation of Omani and international geologists with an overview of existing studies, and with a thorough and up-to-date background information of the relevant literature.

Finally, we would like to thank many colleagues from many different institutes for helpful discussions about several topics. They have been thanked in the acknowledgements.

The authors
Muscat, Oman, March 2020

Content summary

We present a compiled geological/tectonic map of the Jabal Akhdar and Saih Hatat domes and their surroundings, accompanied with crustal-scale cross-sections, kinematic plate reconstructions, and a comprehensive tectono-stratigraphic chart. Our work is largely based on reviewing the literature and complemented by our own investigations, observations and ideas.

The Neoproterozoic is the oldest geological era recorded in the rocks of the study area, and probably witnessed four orogenies, two during the early Cambrian (Cadomian and Angudan), one Late Paleozoic (the enigmatic 'Hercynian Orogeny') and one during the Late Cretaceous (Semail Ophiolite obduction and related processes, referred to as the 'Semail Orogeny'). Furthermore, the study area was affected by at least three Neoproterozoic to earliest Cambrian rifting events (Abu Mahara, Nafun and Fara), a Mid-Ordovician rift event, a latest Paleozoic rifting event (breakup of Pangaea), a late Jurassic rifting event associated with opening of an ocean basin east of Oman and latest Cretaceous to Cenozoic extensional events (orogenic collapse of the Oman Mountains).

The Neoproterozoic to earliest Cambrian extensional events are locally recorded by associated volcanic rocks and extensional structures. The Cadomian Orogeny with an NE–SW-directed stress field has been recently recognized and is associated with tight folds in the Neoproterozoic formations of the western Jabal Akhdar Dome. The Angudan event is recorded by map-scale folds and thrusts of the Jabal Akhdar Dome, indicating a SE–NW direction of compression. The origin/age of these folds has been reinterpreted from 'Hercynian' in the past to Angudan at present. The 'Hercynian' event is characterized mainly by large-scale arch and basin formation as well as block faulting. The latest Paleozoic breakup of Gondwana is well documented by the occurrence of large mafic bodies and volcaniclastics in pre-Permian formations and the Permian Saiq Formation of the Saih Hatat Dome. Oman was situated at an overall passive margin during the Permo-Mesozoic until the late Cretaceous. As an exception, an ocean basin formed between Oman and India during the late Jurassic, leading to some uplift and deformation in northern Oman.

The passive margin changed into an active margin during late Cretaceous. During the Cretaceous, the latest orogeny started probably with Albian intra-oceanic subduction within the Neo-Tethys Ocean, proximal to the Arabian margin. This subduction zone matured, and a self-sustained subduction zone formed during the Cenomanian. This event created new oceanic lithosphere along and above a subduction zone. The supra-subduction spreading center with the entire ophiolite rotated clockwise during the obduction. SW-directed thrusting led to obduction of the Semail Ophiolite which advanced over a distance of some *c.* 550–600 km over Arabia's passive margin (Hawasina Basin) and the Arabian Platform until the Maastrichtian. A possibly migrating foreland basin (Aruma Basin) in front of the advancing ophiolite formed within the Arabian Platform since the Turonian. During obduction, the extended Arabian continental crust and its attached Neo-Tethyan oceanic lithosphere were subducted northeastward to greater depths (minimum of 80 km), and parts of the continental rocks were metamorphosed to eclogite-facies conditions. Once the allochthonous rocks reached the Arabian Platform during the Late Cretaceous, the oceanic lithosphere separated from the continental lithosphere (possible slab breakoff), resulting in rapid exhumation of the continental Arabian Crust of the Saih Hatat area. Important postobduction tectonics occurred during the Late Eocene to Early Miocene with the final exhumation of the Jabal Akhdar and Saih Hatat domes. The onset of this process precedes the continent-continent collision in the Zagros Mountain Belt. The two domes exhibit Permo-Mesozoic sedimentary rocks of the Arabian Platform on their flanks and Neoproterozoic metasedimentary rocks in their cores. Minor sub-recent to recent tectonics in the study area are reflected by earthquakes, uplift of a series of marine terraces, hot springs, and significant present-day wadi incisions.

References

Béchennec, F., Roger, J., Le Métour, J. and Wyns, R. 1992. Geological map of Seeb, sheet NF 40-03, scale 1:250 000, with explanatory notes. *Directorate General of Minerals, Oman Ministry of Petroleum and Minerals.*

Lees, G.M. 1928. The geology and tectonics of Oman and parts of southeastern Arabia. *Quarterly Journal of the Geological Society of London,* **84**, 585–670, https://doi.org/10.1144/GSL.JGS.1928.084.01-04.24

Le Métour, J.F., Béchennec, F., Roger, J. and Wyns, R. 1992. Geological map of Muscat, sheet NF40-04, scale 1:250 000, with explanatory notes. *Directorate General of Minerals, Oman Ministry of Petroleum and Minerals.*

Scharf, A., Mattern, F., Al-Wardi, M., Frijia, G., Moraetis, D., Pracejus, B., Bauer, W. and Cellegari, I. 2021. Appendices to: The Geology and Tectonics of the Jabal Akhdar and Saih Hatat Domes, Oman Mountains. *Geological Society, London, Memoirs*, **54**, 113–117; foldout, https://doi.org/10.1144/M54.7

Wyns, R., Béchennec, F., Le Métour, J. and Roger, J. 1992. Geological map of Tiwi, sheet NF40-8B, scale 1:100 000, with explanatory notes. *Directorate General of Minerals, Oman Ministry of Petroleum and Minerals.*

Acknowledgements

We are indebted to many colleagues for discussions, especially Iftikhar Ahmed, Sobhi Nasir, and Musaab Al-Sarmi, all from Sultan Qaboos University (Oman), Michaela Bernecker, Rajat Mazumder, Andre Pinto, Heninsara Rarivoarison and Katharina Scharf, all from the German University of Technology (GUtech, Oman), Ralf Schuster (Geologische Bundesanstalt in Vienna, Austria), Henk Droste and Mia van Steenwinkel (RockIT Petroleum Geological Training and Consultancy), Loïc Bazalgette and Irene Gomez-Perez (Petroleum Development Oman), Christoph v. Hagke (RWTH Aachen University, Germany), Luca Aldega and Eugenio Carminati (La Sapienza Rome, Italy), Peter Kelemen (Columbia University, USA) and his team from the Oman Drilling Project, Christopher Bailey and his students (College of William & Mary, USA), Pascal Richard (Shell, Global Solutions International BV, The Netherlands), Uwe Ring and Reuben Hansman (University of Stockholm, Sweden), and Kosmas Pavlopoulos (Paris Sorbonne University, Abu Dhabi, UAE). Talal Al-Hosni (Sultan Qaboos University, Oman) kindly provided us with a list of hydrothermal springs from northeastern Oman. Mike Simmons (Neftex Petroleum Consultants, UK) is greatly acknowledged for his help with drawing the sequence stratigraphic scheme of the Arabian Peninsula (fig. A.2 (Scharf *et al.* 2021, Chapter 7, this Memoir)).

The manuscript benefitted from the critical comments of two anonymous reviewers, Arne Grobe (RWTH Aachen University, Germany), Mohammed Al-Kindy (CEO at Geo Consultancy Centre, Oman), Alan Heward (Malvern, UK) and Mike Searle (University of Oxford, UK) from earlier drafts. Furthermore, the reviews by Henk Droste (RockIT Petroleum Geological Training and Consultancy) and Philippe Agard (UPMC, Sorbonne University, Paris, France) are gratefully acknowledged. The latter two reviewers were invited/suggested by Maggie Simmons, Bethan Phillips and Rick Law (all Geological Society of London). Finally, we want to express our appreciation of the efforts of H.E. Eng. Hilal Al-Busaidi, CEO of the Public Authority of Mining (PAM) in Oman, to grant permission for publication of the geological map (fig. A.1 (Scharf *et al.* 2021, Chapter 7, this Memoir)).

Reference

Scharf, A., Mattern, F., Al-Wardi, M., Frijia, G., Moraetis, D., Pracejus, B., Bauer, W. and Cellegari, I. 2021. Appendices to: The Geology and Tectonics of the Jabal Akhdar and Saih Hatat Domes, Oman Mountains. *Geological Society, London, Memoirs*, **54**, 113–117; foldout, https://doi.org/10.1144/M54.7

Chapter 1

Introduction and tectonic framework

Andreas Scharf[1]*, Frank Mattern[1], Mohammed Al-Wardi[1], Gianluca Frijia[2], Daniel Moraetis[3], Bernhard Pracejus[1], Wilfried Bauer[4] and Ivan Callegari[4]

[1]Department of Earth Sciences, College of Science, Sultan Qaboos University, PO Box 36, PC 123, Al-Khod, Muscat, Sultanate of Oman

[2]Department of Physics and Earth Sciences, University of Ferrara, Via Saragat 1, 44122, Ferrara, Italy

[3]Department of Applied Physics and Astronomy, College of Sciences, University of Sharjah, PO Box 27272, Sharjah, United Arab Emirates

[4]Department of Applied Geosciences, German University of Technology GUtech, PO Box 1816, PC 130, Halban, Sultanate of Oman

*Correspondence: scharfa@squ.edu.om

Abstract: The extraordinary outcrop conditions provide a unique opportunity to study the geology and tectonics of the Oman Mountains, which record a geological history of more than 800 million years. We provide a summary of the geological evolution of the Oman Mountains with the emphasis on the Jabal Akhdar and Saih Hatat domes. This Memoir comprises seven chapters. This first chapter summarizes the former studies and the tectonic framework. This is followed by a comprehensive description of all geological formations/rock units (**Scharf *et al.* 2021*a***, Chapter 2, this Memoir) including the famous Semail Ophiolite, the fault and fold pattern (**Scharf *et al.* 2021*b***, Chapter 3, this Memoir) and the overall structure (**Scharf *et al.* 2021c**, Chapter 4, this Memoir). Chapter 5 (**Scharf *et al.* 2021*d***) explains the varied tectonic evolution of the study area, ranging from the Neoproterozoic until present, while Chapter 6 (**Scharf *et al.* 2021*e***) contains the conclusions and a catalogue of open questions. Finally, Chapter 7 (**Scharf *et al.* 2021*f***) provides two over-sized geological maps (1 : 250 000 version available online) and a correlation chart, providing an overview of the geological units/formations. This volume is of interest for all geoscientists, geoscience students and professionals studying the Oman Mountains on the surface as well as in the subsurface because it represents a comprehensive and detailed reference.

The Oman Mountains at the northeastern margin of the Arabian Plate display an impressive geological record in terms of phenomenological diversity and significance since the Precambrian. This record includes (1) the 'Snowball Earth' glaciation (e.g. Leather *et al.* 2002; Kilner *et al.* 2005), (2) the Ediacaran 'Shuram Excursion' (e.g. Burns and Matter 1993; Le Guerroué *et al.* 2006*a*, *b*), (3) the enigmatic Hercynian Unconformity ('Hercynian Orogeny'; e.g. Glennie *et al.* 1974; Beurrier *et al.* 1986*b*; Rabu *et al.* 1986; Faqira *et al.* 2009), (4) the Permo-Carboniferous glaciation (e.g. Heward and Penney 2014), (5) the Pangaea rifting and spreading of the Neo-Tethys Ocean during the Late Paleozoic (e.g. Blendinger *et al.* 1990; Béchennec *et al.* 1991), (6) the iconic obduction of the Semail Ophiolite (e.g. Glennie *et al.* 1974; Searle and Malpas 1980; Lippard *et al.* 1986; Goffé *et al.* 1988; Hacker and Mosenfelder 1996; Hacker *et al.* 1996; Cowan *et al.* 2014; Rioux *et al.* 2012, 2013, 2016), (7) high-pressure/low-temperature metamorphism of the continental margin (e.g. Lippard 1983; Le Métour *et al.* 1986*b*; Goffé *et al.* 1988; El-Shazly *et al.* 1990; Warren *et al.* 2003, 2005), (8) formation and exhumation of the large Jabal Akhdar Dome (JAD) and Saih Hatat Dome (SHD) which shape the Oman Mountains (e.g. Poupeau *et al.* 1998; Saddiqi *et al.* 2006; Hansman *et al.* 2017; Grobe *et al.* 2019), (9) Upper Cretaceous rudist limestones (e.g. Skelton *et al.* 1990; Wyns *et al.* 1992; Philip *et al.* 1995) and organic-rich intrashelf basin environments (e.g. van Buchem *et al.* 2002), (10) surface analogues of subsurface petroleum source and reservoirs rocks (Grantham *et al.* 1988; Koehrer *et al.* 2011; Pöppelreiter *et al.* 2011; Bendias *et al.* 2013; Heward and Penney 2014; Grobe *et al.* 2016), (11) Neogene to active tectonics (e.g. Fournier *et al.* 2006; Mattern *et al.* 2018; Moraetis *et al.* 2018) and (12) a palaeoclimate record for the past 325 ka based on speleothems from cave systems developed in Mesozoic carbonate formations (e.g. Burns *et al.* 1998, 2001). The Semail Ophiolite, as the world's largest and best preserved ophiolite (e.g. Searle and Cox 1999), and the Oman Mountains, as the unique case of an obduction orogen, have been the focus of much research interest from the international geoscientific community during the past decades (e.g. Nicolas *et al.* 1996, 2000).

Due to the massive improvement of Oman's road system during the last few decades and the excellent outcrops, the country became a field laboratory for tectonometamorphic, sedimentological and stratigraphic studies (e.g. Glennie *et al.* 1973, 1974; Robertson *et al.* 1990; Hacker *et al.* 1996; Saddiqi *et al.* 2006; Rollinson *et al.* 2014) with a research focus on ophiolites and ophiolite obduction (e.g. Lippard *et al.* 1986; Nicolas *et al.* 1996; Gregory *et al.* 1998; Searle and Cox 1999, 2002). We compiled these and many other studies to put forward an overview of the geology, tectonostratigraphy and tectonic evolution of the Oman Mountains, as well as a description of their large-scale structures. We concentrate on the eastern part of the Oman Mountains from the western tip of the JAD to the eastern end of the SHD (fig. A.1 (**Scharf *et al.* 2021*f***, Chapter 7, this Memoir)). To assist the international research community, we provide a newly compiled geological map of the eastern part of the Oman Mountains with the JAD and SHD in order to standardize the compiled tectonic units within a single map. This map is largely based on the geological maps of Oman at the scale of 1:250 000 (Béchennec *et al.* 1992*a*, *b*; Le Métour *et al.* 1992; Wyns *et al.* 1992). The map is accompanied by text which is an update on the geological knowledge of the Oman Mountains with some emphasis on the region's geodynamic history. Moreover, we provide a detailed topographic map and several simplified geological cross-sections, as well as an extensive tectonostratigraphic chart (fig. A.2 (**Scharf *et al.* 2021*f***, Chapter 7, this Memoir)).

The JAD consists of autochthonous rocks and is divided into the WNW-striking main dome (*c.* 85 km × 35 km) and NNE-oriented subdome attached to the NE of the main dome (25 km × 10 km; fig. 2.1 (**Scharf *et al.* 2021*a***,

From: Scharf, A., Mattern, F., Al-Wardi, M., Frijia, G., Moraetis, D., Pracejus, B., Bauer, W. and Cellegari, I. 2021. *The Geology and Tectonics of the Jabal Akhdar and Saih Hatat Domes, Oman Mountains*. Geological Society, London, Memoirs, **54**, 1–10, https://doi.org/10.1144/M54.1

Chapter 2, this Memoir)). This NNE-oriented subdome is parallel to the Semail Gap and is known as the Nakhl Subdome (fig. A.1 in **Scharf *et al.* 2021*f***, Chapter 7, this Memoir). In the NE, the Nakhl Subdome sharply bends towards the ESE and extends for *c.* 8 km until Fanja. The overall shape of the JAD is not circular, as expected for a tectonic dome, and may be attributed to the reactivation of faults within the subsurface during doming (see 'Post-obduction deformation' in Chapter 5 (**Scharf *et al.* 2021*d***)).

The SHD has an overall elongated shape and is aligned NW–SE from Fanja for *c.* 95 km towards the SE. The NE–SW extent ranges between 25 and 50 km. We define the SHD as the area consisting of (par-)autochthonous rocks that are surrounded by the Hawasina nappes and/or Semail Ophiolite to the NW and SW, as well as the ocean to the NE. Its southeastern margin is blanketed by Cenozoic shallow-marine sedimentary rocks near Tiwi (fig. A.1). The southeastern border or margin of the SHD is therefore not well defined. Between both domes near Fanja, a WNW–ESE-oriented structural and topographic depression is exposed ('Fanjah Saddle' of Coffield 1990). The autochthonous rocks west and east of the saddle rise 1.3 km above the saddle surface, which consists of allochthonous rocks. The fold axes of the JAD and SHD plunge below the 10-km-wide Fanjah Saddle at an angle of 20–40° (Coffield 1990). Roughly N–S-striking high-angle normal faults with dip directions towards the centre of the saddle cut the flanks of the JAD and SHD (Coffield 1990).

Lees (1928) introduced the Semail Nappe and recognized the allochthonous nature of the Hawasina nappes (see Heward 2016 for a summary of the pioneering expedition of Lees and others during the 1920s). The first descriptions of the Semail Ophiolite after the introduction of plate-tectonic theory were by Reinhardt (1969) and Allemann and Peters (1972).

The entire Oman Mountains were first mapped at a scale of 1:500 000 by Glennie *et al.* (1974). They introduced the main rock units of the Oman Mountains, the pre-Permian metasedimentary successions, the Arabian Platform sediments, the Hawasina nappes, the Semail Ophiolite and the post-obduction sedimentary rocks. Further mapping was carried out by the United States Geological Survey (USGS) in the vicinity of the SHD (Bailey 1981). Geologists of the Open University (UK) mapped the Semail Ophiolite (Lippard *et al.* 1986). French workers of the Bureau de Recherches Géologiques et Minières (BRGM) mapped the Central and Southeastern Oman Mountains at a scale of 1:100 000 (Béchennec *et al.* 1986*a*, *b*; Beurrier *et al.* 1986*a*, *b*; de Gramont *et al.* 1986; Hutin *et al.* 1986; Janjou *et al.* 1986; Le Métour *et al.* 1986*a*, *b*; Rabu *et al.* 1986; Villey *et al.* 1986*a*, *b*, *c*; Roger *et al.* 1991; Wyns *et al.* 1992). This map series was extended by the sheet of Ibra (Southeastern Oman Mountains) at a scale of 1:100 000 (Peters *et al.* 2005). Detailed mapping of the entire Semail Ophiolite was carried out and summarized by Nicolas *et al.* (2000). Eight geological maps of the Muscat-Seeb area have been published at a scale of 1:25 000 (Kajima and Ishii 2012; Kajima and Otake 2012; Kajima *et al.* 2012*a*, *b*, *c*, *d*, *e*, *f*).

Two important compilations of the subsurface stratigraphy and nomenclature were produced by Hughes Clarke (1988) and Forbes *et al.* (2010). The latter provided a comprehensive chart showing the subsurface stratigraphy of Oman (Forbes *et al.* 2010, their enclosure 1). Two Special Publications of the Geological Society of London (Robertson *et al.* 1990; Rollinson *et al.* 2014) represent significant advancements to the understanding of the geology and tectonics of the Oman Mountains. A summary booklet describing the geology and origin of the Oman Mountains was published by Glennie (1995). His work was the first step in understanding the geology of northern Oman for geologists, non-geologist scientists and engineers from the oil industry. This booklet became so famous and in such demand that a second edition was released (Glennie 2005). Searle (2007) wrote a review article about the Permo-Mesozoic evolution of the eastern Arabian Platform and margin with the adjacent Hawasina Basin. This work also targets the obduction of the Semail Ophiolite as well as the post-obduction evolution of parts of the Oman Mountains.

Many informative popular science and geoguide books, including the most recent and well-illustrated work of Al-Kindi (2018) and Searle (2019), have been published during the past decades, outlining the unique and spectacular geology of the Sultanate. With its description of numerous fascinating outcrops, the geological field guide of Hoffmann *et al.* (2016) could be used as a companion document to this book.

The formations at the surface in our study area may have different names to coeval deposits in the subsurface. The latter are commonly used by researchers in the oil industry (e.g. Hughes Clarke 1988; Sharland *et al.* 2004; Forbes *et al.* 2010). We use the surface names in our manuscript, but will also provide the subsurface names for clarity (Chapter 2 and fig. A.2 (**Scharf *et al.* 2021*a*,*f***, Chapters 2 and 7, this Memoir)).

We first describe in detail all major geological formations and tectonic units (**Scharf *et al.* 2021*a***, Chapter 2, this Memoir), provide the regional fault and fold pattern (**Scharf *et al.* 2021*b***, Chapter 3, this Memoir) and characterize the large-scale structure of the study area (**Scharf *et al.* 2021*c***, Chapter 4, this Memoir). This is followed by a comprehensive description of the entire tectonic evolution of the Oman Mountains from the Neoproterozoic until present (**Scharf *et al.* 2021*d***, Chapter 5, this Memoir). This chapter includes a lithospheric scale series of cross-sections, depicting the kinematics and dynamics of the different deformation events with a tectonometamorphic synthesis. The conclusions and a catalogue of open questions, based on the available literature complete this memoir (**Scharf *et al.* 2021*e***, Chapter 6, this Memoir). A compiled geological map at a scale of 1:250 000 and a tectonostratigraphic chart are enclosed in the appendix, providing a handy comprehensive overview (figs A.1 and A.2 (**Scharf *et al.* 2021*f***, Chapter 7, this Memoir). Note that in the literature the transcription of proper Arabic names into English commonly results in different spellings of the same feature; for example, the vowels in the name 'Semail' as in 'Semail Ophiolite' may be transcribed as 'Samail' or 'Sumail'.

Tectonic framework

The eastern part of the Arabian Peninsula (Fig. 1.1) was the site of several tectonic events. Recently, Callegari *et al.* (2020) documented the first evidence of the Cadomian Orogeny in northern Oman. This event is documented by tight to isoclinal folds with an amplitude of a few to tens of metres in the Hajir Formation of the JAD. The shortening direction during this event was NE–SW-aligned (in present coordinates). The age of deformation was at some time after deposition of the Fara Formation and before the Angudan event, that is, between *c.* 540 and 525 ± 5 Ma (Callegari *et al.* 2020; section 'F1 folds' in **Scharf *et al.* 2021*b***, Chapter 3, this Memoir). The Cadomian Orogeny was followed (and probably overlapped) by the Neoproterozoic–Early Cambrian East African Orogeny (*sensu* Stern 1994), resulting in the Angudan Unconformity that coincides with the final stage of the East African Orogeny (*c.* 550–510 Ma; Loosveld *et al.* 1996; Al-Husseini 2000; Immerz *et al.* 2000; Koopman *et al.* 2007; Forbes *et al.* 2010; Al-Kindi and Richard 2014; Droste 2014, fig. 6a). This orogeny is associated with the continent–continent collision of East and West Gondwana. Some authors use the

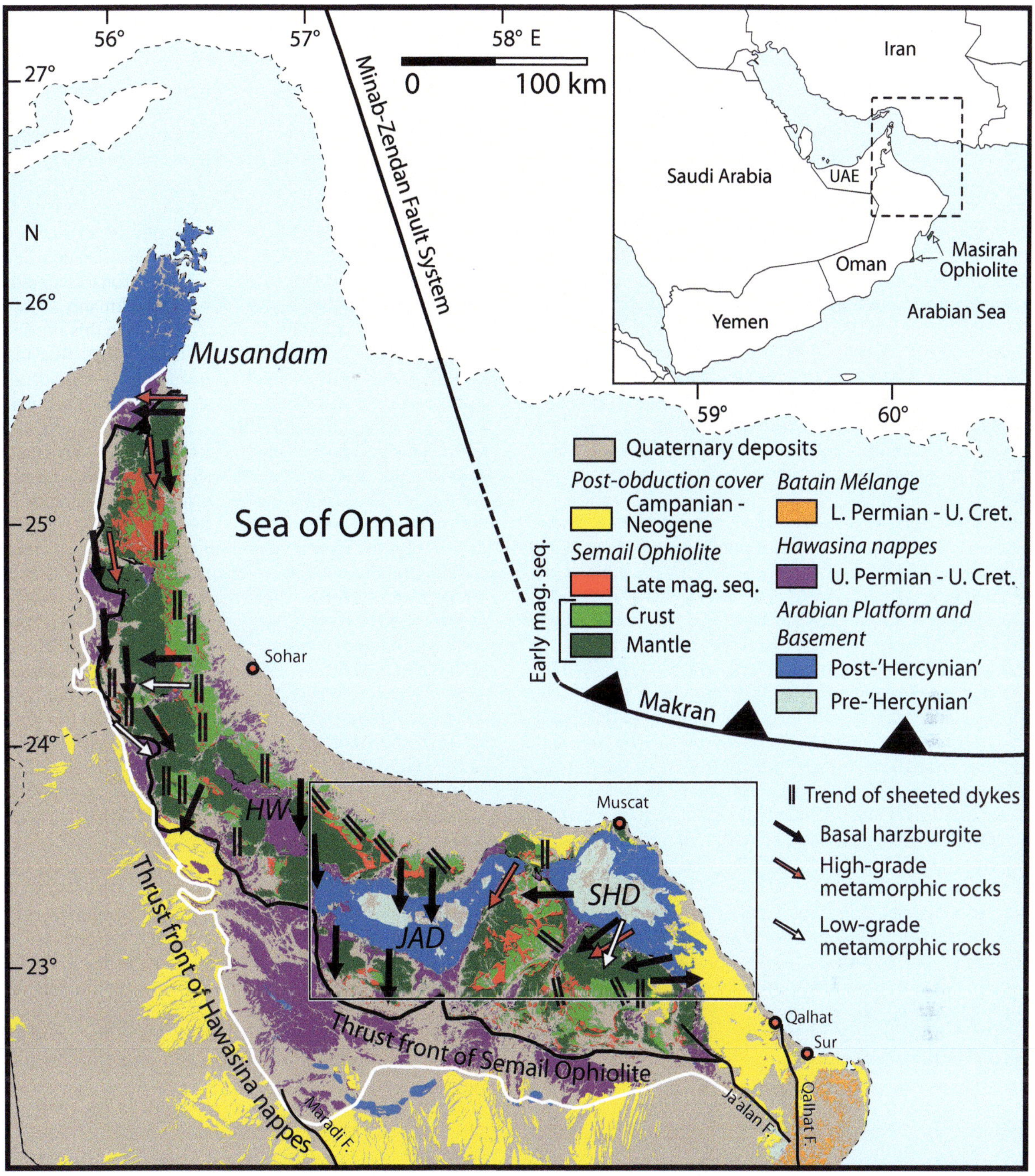

Fig. 1.1. Tectonic overview map of the northeastern Arabian Peninsula. Because of the scale, the metamorphic sole (black) below the ophiolite is only visible in certain areas. All arrows indicate the movement direction of the Semail Ophiolite during the Cenomanian–Maastrichtian. HW, Hawasina Window; JAD, Jabal Akhdar Dome; SHD, Saih Hatat Dome. Map modified after the Geological Map of Oman, 1:1 000 000, Ministry of Petroleum and Minerals (Béchennec *et al.* 1993). Batain Mélange after Shackleton *et al.* (1990), Schreurs and Immenhauser (1999) and Immenhauser *et al.* (2000). Orientation of sheeted dykes and movement direction of the Semail Ophiolite is after Hacker *et al.* (1996). The Makran– and Minab–Zendan thrusts are from Regard *et al.* (2005). The Ja'alan and Qalhat faults are drawn after Fournier *et al.* (2006) and the Maradi Fault is drawn after Filbrandt *et al.* (2006). Parts of the northern Oman Mountains are drawn after the geological map from the United Arab Emirates (UAE) (British Geological Survey 2006). The box outlines the main figure A.1 (**Scharf et al. 2021*f***, Chapter 7, this Memoir). Inset shows the greater study realm and the occurrence of the Masirah Ophiolite.

term 'Malagasy Orogeny' in Arabia as a synonym for the latest part of the East African Orogeny (e.g. Collins and Pisarevsky 2005). We use the phrase 'East African Orogeny'. We also use the name 'Angudan Unconformity' throughout the text.

The Hercynian Orogeny in Europe was caused by the convergence of Gondwana and Laurasia. The 'Hercynian Orogeny' in the northwestern Arabian Plate is manifested in the regional pre-Carboniferous uplift and erosion as well as arch formation (see Faqira *et al.* 2009). Additionally, Abbo *et al.* (2018) concluded that the 'Hercynian' event in the Arabian Plate was not a significant deformation event, but rather a thermal event. The same authors suggested that the large-scale Paleozoic arches and basins across NE Africa and Arabia formed due to a mantle disturbance. Terminologically, we are following Faqira *et al.* (2009) and use the term 'Hercynian' (in quotes), but we are not suggesting that the Hercynian Orogeny (*sensu lato*) and related deformation and/or folding took place in Oman; rather, it seems to have been a thermal event (Abbo *et al.* 2018). In the subsurface of Oman, a significant truncation of Lower Paleozoic strata below the mid-Carboniferous exists. This truncation is related to salt movements and reactivation of basement trends (Droste 1997, fig. 3; Konert *et al.* 2001; Svendsen 2004). The 'Hercynian' event in Oman is only evidenced by 'Hercynian' K/Ar crystallization ages in pre-Permian formations of the SHD at 327 ± 16 Ma (Glennie *et al.* 1974) and of the JAD at 329 ± 11 and 321 ± 10 Ma of chlorite samples from the Mu'aydin Formation. (Beurrier *et al.* 1986*b*). Moreover, pre-'Hercynian' rocks commonly display NE-trending fold axes in the JAD (Mann and Hanna 1990). The NE-trending folds are unconformably overlain by sedimentary rocks of the Haima Supergroup in the subsurface south of the Oman Mountains. The deformation is therefore Early Cambrian in age and not related to an 'Hercynian' event (Droste, H. 2018, pers. comm.). A pre-Amdeh age for the folding and thrusting in the Huqf area has been suggested by Le Métour *et al.* (1995). There is a major erosional gap with an angular unconformity between the pre-Permian rocks and the Permian Saiq Formation in the JAD and SHD (Le Métour *et al.* 1986*b*; Villey *et al.* 1986*c*). The age of the youngest rocks below the unconformity of the JAD and SHD is *c.* 540 Ma (Fara Formation) and *c.* 460 Ma (Amdeh Formation), respectively. These two aspects have been linked to the 'Hercynian' movements in the JAD area (Beurrier *et al.* 1986*b*, fig. 14; Rabu *et al.* 1986, fig. 16). Furthermore, block faulting has been attributed to the 'Hercynian' movements by Beurrier *et al.* (1986*b*), Le Métour *et al.* (1986*b*), Rabu *et al.* (1986) and Villey *et al.* (1986*c*). The trend of the aforementioned fold axes is in line with the trend of 'Hercynian' arches (Faqira *et al.* 2009, fig. 2). The 'Hercynian Unconformity' is also known as the 'pre-Permian unconformity' (e.g. Chauvet *et al.* 2009), the 'mid-Carboniferous tectonic event' (Al-Husseini 2004) and the 'Pre-Unayzah unconformity' in Saudi Arabia (Al-Husseini 2004; Fig. 1.1).

The 'Hercynian' event was followed by the break-up of Gondwana between Arabia and the Cimmerian continental blocks as well as between Arabian and India (e.g. Blendinger *et al.* 1990; Robertson and Searle 1990; Stampfli and Borel 2002; Chauvet *et al.* 2009; D4 in fig. A.2; fig. 5.18c). The timing of continental rifting is Late Carboniferous–Permian (see section on 'Pangaea rifting and formation of the Neo-tethys Ocean and Hawasina Basin' in Chapter 5 (**Scharf *et al.* 2021*d***)). The break-up of Gondwana resulted in the formation of the Batain Basin SE of Oman and the Hawasina Basin NE of Oman during the Early and mid-Permian, respectively (e.g. Béchennec *et al.* 1992*a*, *b*; Immenhauser *et al.* 2000; see also sections on 'Hawasina nappes' in Chapter 2 (**Scharf *et al.* 2021*a***) and 'Pangaea rifting and formation of the Neo-Tethys Ocean and Hawasina Basin' in Chapter 5 (**Scharf *et al.* 2021*d***)). Divergence between Arabia and the Cimmerian continental blocks continued to form the Neo-Tethys Ocean, while extension between Arabia and India aborted. Mostly deep-sea sediments with some limestone were deposited in the Hawasina Basin from the mid-Permian to the Late Cretaceous (e.g. Glennie *et al.* 1974). These rocks were thrust onto the Arabian margin and platform and were themselves overthrust by the Semail Ophiolite during the Late Cretaceous (e.g. Glennie *et al.* 1974; section 'Intraoceanic thrusting, formation and obduction of the Semail Ophiolite' in Chapter 5 of this Memoir (**Scharf *et al.* 2021*d***)).

Large-scale obduction of the prototypical Semail Ophiolite onto the Arabian passive margin during the Late Cretaceous, and the ophiolite sequence itself, have been well researched (e.g. Allemann and Peters 1972; Glennie *et al.* 1974; Searle and Malpas 1980; Coleman and Hopson 1981; Hacker *et al.* 1996; Searle and Cox 1999, 2002; Searle *et al.* 2004; Rioux *et al.* 2016; Soret *et al.* 2017; Guilmette *et al.* 2018). The recent analysis of Guilmette *et al.* (2018) advocates that initial intra-oceanic subduction was induced at *c.* 104 Ma. At *c.* 96 Ma the subduction zone matured and a self-sustained subduction started with a supra-subduction zone spreading centre (e.g. Rioux *et al.* 2016). The newly formed oceanic lithosphere was coevally thrust (obducted) onto Arabian lithosphere. Thrusting lasted until *c.* 68 Ma (Lippard 1983; Goffé *et al.* 1988; Hacker *et al.* 1996; Searle and Cox 1999; El-Shazly *et al.* 2001; Warren *et al.* 2003, 2005; Rioux *et al.* 2016; section 'Intraoceanic thrusting, formation and obduction of the Semail Ophiolite' in Chapter 5 of this Memoir (**Scharf *et al.* 2021*d***)). More than 400–450 km of obduction are considered for the extended part of the Arabian Crust (Cooper 1988; Searle 2007) and another 150 km for the Arabian Platform of more regular thickness as quantified by map analysis. The entire thrust range therefore amounts to > 550–600 km. Thrusting of the ophiolite ensued to the SW (Hacker *et al.* 1996). Today, the ophiolite covers an area of > 10 000 km^2 and is *c.* 550 km long and *c.* 150 km wide (Searle and Cox 1999). The ophiolite measures several kilometres in thickness, which decreases towards the south (Al-Lazki *et al.* 2002; Aldega *et al.* 2017; Carminati *et al.* 2019). According to Reinhardt (1969), Glennie *et al.* (1974), Coleman (1981), Lippard *et al.* (1986), Nicolas (1989) and Searle and Cox (1999), the Semail Ophiolite is composed of *c.* 8–12 km of upper mantle peridotites (mainly harzburgite) and *c.* 2.5–8 km of oceanic crustal rocks; according to Rabu *et al.* (1986) and Villey *et al.* (1986*b*, *c*), the entire ophiolite thickness measures 8–9 km. Hacker and Mosenfelder (1996) provided an inferred pre-emplacement thickness of the Semail Ophiolite of 15–20 km.

Obduction is linked to plate convergence between Arabia and Eurasia and the closure of the intervening oceanic lithosphere (Dercourt *et al.* 1993; Stampfli and Borel 2002; Stampfli and Kozur 2006). Convergence was not only expressed by the obduction of the Semail Ophiolite. While the mantle and crust of the future Semail Ophiolite still formed in a supra-subduction zone spreading centre, the northeastern part of the Arabian passive margin was subducted to the NE beneath the future ophiolite (e.g. Lippard *et al.* 1986; El-Shazly and Coleman 1990; Searle *et al.* 1994; Searle and Cox 1999; Searle *et al.* 2004; Saddiqi *et al.* 2006; Rioux *et al.* 2012, 2013, 2016; Cowan *et al.* 2014; for more details on the direction of obduction and rotation during thrusting see van Hinsbergen *et al.* 2019 and section on 'Intraoceanic thrusting, formation and obduction of the Semail Ophiolite' in Chapter 5 of this Memoir (**Scharf *et al.* 2021*d***)). Eventually, the subducted plate segment was metamorphosed under high-pressure/low-temperature conditions (e.g. Lippard 1983; Le Métour *et al.* 1986*b*; Goffé *et al.* 1988; El-Shazly *et al.* 1990; Searle *et al.* 1994, 2004; Warren *et al.* 2003, 2005), and the Semail Ophiolite was thrust on the Arabian Plate (e.g. Searle *et al.* 2004). A slab break-off may have triggered rapid exhumation of the Arabian crustal slab with its blueschist and eclogite facies, exposed in the northeastern SHD (e.g. Searle *et al.* 2004).

A second ophiolite obduction affected the southeastern part of the Oman Mountains. The Masirah Ophiolite (or 'Eastern Ophiolite Belt' of Gnos *et al.* 1997; Immenhauser *et al.* 1998) is a fragment of oceanic lithosphere thrust onto the easternmost margin of Oman (e.g. Moseley and Abbotts 1979; Mountain and Prell 1990; Ries and Shackleton 1990; Shackleton and Ries 1990; Marquer *et al.* 1995; Schreurs and Immenhauser 1999; Immenhauser *et al.* 2000; Fig. 1.1). The sole of the ophiolite is not exposed, which makes it difficult to tectonically interpret its history of thrusting, emplacement and uplift. The Masirah Ophiolite derived from oceanic lithosphere of the 'Somalia-Mozambique-Madagascar (SoMoMa) Ocean' with a formation age of latest Jurassic to earliest Cretaceous (e.g. Smewing *et al.* 1991; Immenhauser 1995; Peters *et al.* 1995; Gnos *et al.* 1997; Marquer *et al.* 1998; Schreurs and Immenhauser 1999). Latest Jurassic ages (*c.* 150 Ma) are concluded based on U–Pb zircon studies and Ar/Ar dating on plagioclase of gabbro dykes (Peters *et al.* 1995), K–Ar ages on gabbros (Smewing *et al.* 1991) and dating of radiolarian cherts

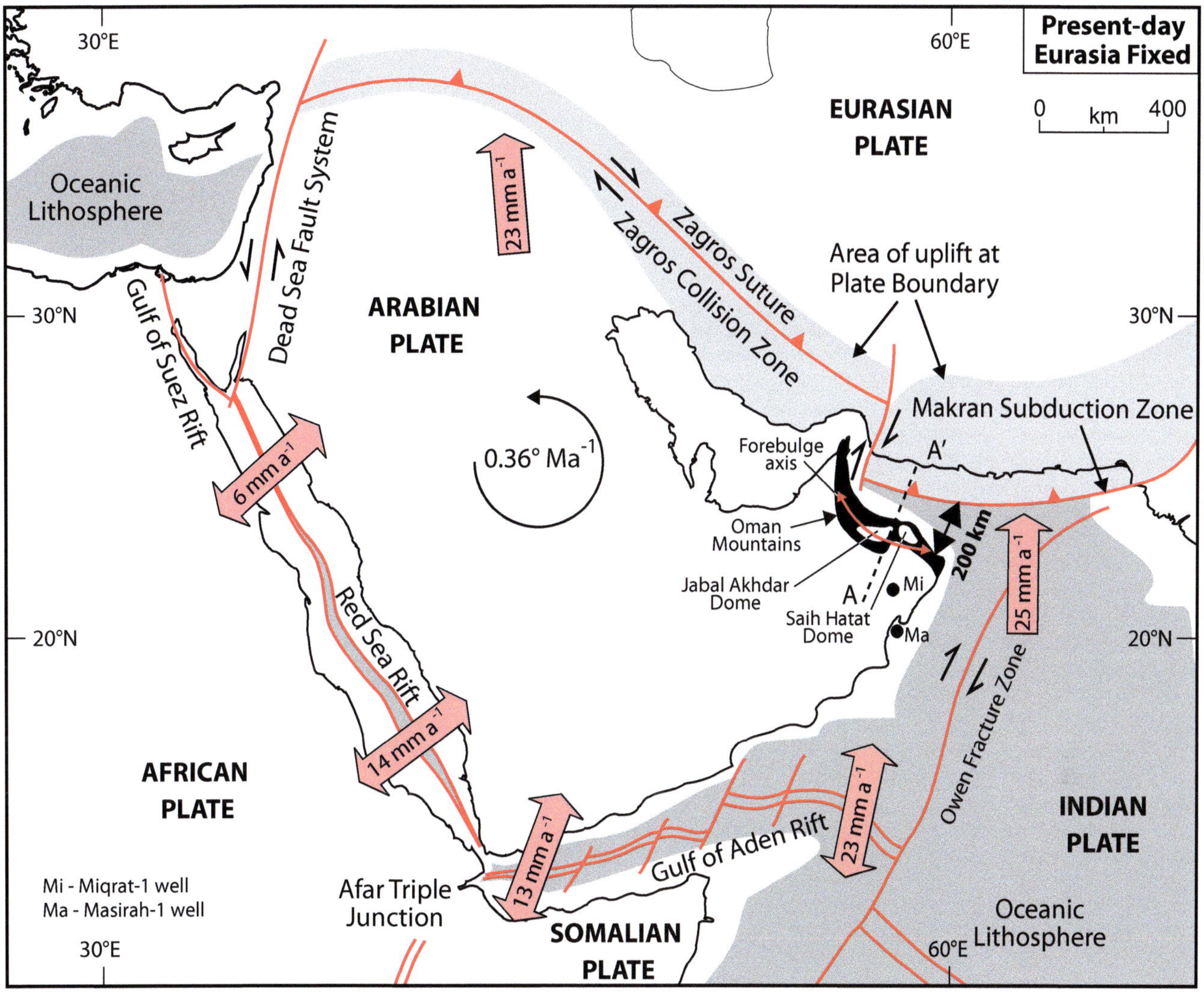

Fig. 1.2. Present tectonic frame of the Arabian Peninsula slightly modified after Hansman *et al.* (2017). Plate velocity from GPS information (Sella *et al.* 2002; McClusky *et al.* 2003; Reilinger *et al.* 2006; Vigny *et al.* 2006; ArRajehi *et al.* 2010). Forebulge axis in the Oman Mountains is after Rodgers and Gunatilaka (2002). Cross section for the trace (A–A′) is depicted in figure 5.14 (**Scharf *et al.* 2021*d***, Chapter 5, this Memoir).

(Moseley 1990). The SoMoMa Ocean separated Gondwana into western (Africa and Arabia) and eastern (Madagascar and India) segments (fig. 5.18f). Formation of the ocean resulted in the gentle extension and uplift of the eastern Arabian margin and the Hawasina Basin (D5 in fig. A.2; Rabu 1987; Béchennec *et al.* 1992*a*, *b*; section 'Arabian Platform' in Chapter 5 (**Scharf *et al.* 2021*d***)). Geochemical analyses reveal that the ophiolite represents a young mid-ocean ridge that had formed during early drift between eastern and western Gondwana (Rollinson 2017). Later, the ocean basin underwent transpression as a result of convergence between India and Arabia, and was obducted and/or emplaced to the WNW for a few tens of kilometres as the 'Masirah Ophiolite' onto the easternmost tip of the Arabian Platform at the Cretaceous–Cenozoic boundary until the early Eocene (*c.* 65–55 Ma; Smewing *et al.* 1991; Peters *et al.* 1995; Gnos *et al.* 1997; Peters and Mercolli 1997; Marquer *et al.* 1998; Schreurs and Immenhauser 1999; Immenhauser *et al.* 2000; Gaina *et al.* 2015; fig. 5.18i). The Masirah Ophiolite therefore remained a part of the oceanic lithosphere for *c.* 80 Ma from its formation to its obduction (e.g. Rollinson 2017). The Hawasina and Haybi Complex were thrust along with the Semail Ophiolite during the Coniacian–Campanian from NE to SW. By contrast, the Masirah Ophiolite was thrust along with the Batain Mélange onto the eastern Oman margin during the Cretaceous–Cenozoic (Schreurs and Immenhauser 1999; Fig. 1.1). The Lower Permian to Upper Cretaceous sediments of the Batain Mélange derived from the Batain Basin, located east of Oman as part of the SoMoMa Ocean (Schreurs and Immenhauser 1999; Immenhauser *et al.* 2000). The Batain Basin is older (latest Carboniferous to Early Permian; Immenhauser *et al.* 2000) than the Hawasina Basin (Late Permian). Although the sediments of the Hawasina nappes and the Batain Mélange derived from different basins, the stratigraphy is similar (see Immenhauser *et al.* 1998, 2000).

The Semail and Masirah ophiolites differ in origin, tectonic setting, age of oceanic lithosphere formation (Neo-Tethys Ocean during the Cenomanian compared with SoMoMa Ocean during the latest Jurassic–earliest Cretaceous), time of obduction onto the Arabian Margin (late Cretaceous compared with latest Cretaceous–early Eocene) and direction of obduction (SW-ward compared with WNW-ward). Generation of the Semail Ophiolite is associated with a spreading centre in a supra-subduction zone setting with coeval formation and thrusting (e.g. Rioux *et al.* 2016), whereas the Masirah Ophiolite formed at a conventional mid-ocean ridge with its emplacement at *c.* 80 Ma after formation (see Rollinson 2017 for further comparison and discussion).

The Oligocene–Miocene to recent regional tectonic setting of the Arabian Plate is summarized in Figure 1.2. The north-eastern plate margin corresponds to the Arabia–Eurasia collision zone, forming the Zagros and Makran mountains (e.g.

Glennie *et al.* 1974; Robertson and Searle 1990; Loosveld *et al.* 1996; Immenhauser *et al.* 2000). Deformation in the Zagros Mountain Belt started before *c.* 25–23 Ma (e.g. Agard *et al.* 2011), amounts to 440 km of shortening (Monthereau 2011) and is still active (Vernant *et al.* 2004). Arabia is bounded in the west by the left-lateral Dead Sea Transform Fault that accommodated a displacement of 100–105 km (Walley 1998). The age of the fault is > 14 Ma (Bayer *et al.* 1988; Makris and Rihm 1991). The southwestern and southern margins of the Arabian Plate are defined by the spreading axes of the Red Sea and Gulf of Aden, respectively. Continental extension in the southern Red Sea area started during *c.* 27.5–23.8 Ma (Bosworth *et al.* 2005; Reilinger and McClusky 2011). Seafloor spreading between the Arabian and African plates (i.e. the Gulf of Aden) commenced at 20 Ma (Fournier *et al.* 2010). The southeastern boundary of Arabia is marked by the 8-Ma-old Owen Transform Fault. Dextral motion along this fault ensues at a rate of 3 ± 1 mm a^{-1} (Fournier *et al.* 2008).

Author contributions **AS**: Conceptualization (lead), Writing original draft (lead), Visualization (lead), Supervision (lead), Investigation (lead); **FM**: Conceptualization (equal), Writing – review and editing (equal), Visualization (equal), Investigation (equal); **MAW**: Writing – review and editing (equal); **GF**: Writing – review and editing (equal); **DM**: Writing – review and editing (equal); **BP**: Writing – review and editing (equal); **WB**: Writing – review and editing (equal); **IC**: Writing – review and editing (equal).

Funding This research received no specific grant from any funding agency in the public, commercial, or not-for-profit sectors.

Data availability Data sharing is not applicable to this article as no datasets were generated or analysed during the current study.

References

Abbo, A., Avigad, D. and Gerdes, A. 2018. The lower crust of the northern broken edge of Gondwana: evidence for sediment subduction and syn-Variscan anorogenic imprint from zircon U-Pb-Hf in granulite xenoliths. *Gondwana Research*, **64**, 84–96, https://doi.org/10.1016/j.gr.2018.08.002

Agard, P., Omrani, J. *et al.* 2011. Zagros orogeny: a subduction-dominated process. *Geological Magazine*, **148**, 692–725, https://doi.org/10.1017/S001675681100046X

Aldega, L., Carminati, E., Scharf, A., Mattern, F. and Al-Wardi, M. 2017. Estimating original thickness and extent of the Semail Ophiolite in the eastern Oman Mountains by paleothermal indicators. *Marine and Petroleum Geology*, **84**, 18–33, https://doi.org/10.1016/j.marpet.geo.2017.03.024

Al-Husseini, M.I. 2000. Origin of the Arabian Plate Structures: Amar Collision and Najd Rift. *GeoArabia*, **5**, 527–542.

Al-Husseini, M.I. 2004. Pre-Unayzah unconformity, Saudi Arabia. Carboniferous, Permian and Early Triassic Arabian Stratigraphy. *GeoArabia Special Publication*, **3**, 15–59.

Al-Kindi, M.H. 2018. *Evolution of Land and Life in Oman: An 800 Million Year Story*. Springer, https://doi.org/10.1007/978-3-319-60152-6

Al-Kindi, M.H. and Richard, P.D. 2014. The main structural styles of the hydrocarbon reservoirs in Oman. *Geological Society, London, Special Publications*, **392**, 409–445, https://doi.org/10.1144/SP392.20

Al-Lazki, A.I., Seber, D., Sandvol, E. and Barazangi, M. 2002. A crustal transect across the Oman Mountains on the eastern margin of Arabia. *GeoArabia*, **7**, 47–78.

Allemann, F. and Peters, T. 1972. The ophiolite-radiolarite belt of the North Oman Mountains. *Eclogae Geologicae Helvetica*, **65**, 657–697.

ArRajehi, A., McClusky, S. *et al.* 2010. Geodetic constraints on present-day motion of the Arabian Plate: implications for Red Sea and Gulf of Aden rifting. *Tectonics*, **29**, TC3011, https://doi.org/10.1029/2009TC002482

Bailey, E.H. 1981. Geological map of Muscat-Ibra area, Sultanate of Oman. *Journal of Geophysical Research*, **86**, 2495–2782, https://doi.org/10.1029/JB086iB04p02495

Bayer, H.-J., Hötzl, H., Jado, A.R., Roscher, B. and Voggenreiter, W. 1988. Sedimentary and structural evolution of the northwest Arabian Sea Margin. *Tectonophysics*, **153**, 137–152, https://doi.org/10.1016/0040-1951(88)90011-X

Béchennec, F., Beurrier, M., Hutin, G. and Rabu, D. 1986*a*. *Geological map of Barka, sheet NF 40-3B, scale 1:100,000, Explanatory Notes*. Directorate General of Minerals, Oman Ministry of Petroleum and Minerals.

Béchennec, F., Beurrier, M., Rabu, D. and Hutin, G. 1986*b*. *Geological map of Bahla, sheet NF 40-07A, scale 1:100,000, with Explanatory Notes*. Directorate General of Minerals, Oman Ministry of Petroleum and Minerals.

Béchennec, F., Tegyey, M., Le Métour, J., Lemière, B., Lescuyer, J.L., Rabu, D. and Milesi, J.P. 1991. Igneous rocks in the Hawasina Nappes and the Hajar Supergroup, Oman Mountains: Their significance in the birth and evolution of the composite extensional Margin of the eastern Tethys. *In*: Peters, T., Nicolas, A. and Coleman, R.J. (eds) *Ophiolite Genesis and Evolution of the Oceanic Lithosphere*. Proceedings of the Ophiolite Conference, Muscat, Oman. Kluwer Academic Publishers, Dordrecht/Boston/London, **1990**, 593–611.

Béchennec, F., Roger, J., Le Métour, J. and Wyns, R. 1992*a*. *Geological map of Seeb, sheet NF 40-03, scale 1:250,000, with Explanatory Notes*. Directorate General of Minerals, Oman Ministry of Petroleum and Minerals.

Béchennec, F., Wyns, R., Roger, J., Le Métour, J. and Chevrel, S. 1992*b*. *Geological map of Nazwa, sheet NF 40-07, scale 1:250,000, with Explanatory Notes*. Directorate General of Minerals, Oman Ministry of Petroleum and Minerals.

Béchennec, F., Le Métour, J., Platel, J.P. and Roger, J. 1993. *Geological map of the Sultanate of Oman, scale 1:1,000,000, with Explanatory Notes*. Directorate General of Minerals, Oman Ministry of Petroleum and Minerals.

Bendias, D., Koehrer, B., Obermaier, M. and Aigner, T. 2013. Mid-Permian Khuff Sequence KS6: paleorelief-influenced facies and sequence patterns in the Lower Khuff time-equivalent strata, Oman Mountains, Sultanate of Oman. *GeoArabia*, **18**, 135–178.

Beurrier, M., Béchennec, F., Hutin, G. and Rabu, D. 1986*a*. *Geological map of As Suwayq, sheet NF 40-03A, scale 1:100,000, with Explanatory notes*. Directorate General of Minerals, Oman Ministry of Petroleum and Minerals.

Beurrier, M., Béchennec, F., Rabu, D. and Hutin, G. 1986*b*. *Geological map of Rustaq, sheet NF 40-03D, scale 1:100,000, with Explanatory notes*. Directorate General of Minerals, Oman Ministry of Petroleum and Minerals.

Blendinger, W., van Vliet, A.T. and Hughes Clarke, M.W. 1990. Updoming, rifting and continental margin development during the Late Paleozoic in northern Oman. *Geological Society, London, Special Publications*, **49**, 27–37.

Bosworth, W., Huchon, P. and McClay, K. 2005. The Red Sea and Gulf of Aden Basins. *Journal of African Earth Sciences*, **43**, 334–378, https://doi.org/10.1016/j.jafrearsci.2005.07.020

British Geological Survey. 2006. *Geological Map of the Northern Emirates, 1:250,000 Scale*. British Geological Survey, Keyworth.

Burns, S.J. and Matter, A. 1993. Carbon isotopic record of the latest Proterozoic from Oman. *Eclogae Geologicae Helvetiae*, **86**, 595–607.

Burns, S.J., Matter, A., Frank, N. and Mangini, A. 1998. Speleothem-based paleoclimate record from northern Oman. *Geology*, **26**, 499–502, https://doi.org/10.1130/0091-7613(1998)026<0499:SBPRFN>2.3.CO;2

Burns, S.J., Fleitmann, D., Matter, A., Neff, U. and Mangini, A. 2001. Speleothem evidence from Oman for continental pluvial events during interglacial periods. *Geology*, **29**, 623–626, https://doi.org/10.1130/0091-7613(2001)029<0623:SEFOFC>2.0.CO;2

Callegari, I., Scharf, A. *et al.* 2020. Gondwana accretion tectonics and implications for the geodynamic evolution of eastern Arabia: first structural evidence of the existence of the Cadomian Orogeny in Oman (Jabal Akhdar Dome, Central Oman Mountains). *Journal of Asian Earth Sciences*, **187**, 104070, https://doi.org/10.1016/j.jseaes.2019.104070

Carminati, E., Aldega, L., Smeraglia, L., Scharf, A. and Mattern, F. 2019. Obduction and collision tectonics in Oman: constraints from structural and thermal analyses. *In*: Rossetti, F., Blanc, A.C., Riguzzi, F., Leroux, E., Pavalopoulos, K., Bellier, O. and Kapsimalis, V. (eds) *The Structural Geology Contribution to the Africa-Eurasia Geology: Basement and Reservoir Structure, Ore Mineralisation and Tectonic Modelling*, Proceedings of the 1st Springer Conference of the Arabian Journal of Geosciences (CAJG-1), Tunisia 2018, IEREK Interdisciplinary Series for Sustainable Development, 319–321, https://doi.org/10.1007/978-3-030-01455-1_70

Chauvet, F., Dumont, T. and Basile, C. 2009. Structures and timing of Permian rifting in the central Oman Mountains (Saih Hatat). *Tectonophysics*, **475**, 563–574, https://doi.org/10.1016/j.tecto.2009.07.008

Coffield, D.Q. 1990. Structures associated with nappe emplacement and culmination collapse in the Central Oman Mountains. *Geological Society, London, Special Publications*, **49**, 447–458.

Coleman, R.G. 1981. Tectonic setting for ophiolite obduction in Oman. *Journal of Geophysical Research*, **86**, 2497–2508, https://doi.org/10.1029/JB086iB04p02497

Coleman, R.G. and Hopson, C.A. 1981. Introduction to the Oman Ophiolite Special Issue. *Journal of Geophysical Research*, **86**, 2495–2496, https://doi.org/10.1029/JB086iB04p02495

Collins, A.S. and Pisarevsky, S.A. 2005. Amalgamating eastern Gondwana: the evolution of the Circum-Indian Orogens. *Earth-Science Reviews*, **71**, 229–270, https://doi.org/10.1016/j.earscirev.2005.02.004

Cooper, D.J.W. 1988. Structure and sequence of thrusting in deep-water sediments during ophiolite emplacement in the south-central Oman Mountains. *Journal of Structural Geology*, **10**, 437–485, https://doi.org/10.1016/0191-8141(88)90031-4

Cowan, R.J., Searle, M.P. and Waters, D.J. 2014. Structure of the metamorphic sole to the Oman Ophiolite, Sumeini Window and Wadi Tayyin: implications for ophiolite obduction processes. *Geological Society, London, Special Publications*, **392**, 155–175, https://doi.org/10.1144/SP392.8

de Gramont, X., Le Métour, J. and Villey, M. 1986. *Geological map of Samad, sheet NF 40-07C, scale 1:100,000, with Explanatory Notes*. Directorate General of Minerals, Oman Ministry of Petroleum and Minerals.

Dercourt, J., Ricou, L.E. and Vrielynck, B. 1993. *Atlas Tethys Palaeoenvironmental Maps*. Gauthier-Villars, Paris, 307.

Droste, H. 1997. Stratigraphy of the Lower Paleozoic Haima Supergroup of Oman. *GeoArabia*, **2**, 419–472.

Droste, H. 2014. Petroleum geology of the Sultanate of Oman. *American Association of Petroleum Geologists, Memoirs*, **106**, 713–755, https://doi.org/10.1036/12431870M1063039

El-Shazly, A.K. and Coleman, R.G. 1990. Metamorphism in the Oman Mountains in relation to the Semail ophiolite emplacement. *Geological Society, London, Special Publications*, **49**, 473–493.

El-Shazly, A.K., Coleman, R.G. and Liou, J.G. 1990. Eclogites and blueschists from Northeastern Oman: petrology and P-T evolution. *Journal of Petrology*, **31**, 629–666, https://doi.org/10.1093/petrology/31.3.629

El-Shazly, A.K., Broker, W., Hacker, B. and Calvert, A. 2001. Formation and exhumation of blueshists and eclogites from NE Oman: new perspectives from Rb-Sr and $^{40}Ar/^{39}Ar$ dating. *Journal of Metamorphic Geology*, **19**, 233–248, https://doi.org/10.1046/j.1525-1314.2001.00309.x

Faqira, M., Rademakers, M. and Afifi, A. 2009. New insights into the Hercynian Orogeny, and their implications for the Paleozoic Hydrocarbon System in the Arabian Plate. *GeoArabia*, **14**, 199–228.

Filbrandt, J.B., Al-Dhahab, S. *et al.* 2006. Kinematic interpretation and structural evolution of North Oman, Block 6, since the Late Cretaceous and implications for timing of hydrocarbon migration into Cretaceous reservoirs. *GeoArabia*, **11**, 97–140.

Forbes, G.A., Jansen, H.S.M. and Schreurs, J. 2010. *Lexicon of Oman Subsurface Stratigraphy. Reference Guide to the Stratigraphy of Oman's Hydrocarbon Basins*. Gulf Petrolink, GeoArabia, Special Publication, **5**.

Fournier, M., Lepvrier, C., Razin, P. and Jolivet, L. 2006. Late Cretaceous to Paleogene Post-obduction extension and subsequent Neogene compression in the Oman Mountains. *GeoArabia*, **11**, 17–40.

Fournier, M., Chamot-Rooke, N., Petit, C., Fabbri, O., Huchon, P., Maillot, B. and Lepvrier, C. 2008. In situ evidence for dextral active motion at the Arabia-India plate boundary. *Nature Geoscience*, **1**, 54–58, https://doi.org/10.1038/ngeo.2007.24

Fournier, M., Chamot-Rooke, N. *et al.* 2010. Somalia plate kinematics, evolution of the Aden – Owen – Carlsberg triple junction, and opening of the Gulf of Aden. *Journal of Geophysical Research*, **115**, B04102, https://doi.org/10.1029/2008JB006257

Gaina, C., van Hinsbergen, D.J.J. and Spakman, W. 2015. Tectonic interactions between India and Arabia since the Jurassic reconstructed from marine geophysics, ophiolite geology, and seismic tomography. *Tectonics*, **34**, 875–906, https://doi.org/10.1002/2014TC003780

Glennie, K.W. 1995. *The Geology of the Oman Mountains: An Outline of their Origin*. 1st ed. Scientific Press, Beaconsfield.

Glennie, K.W. 2005. *The Geology of the Oman Mountains: An Outline of their Origin*. 2nd ed. Scientific Press, Buckinghamshire.

Glennie, K.W., Boeuf, M.G.A., Hughes Clarke, M.W., Moody-Stuart, M., Pilaar, W.F.H. and Reinhardt, B.M. 1973. Late Cretaceous nappes in Oman Mountains and their geological evolution. *American Association of Petroleum Geologists Bulletin*, **57**, 5–27.

Glennie, K.W., Boeuf, M.G.A., Hughes Clarke, M.W., Moody-Stuart, M., Pilaar, W. and Reinhardt, B.M. 1974. Geology of the Oman Mountains *Koninklijk Nederlands Geologisch en Mijnbouwkundig Genootschap Transactions*, **31**, 423.

Gnos, E., Immenhauser, A. and Peters, T. 1997. Late Cretaceous/early Tertiary convergence between the Indian and Arabian plates recorded in ophiolites and related sediments. *Tectonophysics*, **271**, 1–19, https://doi.org/10.1016/S0040-1951(96)00249-1

Goffé, B., Michard, A., Kienast, J.R. and le Mer, O. 1988. A case of obduction related high P, low T metamorphism in upper crustal nappes, Arabian continental margin, Oman: P-T paths and kinematic interpretation. *Tectonophysics*, **151**, 363–386, https://doi.org/10.1016/0040-1951(88)90253-3

Grantham, P.J., Lijmbach, G.W.M., Posthuma, J., Clarke, M.W.H. and Willnik, R.J. 1988. Origin of crude oils in Oman. *Journal of Petroleum Geology*, **11**, 61–80, https://doi.org/10.1111/j.1747-5457.1988.tb00801.x

Gregory, R.T., Gray, D.R. and Miller, J.M. 1998. Tectonics of the Arabian margin associated with the emplacement of the Oman margin along the Ibra transect: new evidence from northeast Saih Hatat. *Tectonics*, **17**, 657–670, https://doi.org/10.1029/98TC02206

Grobe, A., Urai, J.L., Littke, R. and Lünsdorf, N.K. 2016. Hydrocarbon generation and migration under a large overthrust: the carbonate platform under the Semail Ophiolite, Jebel Akhdar, Oman. *International Journal of Coal Geology*, **168**, 3–19, https://doi.org/10.1016/j.coal.2016.02.007

Grobe, A., von Hagke, C., Littke, R., Dunkl, I., Wübbeler, F., Muchez, P. and Urai, J.L. 2019. Tectono-thermal evolution of Oman's Mesozoic passive continental margin under the obducting Semail Ophiolite: a case study of Jebel Akhdar, Oman. *Solid Earth*, **10**, 149–175, https://doi.org/10.5194/se-10-149-2019

Guilmette, C., Smit, M.A. *et al.* 2018. Forced subduction initiation recorded in the sole and crust of the Semail Ophiolite of Oman. *Nature Geoscience*, **11**, 688–695, https://doi.org/10.1038/s41561-018-0209-2

Hacker, B.R. and Mosenfelder, J.L. 1996. Metamorphism and deformation along the emplacement thrust of the Semail ophiolite, Oman. *Earth and Planetary Science Letters*, **144**, 435–451, https://doi.org/10.1016/S0012-821X(96)00186-0

Hacker, B.R., Mosenfelder, J.L. and Gnos, E. 1996. Rapid emplacement of the Oman ophiolite: thermal and geochronical constrains. *Tectonics*, **15**, 1230–1247, https://doi.org/10.1029/96TC01973

Hansman, R.J., Ring, U., Thomson, S.N., den Brock, B. and Stübner, K. 2017. Late Eocene uplift of the Al Hajar Mountains, Oman, supported by stratigraphic and low-temperature thermochronology. *Tectonics*, **36**, 3081–3109, https://doi.org/10.1002/2017TC004672

Heward, A.P. 2016. G.M. Lees (Abu'l Jabal): a pioneering geologist in Oman. *Al Hajar*, **22**, 15–27.

Heward, A.P. and Penney, R.A. 2014. Al Khlata glacial deposits in the Oman Mountains and their implications. *Geological Society, London, Special Publications*, **392**, 279–301, https://doi.org/10.1144/SP392.15

Hoffmann, G., Meschede, M., Zacke, A. and Al Kindi, M. 2016. *Field Guide to the Geology of Northeastern Oman*. Schweizerbart, Geological Field Guides, **110**, 283.

Hughes Clarke, M.W. 1988. Stratigraphy and rock unit nomenclature in the oil-producing area of interior Oman. *Journal of Petroleum Geology*, **11**, 5–60, https://doi.org/10.1111/j.1747-5457.1988.tb00800.x

Hutin, G., Béchennec, F., Beurrier, M. and Rabu, D. 1986. *Geological map of Birkat Al Mawz, sheet NF 40-07B, scale 1:100,000, with Explanatory Notes*. Directorate General of Minerals, Oman Ministry of Petroleum and Minerals.

Immenhauser, A. 1995. *The Autochthonous Mesozoic Sediment Record on the Masirah Island Ophiolite (Sultanate of Oman)*. PhD thesis, University of Bern, Bern, Switzerland.

Immenhauser, A., Schreurs, G., Peters, T., Matter, A., Hauser, M. and Dumitrica, P. 1998. Stratigraphy, sedimentology and depositional environment of the Permian to uppermost Cretaceous Batain Group, eastern Oman. *Eclogae Geolgica Helvetica*, **91**, 217–235.

Immenhauser, A., Schreurs, G., Gnos, E., Oterdoom, W. and Hartmann, B. 2000. Late Paleozoic to Neogene geodynamic evolution of the northeastern Oman margin. *Geological Magazine*, **137**, 1–18, https://doi.org/10.1017/S0016756800003526

Immerz, P.W., Oterdoom, H. and El-Tonbary, M. 2000. The Huqf/Haima hydrocarbon system of Oman and the terminal phase of the Pan-African Orogeny: evaporite depositions in a compressive setting. 4th Middle East Geosciences Conference, GEO 2000, GeoArabia, Abstract, **5**, 113–114.

Janjou, D., Minoux, L., Beurrier, M., de Gramont, X., Le Métour, J. and Villey, M. 1986. *Geological map of Ibri, sheet NF 40-02F, scale 1:100,000, with Explanatory Notes*. Directorate General of Minerals, Oman Ministry of Petroleum and Minerals.

Kajima, M. and Ishii, G. 2012. *Geological map of Al Ansab, sheet NF 40-3C4/C, scale 1:25,000, with Explanatory Notes*. Directorate General of Minerals, Oman Ministry of Commerce and Industry.

Kajima, M. and Otake, M. 2012. *Geological map of Yiti, sheet NF 40-4A3/D, scale 1:25,000, with Explanatory Notes*. Directorate General of Minerals, Oman Ministry of Commerce and Industry.

Kajima, M., Goto, M., Otake, M. and Ishii, G. 2012*a*. *Geological map of Halban, sheet NF 40-3C3/C, scale 1:25,000, with Explanatory Notes*. Directorate General of Minerals, Oman Ministry of Commerce and Industry.

Kajima, M., Ishii, G. and Goto, M. 2012*b*. *Geological map of Ar Rusayl, sheet NF 40-3C3/C, scale 1:25,000, with Explanatory Notes*. Directorate General of Minerals, Oman Ministry of Commerce and Industry.

Kajima, M., Ishii, G., Goto, M. and Otake, M. 2012*c*. *Geological map of Muscat, sheet NF 40-4A3/C, scale 1:25,000, with Explanatory Notes*. Directorate General of Minerals, Oman Ministry of Commerce and Industry.

Kajima, M., Ishii, G., Otake, M. and Goto, M. 2012*d*. *Geological map of Al Khuwayr, sheet NF 40-3C4/D, scale 1:25,000, with Explanatory Notes*. Directorate General of Minerals, Oman Ministry of Commerce and Industry.

Kajima, M. *et al.* 2012*e*. *Geological map of Al Manumah, sheet NF 40-3C3/A, scale 1:25,000, with Explanatory Notes*. Directorate General of Minerals, Oman Ministry of Commerce and Industry.

Kajima, M. *et al.* 2012*f*. *Geological map of As Seeb, sheet NF 40-3C3/B, scale 1:25,000, with Explanatory Notes*. Directorate General of Minerals, Oman Ministry of Commerce and Industry.

Kilner, B., Mac Niocaill, C. and Brasier, M. 2005. Low-latitude glaciation in the Neoproterozoic of Oman. *Geology*, **33**, 413–416, https://doi.org/10.1130/G21227.1

Koehrer, B., Aigner, T. and Pöppelreiter, M. 2011. Field-scale geometries of Upper Khuff reservoir geobodies in an outcrop analogue (Oman Mountains, Sultanate of Oman). *Petroleum Geoscience*, **17**, 3–16, https://doi.org/10.1144/1354-079310-009

Konert, G., Afifi, A.M., Al-Hajri, S.A. and Droste, H.J. 2001. Paleozoic stratigraphy and hydrocarbon habitat of the Arabian Plate. *GeoArabia*, **6**, 407–442.

Koopman, A., van der Berg, M., Romine, K. and Teasdale, J. 2007. Proterozoic to Cambrian plate-tectonics and its control on the structural evolution of the Ara Salt-Basin in Oman. Abstract AAPG European Region Conference, Athens, Greece: AAPG Search and Discovery Article #90072.

Leather, J., Allen, P.A., Braiser, M.D. and Cozzi, A. 2002. A Neoproterozoic Snowball Earth under scrutiny: evidence from the Fiq glaciation of Oman. *Geology*, **30**, 891–894, https://doi.org/10.1130/0091-7613(2002)0302.0.CO.2

Lees, G.M. 1928. The geology and tectonics of Oman and parts of southeastern Arabia. *Quarterly Journal of the Geological Society of London*, **84**, 585–670, https://doi.org/10.1144/GSL.JGS.1928.084.01-04.24

Le Guerroué, E., Allen, P.A. and Cozzi, A. 2006*a*. Chemostratigraphic and sedimentological framework of the largest negative carbon isotopic excursion in Earth history: the Neoproterozoic Shuram Formation (Nafun Group, Oman). *Precambrian Research*, **144**, 68–92, https://doi.org/10.1016/j.precamres.2006.01.007

Le Guerroué, E., Allen, P., Cozzi, A., Etienne, J.L. and Fanning, M. 2006*b*. 50 Myr recovery from the largest negative δ^{13}C excursion in the Ediacaran ocean. *Terra Nova*, **18**, 147–153, https://doi.org/10.1111/j.1365-3121.2006.00674.x

Le Métour, J., Villey, M. and de Gramont, X. 1986*a*. *Geological map of Masqat, sheet NF 40-4A, scale 1:100,000, Explanatory notes*. Directorate General of Minerals, Oman Ministry of Petroleum and Minerals.

Le Métour, J., Villey, M. and de Gramont, X. 1986*b*. *Geological map of Quryat, sheet NF 40-4D, scale 1:100,000, Explanatory notes*. Directorate General of Minerals, Oman Ministry of Petroleum and Minerals.

Le Métour, J.F., Béchennec, F., Roger, J. and Wyns, R. 1992. *Geological map of Muscat, sheet NF40-04, scale 1:250,000, with Explanatory notes*. Directorate General of Minerals, Oman Ministry of Petroleum and Minerals.

Le Métour, J.F., Michel, J.C., Béchennec, F., Platel, J.P. and Roger, J. 1995. *Geology and mineral wealth of the Sultanate of Oman*. Directorate General of Minerals. Oman Ministry of Petroleum and Mineral.

Lippard, S.J. 1983. Cretaceous high pressure metamorphism in NE Oman and its relationship to subduction and ophiolite nappe emplacement. *Journal of Geological Society London*, **140**, 97–104, https://doi.org/10.1144/gsjgs.140.1.0097

Lippard, S.J., Shelton, A.W. and Gass, I.G. 1986. The ophiolite of northern Oman. *Geological Society, London, Memoirs*, **11**, 178.

Loosveld, R.J.H., Bell, A. and Terken, J.J.M. 1996. The tectonic evolution of interior Oman. *GeoArabia*, **1**, 28–51.

Makris, J. and Rihm, R. 1991. Shear controlled evolution of Red Sea: pull apart model. *Tectonophysics*, **198**, 441–446, https://doi.org/10.1016/0040-1951(91)90166-P

Mann, A. and Hanna, S.S. 1990. The tectonic evolution of pre-Permian rocks, Central and Southeastern Oman Mountains. *Geological Society, London, Special Publications*, **49**, 307–325.

Marquer, D., Peters, T. and Gnos, E. 1995. A new structural interpretation for the emplacement of the Masirah Ophiolites (Oman): a main Paleocene intra-oceanic thrust. *Geodinimica Acta*, **8**, 13–19, https://doi.org/10.1080/09853111.1995.11105269

Marquer, D., Mercolli, I. and Peters, T. 1998. Early Cretaceous intra-oceanic rifting in the Proto-Indian Ocean recorded in the

Masirah Ophiolite, Sultanate of Oman. *Tectonophysics*, **292**, 1–16, https://doi.org/10.1016/S0040-1951(98)00063-8

Mattern, F., Moraetis, D. *et al.* 2018. Coastal dynamics of uplifted and emerged Late Pleistocene near-shore coral patch reefs at Fins (eastern coastal Oman, Gulf of Oman). *Journal of African Earth Sciences*, **138**, 192–200, https://doi.org/10.1016/j.afrearsci.2017.11.018

McClusky, S., Reilinger, R., Mahmoud, S., Ben Sari, D. and Tealeb, A. 2003. GPS constraints on Africa (Nubia) and Arabia plate motions. *Geophysical Journal International*, **155**, 126–138, https://doi.org/10.1046/j.1365-246X.2003.02023.x

Monthereau, F. 2011. Timing of uplift in the Zagros belt/Iranian plateau and accommodation of late Cenozoic Arabia-Eurasia convergence. *Geological Magazine*, **148**, 726–738, https://doi.org/10.1017/S0016756811000306

Moraetis, D., Mattern, F., Scharf, A., Frijia, G., Kusky, T.M., Yuan, Y. and Hussain, I.L. 2018. Neogene to Quaternary uplift history along the passive margin of the northeastern Arabian Peninsula eastern Hajar Mountains, Oman. *Quaternary Research*, **90**, 418–434, https://doi.org/10.1017/qua.2018.51

Moseley, F. 1990. The structure of Masirah Island, Oman. *Geological Society, London, Special Publications*, **49**, 665–671.

Moseley, F. and Abbotts, I.L. 1979. The ophiolite mélange of Masirah, Oman. *Journal of the Geological Society*, **136**, 713–724, https://doi.org/10.1144/gsjgs.136.6.0713

Mountain, G.S. and Prell, W.L. 1990. A multiphase plate tectonic history of the southeast continental margin of Oman. *Geological Society, London, Special Publications*, **49**, 725–743.

Nicolas, A. 1989. *Structures of Ophiolites and Dynamics of Oceanic Lithosphere*. Netherlands, Kluwer, p. **367**.

Nicolas, A., Boudier, F. and Ildefonse, B. 1996. Variable crustal thickness in the Oman ophiolite: implications for oceanic crust. *Journal of Geophysical Research*, **101**, 17,941–17,950, https://doi.org/10.1029/96JB00195

Nicolas, A., Boudier, F., Ildefonse, B. and Ball, E. 2000. Accretion of Oman and United Arab Emirates ophiolite: discussion of a new structural map. *Marine Geophysical Researches*, **21**, 147–179, https://doi.org/10.1023/A:1026769727917

Peters, T. and Mercolli, I. 1997. Formation and evolution of the Masirah Ophiolite (Sultanate of Oman). *Ofioliti*, **22**, 15–34.

Peters, T., Immenhauser, A., Mercolli, I. and Meyer, J. 1995. *Geological map of Masirah North and Masirah South*. Explanatory notes, scale 1:50,000, sheet K768 North and K768 South. Directorate General of Minerals, Oman Ministry of Petroleum and Minerals, Muscat, Oman.

Peters, T., Blechschmidt, I., Krystyn, L., Dumitrica, P., Mercolli, I., El Amin, O. and Al Towaya, A. 2005. *Geological map of Ibra. Explanatory notes, sheet NF 40-08A, scale 1:100,000*. Directorate General of Minerals, Ministry of Commerce and Industry of Oman.

Philip, J., Borgomano, J. and Al-Maskiry, S. 1995. Cenomanian–Early Turonian carbonate platform of Northern Oman: stratigraphy and palaeo-environments. *Palaeogeography, Palaeoclimatology, Palaeoecology*, **119**, 77–92, https://doi.org/10.1016/0031-0182(95)00061-5

Pöppelreiter, M.C., Schneider, C.J., Obermaier, M., Forke, H.C., Koehrer, B. and Aigner, T. 2011. Seal turns into reservoir: Sudair equivalents in outcrop, Al Jabal al-Akhdar, Sultanate of Oman. *GeoArabia*, **16**, 69–108.

Poupeau, G., Saddiqi, O., Michard, A., Goffé, B. and Oberhänsli, R. 1998. Late thermal evolution of the Oman Mountains subophiolitic windows: apatite fission-track thermochronology. *Geology*, **26**, 1139–1142, https://doi.org/10.1130/0091-7613(1998)026<1139:LTEOTO>2.3.CO;2

Rabu, D. 1987. *Géologie de l'Autochtone des Montagnes d'Oman: la fenêtre du Jabal Akhdar*. La semelle métamorphique de la Nappe ophiolitique de Samail dans les parties orientaler et central des Montagnes d'Oman: une revue. PhD thesis, Pierre and Marie University, Paris 6, France, Bureau de Recherches Géologiques et Minières, Document No. 130.

Rabu, D., Béchennec, F., Beurrier, M. and Hutin, G. 1986. *Geological map of Nakhl, sheet NF40-3E, scale: 1:100,000, with Explanatory Notes*. Directorate General of Minerals, Oman Ministry of Petroleum and Minerals.

Regard, V., Bellier, O. *et al.* 2005. Cumulative right-lateral fault slip rate across the Zagros-Makran transfer zone: role of the Minab-Zendan fault system in accommodating Arabia-Eurasia convergence in southeast Iran. *Geophysical Journal International*, **162**, 177–203, https://doi.org/10.1111/j.1365-246X.2005.02558.x

Reilinger, R., McClusky, S. *et al.* 2006. GPS constraints on continental deformation in the Africa-Arabia-Eurasia continental collision zone and implications for the dynamics of plate interactions. *Journal of Geophysical Research*, **111**, B05411, https://doi.org/10.1029/2005JB004051

Reilinger, R. and McClusky, S. 2011. Nubia-Arabia-Eurasia plate motions and the dynamics of Mediterranean and Middle East tectonics. *Geophysical Journal International*, **186**, 971–979, https://doi.org/10.1111/j.1365-246X.2011.05133.x

Reinhardt, B.M. 1969. On the genesis and emplacement of ophiolites in the Oman Mountains geosyncline. *Schweizerische Mineralogische und Petrographische Mitteilungen*, **49**, 1–30.

Ries, A.C. and Shackleton, R.M. 1990. Structures in the Huqf-Haushi uplift, east central Oman. *Geological Society, London, Special Publications*, **49**, 653–663.

Rioux, M., Bowring, S.A., Kelemen, P.B., Gordon, S., Dudás, F. and Miller, R. 2012. Rapid crustal accretion and magma assimilation in the Oman-U.A.E. ophiolite: high precision U-Pb zircon geochronology of the gabbroic crust. *Journal of Geophysical Research*, **117**, B07201, https://doi.org/10.1029/2012JB009273

Rioux, M., Bowring, S., Kelemen, P., Gordon, S., Miller, R. and Dudás, F. 2013. Tectonic development of the Samail ophiolite: high-precision U-Pb zircon geochronology and Sm-Nd isotopic constraints on crustal growth and emplacement. *Journal of Geophysical Research: Solid Earth*, **118**, 2085–2101, https://doi.org/10.1002/jgrb.50139

Rioux, M., Garber, J., Bauer, A., Bowring, S., Searle, M.P., Kelemen, P. and Hacker, B. 2016. Synchronous formation of the metamorphic sole and igneous crust of the Semail ophiolite: new constraints on the tectonic evolution during ophiolite formation from high-precision U-Pb zircon geochronology. *Earth and Planetary Science Letters*, **451**, 185–195, https://doi.org/10.1016/j.epsl.2016.06.051

Robertson, A.H.F. and Searle, M.P. 1990. The northern Oman Tethyan continental margin: stratigraphy, structure, concepts and controversies. *Geological Society, London, Special Publications*, **49**, 3–25.

Robertson, A.H.F., Searle, M.P. and Ries, A.C. 1990. *The Geology and Tectonics of the Oman Regions*. Geological Society, London, Special Publications, **49**.

Rodgers, D.W. and Gunatilaka, A. 2002. Bajada formation by monsoonal erosion of a subaerial forebulge, Sultanate of Oman. *Sedimentary Geology*, **154**, 127–146, https://doi.org/10.1016/S0037-0738(02)00126-4

Roger, J., Béchennec, F., Janjou, D., Le Métour, J., Wyns, R. and Beurrier, M. 1991. *Geological map of Ja'alan, sheet NF 40-08E, scale:100,000, with Explanatory Notes*. Directorate General of Minerals, Oman Ministry of Petroleum and Minerals.

Rollinson, H. 2017. Masirah – The other Oman ophiolite: a better analogue for mid-ocean ridge processes? *Geoscience Frontiers*, **8**, 1253–1262, https://doi.org/10.1016/j.gsf.2017.04.009

Rollinson, H.R., Searle, M.P., Abbasi, I.A., Al-Lazki, A.I. and Al Kindi, M.H. 2014. Tectonic Evolution of the Oman Mountains. Geological Society, London, Special Publications, **392**.

Saddiqi, O., Michard, A., Goffé, B., Poupeau, G. and Oberhänsli, R. 2006. Fission-track thermochronology of the Oman Mountains continental widows, and current problems of tectonic interpretation. *Bulletin de la Société Géologique de France*, **177**, 127–134, https://doi.org/10.2113/gssgfbull.177.3.127

Scharf, A., Mattern, F. *et al.* 2021*a*. Tectonostratigraphy of the eastern part of the Oman Mountains. *Geological Society, London, Memoirs*, **54**, 11–47, https://doi.org/10.1144/M54.2

Scharf, A., Mattern, F. *et al.* 2021*b*. Thrusts, extensional faults and fold patterns of the major units. *Geological Society, London, Memoirs*, **54**, 49–60, https://doi.org/10.1144/M54.3

Scharf, A., Mattern, F. *et al.* 2021*c*. Large-scale structure of the study area. *Geological Society, London, Memoirs*, **54**, 61–66, https://doi.org/10.1144/M54.4

Scharf, A., Mattern, F. *et al.* 2021*d*. Tectonic evolution of the Oman Mountains. *Geological Society, London, Memoirs*, **54**, 67–104, https://doi.org/10.1144/M54.5

Scharf, A., Mattern, F. *et al.* 2021*e*. Conclusions, differences between the Jabal Akhdar and Saih Hatat domes and unanswered questions. *Geological Society, London, Memoirs*, **54**, 105–111, https://doi.org/10.1144/M54.6

Scharf, A., Mattern, F. *et al.* 2021*f*. Appendices to: The Geology and Tectonics of the Jabal Akhdar and Saih Hatat Domes, Oman Mountains. *Geological Society, London, Memoirs*, **54**, 113–116, https://doi.org/10.1144/M54.7

Schreurs, G. and Immenhauser, A. 1999. West-northwest directed obduction of the Batain Group on the eastern Oman continental margin at the Cretaceous-Tertiary boundary. *Tectonics*, **18**, 148–160, https://doi.org/10.1029/1998TC900020

Searle, M.P. 2007. Structural geometry, style and timing of deformation in the Hawasina Window, Al Jabal al Akhdar and Saih Hatat culminations, Oman Mountains. *GeoArabia*, **12**, 99–130.

Searle, M.P. 2019. *Geology of the Oman Mountains, Eastern Arabia*. Springer.

Searle, M.P. and Cox, J.S. 1999. Tectonic setting, origin and obduction of the Oman ophiolite. *Geological Society of America Bulletin*, **111**, 104–122, https://doi.org/10.1130/0016-7606(1999)111<0104:TSOAOO>2.3.CO;2

Searle, M.P. and Cox, J.S. 2002. Subduction zone metamorphism during formation and emplacement of the Semail Ophiolite in the Oman Mountains. *Geological Magazine*, **139**, 241–255, https://doi.org/10.1017/S0016756802006532

Searle, M.P. and Malpas, J. 1980. Structure and metamorphism of rocks beneath the Semail ophiolite of Oman and their significance in ophiolite obduction. *Earth and Environmental Science, Transactions of the Royal Society of Edinburgh*, **71**, 247–262, https://doi.org/10.1017/S0263593300013614

Searle, M.P., Waters, D.J., Martin, H.N. and Rex, D.C. 1994. Structure and metamorphism of blueschist – eclogite facies rocks from the northeastern Oman Mountains. *Journal of Geological Society, London*, **151**, 555–576, https://doi.org/10.1144/gsjgs.151.3.0555

Searle, M.P., Warren, C.J., Waters, D.J. and Parrish, R.R. 2004. Structural evolution, metamorphism and restoration of the Arabian continental margin, Saih Hatat region, Oman Mountains. *Journal of Structural Geology*, **26**, 451–473, https://doi.org/10.1016/j.jsg.2003.08.005

Sella, G.F., Dixon, T.H. and Mao, A. 2002. Revel: a model for Recent plate velocities from space geodesy. *Journal of Geophysical Research*, **107**, 2018, https://doi.org/10.1029/2000JB000033

Shackleton, R.M. and Ries, A.C. 1990. Tectonics of the Masirah fault zone and eastern Oman. *Geological Society, London, Special Publications*, **49**, 715–724.

Shackleton, R.M., Ries, A.C., Bird, P.R., Filbrandt, J.B., Lee, C.W. and Cunningham, C.C. 1990. The Batain mélange of NE Oman. *Geological Society, London, Special Publications*, **49**, 673–696.

Sharland, P.R., Casey, D.M., Davies, R.B., Simmons, M.D. and Sutcliffe, O.E. 2004. Arabian Plate Sequence Stratigraphy – revisions to SP2. *GeoArabia*, **9**, 199–214.

Skelton, P.W., Nolan, S.C. and Scott, R.W. 1990. The Maastrichtian transgression onto the northwestern flank of the Proto-Oman Mountains: sequences of rudist-bearing beach to open shelf facies. *Geological Society, London, Special Publications*, **49**, 521–547.

Smewing, J.D., Abbotts, I.L., Dunne, L.A. and Rex, D.C. 1991. Formation and emplacement ages of the Masirah ophiolite, Sultanate of Oman. *Geology*, **19**, 453–456, https://doi.org/10.1130/0091-7613(1991)019<0453:FAEAOT>2.3.CO;2

Soret, M., Agard, P., Dubacq, B., Plunder, A. and Yamato, P. 2017. Petrological evidence for stepwise accretion of metamorphic soles during subduction infancy (Semail ophiolite, Oman and UAE). *Journal of Metamorphic Geology*, **35**, 1051–1080, https://doi.org/10.1111/jmg.12267

Stampfli, G.M. and Borel, G.D. 2002. A plate tectonic model for the Paleozoic and Mesozoic constrained by dynamic plate boundaries and restored synthetic oceanic isochrons. *Earth and Planetary Science Letters*, **196**, 17–33, https://doi.org/10.1016/S0012-821X(01)00588-X

Stampfli, G.M. and Kozur, H.W. 2006. Europe from the Variscan to the Alpine cycles. *Geological Society, London, Memoirs*, **32**, 57–82, https://doi.org/10.1144/GSL.MEM.2006.032.01.04

Stern, R.J. 1994. Arc assembly and continental collision in the Neoproterozoic East African Orogen: implications for the consolidation of Gondwanaland. *Annual Reviews Earth and Planetary Science*, **22**, 319–351, https://doi.org/10.1146/annurev.ea.22.050194.001535

Svendsen, N.B. 2004. The Sahmah Formation of Oman: exploration implications for the Rub' Al-Khali Basin. *GeoArabia*, **9**, 119–136.

van Buchem, F.S.P., Razin, P., Homewood, P.W., Oterdoom, W.H. and Philip, J. 2002. Stratigraphic organization of carbonate ramps and organic-rich intrashelf basins: Natih Formation (middle Cretaceous) of northern Oman. *American Association of Petroleum Geologists Bulletin*, **86**, 21–53.

van Hinsbergen, D.J.J., Maffione, M., Koornneef, L.M.T. and Guilmette, C. 2019. Kinematic and paleomagnetic restoration of the Semail ophiolite (Oman) reveals subduction initiation along an ancient Neotethyan fracture zone. *Earth and Planetary Science Letters*, **518**, 183–196, https://doi.org/10.1016/j.epsl.2019.04.038

Vernant, P., Nilforoushan, F. *et al.* 2004. Present-day crustal deformation and plate kinematics in the Middle East constrained by GPS measurements in Iran and northern Oman. *Geophysical Journal International*, **157**, 381–398, https://doi.org/10.1111/j.1365-246X.2004.02222.x

Vigny, C., Huchon, P., Ruegg, J.-C., Khanbari, K. and Asfaw, L.M. 2006. Confirmation of Arabia plate slow motion by new GPS data in Yemen. *Journal of Geophysical Research*, **111**, B02402, https://doi.org/10.1029/2004JB003229

Villey, M., Béchennec, F., Beurrier, M., Le Métour, J. and Rabu, D. 1986*a*. *Geological map of Yanqul, sheet NF 40-2C, scale: 1:100,000, with Explanatory notes*. Directorate General of Minerals, Oman Ministry of Petroleum and Minerals.

Villey, M., de Gramont, X. and Le Métour, J. 1986*b*. *Geological map of Seeb, sheet NF 40-3C, scale: 1:100,000, with Explanatory notes*. Directorate General of Minerals, Oman Ministry of Petroleum and Minerals.

Villey, M., Le Métour, J. and de Gramont, X. 1986*c*. *Geological map of Fanjah, sheet NF 40-3F, scale: 1:100,000, with Explanatory notes*. Directorate General of Minerals, Oman Ministry of Petroleum and Minerals.

Walley, C.D. 1998. Some outstanding issues in the geology of Lebanon and their importance in the tectonic evolution of the Levantine region. *Tectonophysics*, **298**, 37–62, https://doi.org/10.1016/S0040-1951(98)00177-2

Warren, C.J., Parrish, R.R., Searle, M.P. and Waters, D.J. 2003. Dating the subduction of the Arabian continental margin beneath the Semail Ophiolite. *Geology*, **31**, 889–892, https://doi.org/10.1130/G19666.1

Warren, C.J., Parrish, R.R., Waters, D.J. and Searle, M.P. 2005. Dating the geologic history of Oman's Semail ophiolite: insight from U-Pb geochronology. *Contributions to Mineralogy and Petrology*, **150**, 403–422, https://doi.org/10.1007/s00410-005-0028-5

Wyns, R., Béchennec, F., Le Métour, J. and Roger, J. 1992. *Geological map of Tiwi, sheet NF40-8B, scale 1:100,000, with Explanatory notes*. Directorate General of Minerals, Oman Ministry of Petroleum and Minerals.

Chapter 2

Tectonostratigraphy of the eastern part of the Oman Mountains

Andreas Scharf[1]*, Frank Mattern[1], Mohammed Al-Wardi[1], Gianluca Frijia[2], Daniel Moraetis[3], Bernhard Pracejus[1], Wilfried Bauer[4] and Ivan Callegari[4]

[1]Department of Earth Sciences, College of Science, Sultan Qaboos University, PO Box 36, PC 123, Al-Khod, Muscat, Sultanate of Oman

[2]Department of Physics and Earth Sciences, University of Ferrara, Via Saragat 1, 44122, Ferrara, Italy

[3]Department of Applied Physics and Astronomy, College of Sciences, University of Sharjah, PO Box 27272, Sharjah, United Arab Emirates

[4]Department of Applied Geosciences, German University of Technology GUtech, PO Box 1816, PC 130, Halban, Sultanate of Oman

*Correspondence: scharfa@squ.edu.om

Abstract: This chapter provides comprehensive descriptions of 52 numbered formations/rock units of the Southeastern Oman Mountains, based on available literature. The oldest eight siliciclastic and carbonate formations are positioned below the 'Hercynian' Unconformity. The overlying formation (9–16) mostly represent carbonates which accumulated in a passive margin platform setting during or after the opening of the Neo-Tethys Ocean. The passive margin slope and platform collapsed during the late Cretaceous because of the obduction of the Semail Ophiolite along with the deep marine Hawasina sedimentary rocks. The collapsing passive margin interval was recorded within the syn-obductional Aruma Group (17; Muti Formation). Above this formation are the allochthonous units (18–42) of the tectonically lower Hawasina deep-sea basin and the structurally overlying Semail Ophiolite. The former contains Permian to Upper Cretaceous formations, while the latter is Cenomanian in age. Above the allochthonous rocks, the Neo-autochthonous formations were deposited, starting with the post-obductional uppermost Cretaceous Aruma Group (43; Al-Khod Formation) until the Quaternary deposits (52). All these formations/rock units are depicted on an accompanying map and stratigraphic chart.

Autochthonous

Units below the Hercynian Unconformity

The following formation numbers refer to the system in figures A.1 and A.2 (Scharf *et al.* 2021*a*, Chapter 7, this Memoir). Where formation names differ in the subsurface, their subsurface equivalents are included in brackets in figure A.2. The pre-mid-Carboniferous formations are collectively labelled as 'Autochthonous Unit A' (Beurrier *et al.* 1986; Le Métour 1988; Rabu 1988; Le Métour *et al.* 1990; Figs A.1 & A.2).

Formations 1–5 (figs A.1 & A.2) are exposed in the Jabal Akhdar Dome (JAD) and 6–7 in the Saih Hatat Dome (SHD). All five formations within the JAD are part of the Huqf Supergroup and were affected by the Neoproterozoic–Cambrian East African Orogeny, and therefore occur below the Angudan Unconformity.

The ages and equivalents of formations 6 and 7 are not yet clear, and the SHD may share the same geodynamic development. The Huqf Supergroup is divided into the Abu Mahara, Nafun I, Nafun II, Ara and Nimr groups (fig. A.2). The base of Nafun Group I is dated as *c.* 635 Ma and the top of Nafun Group II as *c.* 547 Ma (Table 2.1). These ages are well constrained in Oman. By contrast, the ages of the individual formations within the Nafun groups are not well constrained. The ages differ by a few or some million years, depending on author. Table 2.1 provides a synoptic overview of the different formation ages of the Nafun groups (Al-Husseini 2014). The age differences are due to a lack of geochronological data and rely on, for example, different estimates of sedimentation rates within the formations. For simplicity, we mostly follow the ages of Forbes *et al.* (2010), but highlight that the ages within the Ediacaran Nafun groups (fig. A.2) are uncertain (for further discussion, see Bowring *et al.* 2007, 2009; Le Guerroué *et al.* 2009; Al-Husseini 2014; Table 2.1).

Formation 8 (Amdeh) of the Haima Supergroup and formation 9 (Al Khlata) of the Haushi Group occur only in the SHD. The lithologies of the metasedimentary formations 1–9 in the core of the domes include metamorphic sediments, such as diamictite, greywacke, sandstone, siltstone, shale and limestone/dolostone, which are relatively susceptible to erosion compared with the surrounding younger carbonate formations 10–17. This has resulted in a present-day geomorphology of the domes as 'bowl(s)' in their central parts (units 1–9), surrounded by cliffs (units 10–17). From west to east, these are Sahtan, Bani Kharus, Mu'aydin, Ghubrah or Mistal, Hedak and Jabir bowls in the JAD, and the Hatat Bowl in the SHD (Fig. 2.1). Permanent springs occur at the base of carbonate formations 10–17 where they overlie impermeable shaly units in formations 1–9 and provide the basis for permanent settlements and local agriculture. The latter is cultivated along artificial terraces along a steep slope and with an ingenious water irrigation system (local name is 'aflaj') that has been designated part of the UNESCO World Heritage Site since 2006 (https://whc.unesco.org/en/list/1207). The terraces have been cultivated for thousands of years (e.g. Moraetis *et al.* 2020). The cover picture of this volume depicts the Mu'aydin Bowl with its water irrigation systems. The artificial terraces are within the Mistal Formation and below the 'Hercynian' Unconformity.

Mistal Formation (1). The oldest formation of the JAD is the Mistal Formation (1). A correlation of the Mistal Formation with formations of the Abu Mahara Group in the subsurface is still unclear (see discussion in Forbes *et al.* 2010). Parts of the Mistal Formation may correlate with the Ghadir Manqil Formation in the subsurface (Forbes *et al.* 2010). The Mistal Formation was defined in the Mistal Bowl (Kapp and Llewellyn 1965; Allen 2007; Fig. 2.1) where it measures ≤1000 m in thickness (Rabu *et al.* 1986). The Mistal Formation is even thicker in the Sahtan Bowl, reaching 1500 m (Beurrier *et al.* 1986). The Mistal Formation is divided into four members (Ghubrah, Saqlah, Fiq and Amq; fig. A.2). The subsurface equivalent of the Amq Member are the Masirah Bay and

From: Scharf, A., Mattern, F., Al-Wardi, M., Frijia, G., Moraetis, D., Pracejus, B., Bauer, W. and Cellegari, I. 2021. *The Geology and Tectonics of the Jabal Akhdar and Saih Hatat Domes, Oman Mountains*. Geological Society, London, Memoirs, **54**, 11–47, https://doi.org/10.1144/M54.2

Table 2.1. *Correlation of formations of the Nafun Group and age and thickness data*

Surface and formation: Subsurface and Huqf area	Surface and formation: Jabal Akhdar	Huqf type section thickness (m)	Jabal Akhdar thickness (m)	Miqrat-1 (Fig. 2.1) thickness (Forbes *et al.* 2010) (m)	Masirah-1 (Fig. 2.1) thickness‡ (m)	Bowring *et al.* (2007); Forbes *et al.* (2010) age (Ma)	Le Guerroué *et al.* (2006*a*, *b*) age (Ma)	Al-Husseini (2014) calculated (approximate) age (Ma)	Used in fig. A.2
Top Nafun Group						547	547	547.0 (547)	547
Buah Formation	Kharus Formation	0–310*	150†	250	360			+13.7	
Base Buah Formation	Base Kharus Formation					551.5–552.5	575	560.7 (561)	555§
Shuram Formation	Mu'aydin Formation	230–330*	800–900‡	350	560			+21.4	
Base Shuram Formation	Base Mu'aydin Formation					554–562	600	582.1 (582)	590
Sub-Shuram Unconformity						562–590		+0.0	
Top Khufai	Top Hajir					590	600	582.1 (582)	590
Khufai Formation	Hajir Formation	240–340*	100†	325	250			+9.5	
Base Khufai Formation	Base Hajir Formation					615	610	591.6 (592)	615
Masirah Bay Formation	Amq Member	>250 (5)	170–200‖	111	1138			+43.5	
Top Hadash	Top Amq Member					634.5	634.5	N/A	
Hadash Formation	Amq Member	15	15	8	0			N/A	
Base Nafun Group						635	635	635.0 (635)	635

Modified from Al-Husseini (2014). *Gorin *et al.* (1982); †Glennie *et al.* (1974); ‡Rabu *et al.* (1986); §Cozzi *et al.* (2004); ‖Allen and Leather (2006). Migrat-1 and Masirah-1 are wells south of the Oman Mountains. Their locations are depicted in figure 1.2 (Chapter 1, this Volume).

Hadash formations (fig. A.2). The overall lithologies of the members are sandstone, greywacke, mafic lava, feldspathic wacke, diamictite and dolostone (Beurrier *et al.* 1986; Rabu *et al.* 1986; Villey *et al.* 1986*c*). The assemblage of glaciogenic sedimentary rocks (i.e. diamictite with dropstones) within the Ghubrah and the Fiq members as well as the overlying post-glaciogenic cap carbonates of the Amq Member suggest two cycles of low-latitude Neoproterozoic glaciations (Brasier *et al.* 2000; Leather *et al.* 2002; Kellerhals and Matter 2003; Kilner *et al.* 2005; Le Guerroué *et al.* 2005; Allen *et al.* 2011). The Ghubrah Member has an age of 723 + 16–10 Ma, 713.7 ± 0.5 Ma to 711.8 ± 1.6 Ma (U–Pb zircon age derived from a tuffaceous bed near the base of the formation; Brasier *et al.* 2000; Bowring *et al.* 2007; Allen *et al.* 2011), whereas the top of the Fiq Member has an inferred age of <645–635 Ma (Allen and Leather 2006; Le Guerroué *et al.* 2006*a*; Bowring *et al.* 2007; Rieu *et al.* 2007; Allen *et al.* 2011). These ages are based on a correlation of dated ash from the globally occurring cap carbonates with indicative negative $\delta^{13}C$ values from China, Namibia and Canada. The Ghubrah Member could represent glaciogenic sediments of the Sturtian Glaciation, whereas the Fiq Member represents those of the Marinoan Glaciation (Allen *et al.* 2004). The Saqlah Member occurs at the base of the Fiq Member and marks a hiatus between the older Ghubrah Member. The Saqlah Member consists of submarine pillow basalt, volcanic ash and volcaniclastic rocks (e.g. Allen *et al.* 2011). The thickness of this member measures up to 60 m in the eastern JAD, while it is absent from the western JAD (Beurrier *et al.* 1986; Rabu *et al.* 1986). Detailed studies of the Fiq Member have been carried out by Allen *et al.* (2004, 2011), showing its glaciogenic character. The slates of the Amq Member were deposited in a deep marine basin, whereas synchronous rocks in the Huqf area are coarse siliciclastic coastal deposits (Allen and Leather 2006). Unlike the other three members of the Mistal Formation, which all belong to the Abu Mahara Group, the Amq Member belongs to the Nafun I Group (fig. A.2).

The age of the Sturtian Glaciation has been indicated as *c.* 715–680 Ma (Stern *et al.* 2006). The beginning of the Marinoan Glaciation is constrained between 660 and 650 Ma (Stern *et al.* 2006) and its end at *c.* 635 Ma (635.5 ± 1.2 Ma, ≤635.2 ± 0.6 Ma and 636.41 ± 0.45 Ma according to Hoffmann *et al.* 2004, Condon *et al.* 2005 and Calver *et al.* 2013, respectively). No evidence for the Gaskiers Glaciation (*c.* 585–582 Ma) has been reported from Oman (Stern *et al.* 2006).

Hajir Formation (2). The Hajir Formation (2) or Khufai Formation (subsurface name) of the JAD was first described by Kapp and Llewellyn (1965; Allen 2007) and Glennie *et al.* (1974). The basal age of this formation has been suggested to be 615 Ma (Bowring *et al.* 2007; Forbes *et al.* 2010), 610 Ma (Le Guerroué *et al.* 2006*a*, *b*; Al-Husseini 2014, table 1) or 592 Ma (Al-Husseini 2014). The age of the top of the Hajir Formation is 601 ± 8 Ma, based on interpolation of the subsidence pathway (Le Guerroué 2006, fig. 4.4). Bowring *et al.* (2007) and Forbes *et al.* (2010) indicated an age of 590 Ma for the top of this formation and Al-Husseini (2014) suggested an age of 582 Ma (Table 2.1). The thickness of the formation varies from 100 m in the type locality within the Sahtan Bowl to 0 m where it vanishes towards the NW and west of the JAD (Beurrier *et al.* 1986; Rabu *et al.* 1986). This reduction of thickness is attributed to truncation at the Hercynian Unconformity. The sulfurous, thin to medium-bedded, black dolomitic limestone, is locally rich in organic matter, and interpreted to have been deposited in a restricted lagoon (Beurrier *et al.* 1986; Rabu *et al.* 1986) or to have formed in the distal environment of a prograding

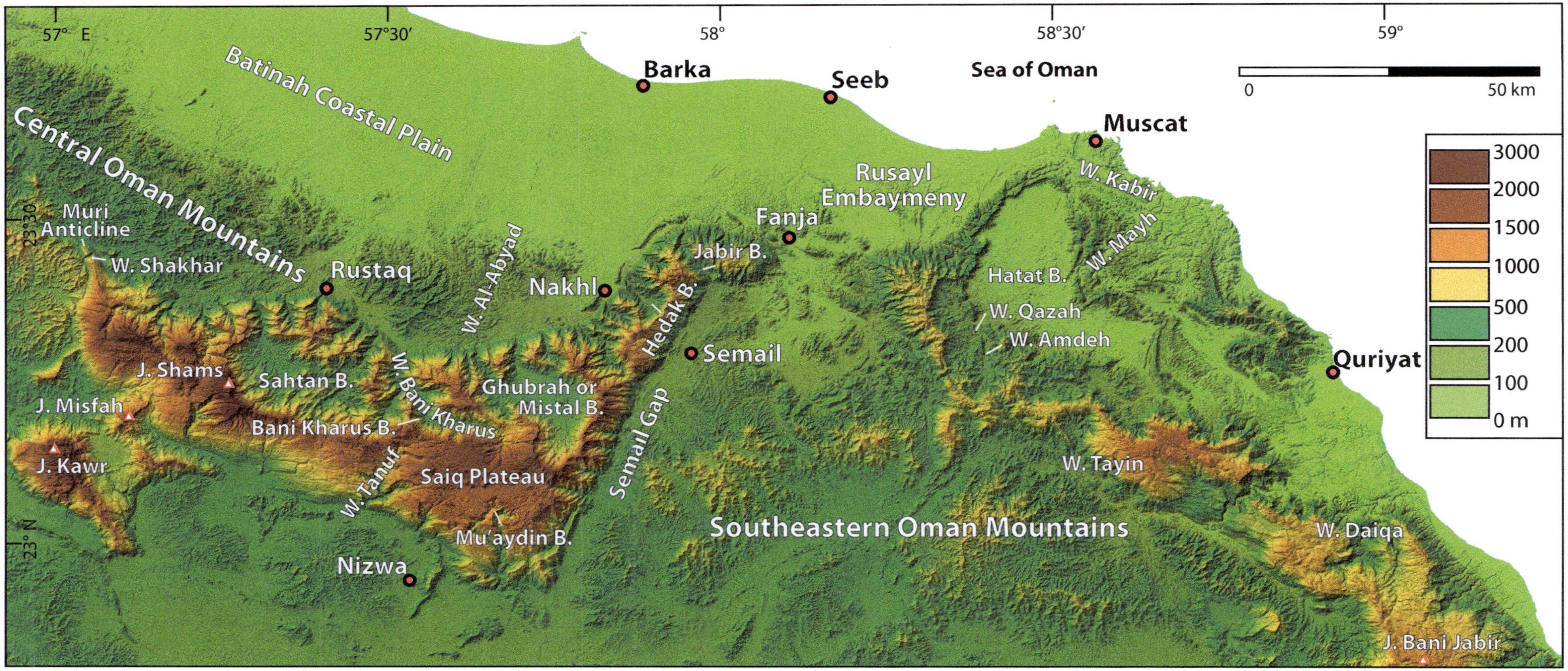

Fig. 2.1. Digital elevation model (DEM) of the study area, showing the main localities discussed in the text and geomorphological features. B, Bowl; J, Jabal; W, Wadi. Map is based on data from Advances Spaceborne Thermal Emission and Reflection Radiometer (ASTER) global DEM (GDEM) (product of METI and National Aeronautics and Space Administration), compiled with SimpleDEMViewer for Mac (version 4.4.9). The Semail Gap divides the Oman Mountains into the Central and Southeastern Oman Mountains. It is an important historical trade route, linking the Batinah Coastal Plain to the villages and towns of Oman's interior (Glennie *et al.* 1974).

ramp (Gorin *et al.* 1982; Wright *et al.* 1990; McCarron 2000; Le Guerroué *et al.* 2006*a*, Osburn *et al.* 2014). The type of contact with the overlaying Mu'aydin Formation is currently unresolved. Sedimentation along this contact in the JAD and Huqf area is either continuous (e.g. Le Guerroué *et al.* 2009) or discontinuous, marked by a major unconformity (McCarron 2000; Bowring *et al.* 2007, 2009; Forbes *et al.* 2010). McCarron (2000; Bowring *et al.* 2009) located a major sequence boundary at this contact. According to Forbes *et al.* (2010), the Khufai–Shuram transition is unconformable in the subsurface. In the JAD this gradational contact is marked by a flooding surface (Osburn *et al.* 2014). Recently, Mattern and Scharf (2019) recognized synsedimentary extensional faults at the Hajir–Mu'aydin boundary in the central JAD. This syntectonic faulting may have been related to extensional movements along a major fault zone, a precursor of the Semail Gap (Mattern and Scharf 2019). We depict the version of Le Guerroué (2006) with continuous sedimentation across the Hajir–Mu'aydin contact in figure A.2, but note that this interpretation is under debate.

Mu'aydin Formation (3). The Mu'aydin Formation (3) or Shuram Formation (subsurface name) overlies the Hajir Formation and in part the Mistal Formation (e.g. in the Mu'aydin Bowl; fig. A.1). As reported by Kapp and Llewellyn (1965; Allen 2007) and Glennie *et al.* (1974) this formation is well-exposed in the western part of the JAD but disappears towards the NE. The 800–900-m-thick formation is a monotonous unit of green and reddish siltstones that can be divided into three members at the 1:250 000 map scale (Rabu *et al.* 1986). The lower member contains a 2–3-m-thick basal dark-coloured quartz sandstone layer, overlain by 40–50 m of alternating thick russet sandstone and mauve siltstone (Rabu *et al.* 1986). The middle member consists of monotonous greenish to mauve micaceous siltstone (Rabu *et al.* 1986). The upper member is represented by micaceous siltstone and interbedded white limestone, forming a transition to the younger Kharus Formation (Rabu *et al.* 1986). According to Rabu *et al.* (1986), the overall sedimentary palaeoenvironment of the Mu'aydin Formation was coastal. Le Guerroué *et al.* (2006*a*) interpreted the Mu'aydin Formation of the JAD as a distal siliciclastic shelf deposit.

The age for the base of the Mu'aydin Formation is highly debated; a possible hiatus between the Hajir and Mu'aydin formations may therefore exist (see above and Bowring *et al.* 2007, 2009; Le Guerroué *et al.* 2009; Forbes *et al.* 2010). The base of the Mu'aydin Formation is probably younger than 601 ± 8 Ma based on interpolation of the subsidence pathway (Le Guerroué 2006) or 562–554 Ma (Bowring *et al.* 2007). Additionally, Al-Husseini (2014) provided an age of 582 Ma for the base of the Mu'aydin Formation (Table 2.1).

The age of the top of the Mu'aydin Formation is also not well constrained. The age may range from 574 ± 7 Ma (Le Guerroué 2006) or 561 Ma (Al-Husseini 2014) to 562–554 Ma (Bowring *et al.* 2007). We chose 555 Ma for figure A.2 because, according to Forbes *et al.* (2010), the range of 562–554 Ma (Bowring *et al.* 2007) is in line with the chronostratigraphic Neoprotrozoic findings. Moreover, Cozzi *et al.* (2004) presented an age for the base of the Buah Formation (Kharus) of 555 Ma (see below: Kharus Formation).

The Mu'aydin Formation is characterized by a dramatic negative carbon isotope shift in δ^{13}C values from +5‰ (or +4‰, depending on author) to −12‰ ('Shuram Excursion' or 'Shuram/Wonoka Excursion'; Burns and Matter 1993; Le Guerroué *et al.* 2006*a*, *b*; Bowring *et al.* 2007; Le Guerroué and Cozzi 2009; Le Guerroué 2010). The negative anomaly of the δ^{13}C values has been reported for different Ediacaran formations around the world (Australia, China, Oman, Scandinavia, Siberia; e.g. Le Guerroué 2010). The age of this global event in Oman is not precisely known because of the lack of radiometric data (e.g. Bowring *et al.* 2007). A possible age of this event was provided by Le Guerroué (2010). This author used the strontium isotope ratio as a reliable geochronometer, and deduced that the excursion in Oman started at *c.*600 Ma and ended at *c.*550 Ma. In Oman, the Shuram Excursion is preserved within a *c.* 800 m thick stratigraphic section (Le Guerroué *et al.* 2006*a*).

The cause of the shift is unclear. It could be related to progressive oxidation of the ocean during the Ediacaran (Rothman *et al.* 2003; Condon *et al.* 2005; Fike *et al.* 2006; Bowring *et al.* 2007; Grotzinger *et al.* 2011). An alternative explanation of the negative δ^{13}C values could be diagenetic alteration following deposition of the carbonates (Grotzinger *et al.* 2011).

Kharus Formation (4). The Kharus Formation (4) or Buah Formation (subsurface name) is only exposed in the Sahtan Bowl and in Wadi Bani Kharus (JAD; Fig. 2.1) with a thickness of *c.* 245 m (Kapp and Llewellyn 1965; Glennie *et al.* 1974; Rabu *et al.* 1986; Béchennec *et al.* 1992*a*; Allen 2007). The Kharus Formation has not yet been precisely dated. However, U–Pb zircon ages of volcaniclastic rocks, dating the overlying Fara Formation in the JAD, prove that the top of the Kharus Formation must be of age >547–544 Ma (Bowring *et al.* 2007). Cozzi *et al.* (2004) constrained the age of the top of the Buah Formation (Kharus) in the Huqf area as 550 Ma. The assumed accumulation rate of this formation is 0.05 m ka^{-1} (Cozzi *et al.* 2004, based on typical accumulation rates of carbonate ramp settings with slow tectonic subsidence *sensu* Burchette and Wright 1992). It would have required *c.* 4–5 Ma to accumulate the 200–250-m-thick formation (Cozzi et al. 2004); the base of the formation may therefore have an age of 555 Ma (Cozzi *et al.* 2004).

Bowring *et al.* (2007; Forbes *et al.* 2010) provided a speculative age for the base of this formation in the subsurface as 551.5–552.5 Ma. Slightly older ages from the surface and subsurface were concluded by Le Guerroué *et al.* (2006*b*, table 1) and Al-Husseini (2014) of 574 ± 7 Ma and 561 Ma, respectively (Table 2.1). Geochronological data from the subsurface suggest a top Buah age of 547 Ma (Amthor *et al.* 2003; Bowring *et al.* 2007; Forbes *et al.* 2010).

The Kharus Formation consists of dark organic-rich limestone and dolomite, possibly sourcing much of Oman's oil (Cozzi and Al-Siyabi 2004; Droste 2014). It also contains silicified stromatolites. Its depositional setting is interpreted as an intertidal to supratidal carbonate platform (Béchennec *et al.* 1992*a*) or as a distally steepened storm-dominated carbonate ramp (Cozzi *et al.* 2004).

Fara Formation (5). The 380-m-thick Fara Formation (5) of the Ara Group (subsurface name) has been divided into three members (Beurrier *et al.* 1986). The type locality of the Fara Formation is in Wadi Bani Awf, east of Rustaq (Fig. 2.1).

Sedimentary rocks of the Ara Group occur mainly in the South Oman Salt Basin, Ghaba Salt Basin and Fahud Salt Basin in the subsurface of interior Oman, south of the thrust front of the Hawasina nappes (e.g. Forbes *et al.* 2010). Some salt domes in the Ghaba Salt Basin are indicative of active halokinesis (Peters *et al.* 2003). Evaporites of the Ghaba Salt Basin may extend north of the thrust front of the Hawasina nappes below the Southeastern Oman Mountains, but still south of the SHD (Mount *et al.* 1998; Heward and Penney 2014, fig. 1). No sedimentary rocks of the Ara Group have been mapped or reported from the SHD. Recent studies confirm that undated anhydrite and gypsum bodies exist within the Hawasina Window and the Northern Oman Mountains at Jabal Qumayrah and Jabal Sumeini near the Oman–UAE border (Csontos *et al.* 2010; Cooper *et al.* 2012, 2013, 2018). The undated evaporite bodies possibly belong to the Ara Group, based on structural considerations, gravity profiles and $^{87}Sr/^{86}Sr$ ratio analyses (Csontos *et al.* 2010; Cooper *et al.* 2012, 2013, 2018; Ali *et al.* 2014). These evaporites may originally represent the Fahud Salt Basin (Cooper *et al.* 2013; Ali *et al.* 2014). According to Ali *et al.* (2014, fig. 17) the evaporites probably migrated towards the surface along deep-rooted basement faults driven by density inversion during the Oligocene. These deep faults may be inverted rift-related normal faults (Ali *et al.* 2014).

The Fara Formation in the JAD is partly time-equivalent to the Ara Group (e.g. Forbes *et al.* 2010). An ignimbrite level within the Fara Formation from the JAD, 200 m above the base of the formation, is 544.5 ± 3.3 Ma in age (U–Pb on zircon; Brasier *et al.* 2000). Bowring *et al.* (2007) produced two additional U–Pb zircon ages for the Fara Formation The two weighted mean ages of 547.23 ± 0.28 Ma and 542.54 ± 0.45 Ma are from the base and top, respectively.

U–Pb zircon dating of tuffites from the subsurface of the lower and middle parts of the Ara Group from the South Oman Salt Basin revealed ages of 546.72 ± 0.21 Ma, 548.9 ± 0.98 Ma (Bowring *et al.* 2007) and 542.6 ± 0.3 (Amthor *et al.* 2003), respectively. The overall age of the Ara Group is *c.* 547–538 Ma (Forbes *et al.* 2010).

The Fara Formation overlies the Kharus Formation in apparent conformity (Béchennec *et al.* 1992*a*) and is itself unconformably overlain by the Saiq Formation. The lower 60-m-thick member consists of black chert interbedded with grey dolomite (Béchennec *et al.* 1992*a*). The top part of the lower member contains metre-thick layers of banded silicic tuff (Béchennec *et al.* 1992*a*). The 200-m-thick middle member contains conglomerate, siltstone, breccia, grit, sandstone, sandy limestone, stromatolites and abundant phosphate grains (Beurrier *et al.* 1986). The 120-m-thick upper member is highly diverse in its lithology. It contains quartzose siltstone, grey-green argillaceous siltstone, green silicic tuffite, greywacke, sandstone and carbonate-cemented conglomerate (Béchennec *et al.* 1992*a*). The change from the stromatolite-bearing limestone deposits of the Kharus Formation to the chert deposits of the Fara Formation reflects a basin deepening (Allen 2007; Bowring *et al.* 2007).

In contrast to the Fara Formation, the partly time-equivalent Ediacaran–Early Cambrian subsurface Ara Group consists of six cyclic evaporite-carbonate sequences with shale, resulting in a high potential for hydrocarbon seals (e.g. Mattes and Conway Morris 1990; Amthor *et al.* 2003; Schröder *et al.* 2003; Forbes *et al.* 2010). The Ara Group was deposited in tectonically active basins under stable arid conditions (Schröder *et al.* 2003). The carbonates and evaporites accumulated during relatively high and low sea-levels, respectively (Mattes and Conway Morris 1990).

Hatat Formation (6). The Hatat Formation (6) is the oldest formation in the SHD and may be the time equivalent of the Mistal Formation of the JAD (e.g. Le Métour *et al.* 1986*b*; fig. A.2). No age data of the Hatat Formation are currently available.

The Hatat Formation has been divided into two members at the 1:250 000 map scale (Béchennec *et al.* 1992*a*). The lower member consists of banded schist, metagreywacke, other metaclastics, greenstone, and some pillowed metabasalt with a total thickness of 2500 m (Béchennec *et al.* 1992*a*). The upper member consists of banded schist, metagreywacke and 80–100-m-thick brown dolomite. Glaciogenic sediments in the upper part of the Hatat Formation have been correlated with those of the JAD (Le Métour *et al.* 1986*b*, fig. 11). However, it is not clear to which glaciation they belong (Sturtian or Marinoan). Because of the reduced thickness of the upper member, we do not subdivide the Hatat Formation (fig. A.1). Besides the abovementioned subdivision, the Hatat Formation has been also subdivided into four members (Saadi-, Jahlut-, Amarat- and Sarin members; Le Métour *et al.* 1986*b*) at the 1:100 000 map scale. The Sarin Member contains diamictites. The Saadi and Jahlut members on one side and the Amarat and Sarin members on the other side correspond to the lower and upper members of the Hatat Formation, respectively (fig. A.1).

Hiyam Formation (7). The Hiyam Formation (7) of the SHD may be the equivalent of the Kharus Formation. It consists of some grey metaclastics, common metalimestone and also metadolostone with oncoids, pisolites and stromatolites. Le Métour *et al.* (1986*b*) subdivided the Hiyam Formation into six units at the map scale of 1:100 000. The first five units are grouped together as the Hi_1 Member and the highest unit as the Hi_2 Member (Béchennec *et al.* 1992*a*). We do not distinguish between the two members in our geological map (fig. A.1). The depositional environment has been interpreted as mainly intertidal to supratidal (Le Métour *et al.* 1986*b*). The formation is exposed along the western, southern and eastern margin of the Hatat Bowl (Kapp and Llewellyn 1965; Glennie *et al.* 1974; Le Métour *et al.* 1986*b*; Béchennec *et al.* 1992*a*). The thickness of the formation is *c.* 400 m (Béchennec *et al.* 1992*a*).

Former correlations considered the Hajir Formation (2) an equivalent of the Hiyam Formation (7) (e.g. Le Métour *et al.* 1986*a*, *b*; Villey *et al.* 1986*b*, *c*). However, Mattes and Conway Morris (1990) and Mann and Hanna (1990, fig. 3), with inconclusive reference to Lovelock *et al.* (1981), proposed that the Kharus Formation is the equivalent of the Hiyam Formation. We point out that both formations (Kharus and Hiyam) are the youngest Neoproterozoic formations in the JAD and SHD, show an overall carbonate composition, contain stromatolites which can be selectively silicified, and display dolomitization and silicification at their tops. Based on these lithological similarities, we consider the Hiyam Formation the equivalent of the Kharus Formation (fig. A.2).

Amdeh Formation (8). The Amdeh Formation (8) overlies the Hiyam Formation at a tectonized unconformity (Lovelock *et al.* 1981). This unconformity represents the Angudan Unconformity. The Amdeh Formation was first described by Lees (1928, ‘Wadi Amdeh Quartzites’), Kapp and Llewellyn (1965), Glennie *et al.* (1974), Lovelock *et al.* (1981) and Le Métour *et al.* (1986*b*). According to Lovelock *et al.* (1981, fig. 2), this formation is divided into five members with a total thickness of 3400 m ± 10% (from the bottom to top): the 240-m-thick Lower Siltstone Member (Am1); the 253-m-thick Lower Quartzite Member (Am2); the 445-m-thick Middle Shale Member (Am3); the 1677-m-thick Upper Quartzite Member (Am4); and the 805-m-thick Upper Siltstone Member (Am5). The Amdeh members Am1, Am3 and Am5 consist mostly of grey, dark-coloured and green to mauvish siltstone and shale/slate layers. The members are interbedded with weakly metamorphosed, rippled and cross-bedded sandstone (Am2, Am4), whereas the three shaly members (Am1, Am3 and Am5) geomorphologically recede the two quartzite members from ridges and hills (Le Métour *et al.* 1986*b*).

While the lower two members do not seem to contain any fossils or trace fossils, the upper three members bear bivalves, brachiopods, trilobites, acritarchs and ichnofacies such as

Fig. 2.2. Angular unconformity between the Neoproterozoic Kharus Formation and the Permian Saiq Formation, signifying pre-Permian tilting of Neoproterozoic rocks that were folded during the East African or 'Hercynian' Orogeny. The unconformity is located *c.* 40 m above the Wadi level. Wadi Bani Kharus, JAD.

Cruziana, *Skolithos* and *Daedalus* (Lovelock *et al.* 1981; Le Métour *et al.* 1986*b*) as well as *Sacabambaspis*, a jawless fish, chitinozoa (Sansom *et al.* 2009), ostracods as well as conodonts (Heward *et al.* 2018*b*). Rickards *et al.* (2010, fig. 2), Heward *et al.* (2018*b*, fig. 2) and Miller *et al.* (2018, fig. 2) correlated the five Amdeh members with clastic rocks of the Haima Supergroup and the Nimr Group in the subsurface of the interior Oman, deriving an age spanning the Cambrian and the Darriwilian (fig. A.2).

Detailed investigations of the Middle Shale Member (Am3) in Wadi Qazha (southwestern SHD) were performed by Heward *et al.* (2018*b*). The Am3 Member is subdivided into a lower shaly unit of 310 m thickness and a upper sandy unit of 232.5 m thickness (Heward *et al.* 2018*b*). The equivalents of the lower and upper units of the Ghaba Salt Basin subsurface may be the Mabrouk and Barakat formations, respectively (Heward *et al.* 2018*b*; fig. A.2). Based on sedimentological investigations, the Am3 Member is thicker (542.5 m, Heward *et al.* 2018*b*) than previous thought (445 m, Lovelock *et al.* 1981). Heward *et al.* (2018*b*) allocated a Tremadocian–Floian age (Early Ordovician) for the 542.5-m-thick Am3 Member. The conodont assemblage is assigned to the Tremadocian (*Tetraprioniodus*, *Drepanoistodus*, *Drepanodus*, *Scolopodus*, *?Tropodus*, *Semiacontiodus* and *Teridontus*). The Floian age is concluded based on the taxa Balognathidaegen. et sp. indet. A and B and aff. *Erraticodon* (Heward *et al.* 2018*b*).

The amount and concentration of quartz in this thick formation is impressive. The original sandstones are texturally and compositionally mature (well sorted, more than 90% quartz). Mattern *et al.* (2018*a*) studied two heavy mineral placers in the Amdeh Member 4 from Wadi Qazah (western SHD;

Fig. 2.1), mainly qualitatively by X-ray diffraction (XRD), X-ray fluorescence (XRF), and reflected and transmitted light microscopy. The dominant heavy mineral is anatase, followed by ilmenite and zircon. Taking the quartz-rich nature of the member into account, these heavy minerals point to acidic plutonic and volcanic rocks, different pegmatites, and mica schists and gneisses as possible sources, but mature quartz-rich sandstone could also be considered.

Heward *et al.* (2018*a*) and Miller *et al.* (2018) studied the 690-m-thick Amdeh Member 5 in an eastern inlier of the SHD. This member consists of quartzitic sandstone with shales and bivalve-rich shell beds. The authors observed trace fossils such as *Cruziana* and *Skolithos*, fragments of the arandaspid fish *Sacabambaspis*, ossicles and moulds of the early disparid crinoid *Iocrinus*, two new genera of conodonts (*Aldridgeognathus manniki* and *Omanognathus daiqaensis*) and the rare trinucleid trilobite *Yinpanolithus*. The constant shallow-marine facies on the one hand and the thickness of this member on the other suggests a stable system of subsidence and sediment supply during the Middle Ordovician.

Formations 1–7 have been deformed by the East African Orogeny during the Early Cambrian. The resulting Angudan Unconformity can be studied between the Neoproterozoic Hiyam and the Paleozoic Amdeh formations in the SHD (Wadi Amdeh, northwestern SHD). Where Proterozoic successions are overlain by the Permian Saiq Formation, the East African Orogeny, the 'Hercynian' event as well as tilting and block faulting during rifting in the course of the Pangea Rifting, may have contributed to the respective angular unconformity. This unconformity is spectacularly exposed in the JAD (Fig. 2.2). Searle and Alsop (2007) showed that in the SHD the unconformity has been inverted by mega sheath folds (section 'F4 folds' in Scharf *et al.* 2021*b*, Chapter 3, this Memoir). Gray *et al.* (2000) and Grobe *et al.* (2018) refer to an overprinting of the original unconformity by later tectonic events (from Late Cretaceous subduction and/or exhumation-related deformation to Cenozoic doming) that affected the SHD. The northeasternmost parts of the Amdeh Formation, together with the formations of the Hajar Supergroup, underwent high-pressure/low-temperature metamorphism during the Late Cretaceous convergence between Arabian and Eurasia (Searle *et al.* 1994, 2004; section 'F3 folds' in Chapter 3).

Formations above the Hercynian Unconformity

The Al Khlata Formation, the Permo-Mesozoic of the Arabian Platform (Hajar Supergroup) and the fill of the Aruma Foreland Basin (Aruma Group). The formations above the Hercynian Unconformity are subdivided into the Haushi, Akhdar, Sahtan, Kahmah, Wasia and Aruma groups (fig. A.2). The Permo-Mesozoic groups (10–17) are part of the Hajar Supergroup (Hanna 1990; fig. A.2) and known as 'Autochthonous Unit B' (Beurrier *et al.* 1986; Le Métour 1988; Rabu 1988; Le Métour *et al.* 1990). We also include the Al Khlata Formation in Autochthonous Unit B (figs A.1 & A.2).

Haushi Group. The base of the Haushi Group *sensu* Hughes Clarke (1988; Forbes *et al.* 2010) represents the earliest sediments above the Hercynian Unconformity. The base of this group is characterized by the glacio-lacustrine Al Khlata Formation (9) in the SHD and the Huqf area (e.g. Heward and Penney 2014), and the upper part is represented by the postglacial Sakmarian–Wordian (Permian) Gharif Formation in the Huqf area (Martin *et al.* 2008). The Gharif Formation may be present in the SHD, but has not yet been recognized there (see Heward and Penney 2014).

The depositional sequences of the Al Khlata Formation have been re-evaluated by Osterloff *et al.* (2004*a*, *b*). A subdivision in production units based on palynological calibration (P9, P5, P1) has been established (Penney *et al.* 2008; Forbes *et al.* 2010). The Al Khlata Formation occurs extensively in the subsurface of southern, central and northern Oman. A thin equivalent, comprising the younger P5 and P1 units, also crops out in the Huqf area (Osterloff *et al.* 2004*b*; Forbes *et al.* 2010). The Al Khlata Formation, particularly the oldest and sandy P9 unit, contains important oil reservoirs in the subsurface (Levell *et al.* 1988; Heward 1990; Al-Abri *et al.* 2018). Further studies were carried out by Al-Belushi *et al.* (1996), Hartmann *et al.* (2000), Penney *et al.* (2008), Stephenson *et al.* (2008), Martin *et al.* (2012) and Al-Abri *et al.* (2018).

The main source of information on the Al Khlata Formation in the SHD is Heward and Penney (2014), and the following is a summary of their work. The Al Khlata Formation was thought to be thin and absent from northern Oman (e.g. Levell *et al.* 1988). However, the recovery of Late Carboniferous spores from a shale and the recognition of diamictite with granitic clasts in exposures in Wadi Daiqa (Fig. 2.1) demonstrated the presence of the formation in the eastern part of the SHD. The glaciogenic deposits of the Al Khlata Formation occur in our study area within a *c.* 250 m-thick succession in Wadi Daiqa (figs. A.1 & Fig. 2.1 southern SHD). Further possible outcrops are located in Wadi Al-Arabiyin (Hayl Al-Quwasim inlier of Heward and Penney 2014) with a thickness of 30 m and 20 km to the NNW of Wadi Daiqa (Wadi Mijlas; Heward and Penney 2014), as a *c.* 50 m thick clastic layer. A 175 m × 145 m patch of diamictite from Wadi Amdeh was also considered as a possible remnant of the Al Khlata Formation, but is now considered, along with other nearby conglomerates, as Mesozoic or Cenozoic slope deposits (Heward 2019). However, petrographical analyses of 30 pebbles from this possible diamictite outcrop did not support an Al Khlata origin because of the lack of igneous components and the discovery of gastropod fossils, which is atypical of the Al Khlata Formation. Furthermore, Bauer (2019) encountered microscopic glaucophane within the rocks and determined the depositional age of the diamictite as pre-Campanian. The metamorphism is probably related to the Late Cretaceous high-pressure/low-temperature event that effected the northeastern SHD (Bauer 2019).

In Wadi Daiqa, the Al Khlata Formation consists of white quartzose sandstone, yellow-brown weathered diamictites and dark-grey shales with orange-pink ironstones. Based on palynology, the Al Khlata Formation has been dated in the SHD as Late Carboniferous (P9, ?Late Moscovian–Gzhelian; Heward and Penney 2014). These are the only outcrops of the sandy Al Khlata P9 unit currently known in Oman.

Provenance studies on detrital zircons from the Al Khlata Formation showed age peaks of *c.* 525 Ma (Angudan Orogeny), *c.* 620 Ma (amalgamation of the Arabian-Nubian Shield) and *c.* 800 Ma (juvenile basement of the Arabian peninsula), indicating local sources of the detritus. However, exotic components with ages of *c.* 1000 Ma and of 2500–3000 Ma, typical for African sources (but not from India), were also recorded (Craddock *et al.* 2019), indicating glacial transport from mainland Africa in broadly northern directions.

The Gharif Formation has a constant thickness of 230 m in central Oman (Hughes Clarke 1988; Forbes *et al.* 2010). The sedimentary rocks of this formation probably derived from the varied material supplied by the Al Khlata glacial interval (Forbes *et al.* 2010). Heward and Penney (2014) pointed out that in Wadi Daiqa and the Wadi Mijlas Gorge, and above the Al Khlata Formation, a horizon of coarse-grained, cross-bedded feldspatic sandstones and purple siltstones exists. In addition, this horizon contains fossil wood. This undated horizon may be the equivalent of the uppermost Gharif Formation of the

Huqf area. This interpretation considers that the middle and upper Gharif sandstones in the Huqf area contain more feldspar than the Al Khlata Formation (Hartmann *et al.* 2000) and that fossil wood is common in the uppermost Gharif Formation of the Huqf area (Heward and Penney 2014).

Akhdar Group. The Akhdar Group of Glennie *et al.* (1974) consists of two formations and is conformably overlying the Haushi Group in the subsurface (see Forbes *et al.* 2010). The base forms a major transgression. Thinning and/or absence of the Gharif Formation in the Oman Mountains could be related to intra-Gharif unconformities and possibly related to Permian rifting (see Blendinger *et al.* 1990). In the JAD, the Akhdar Group unconformably overlies pre-Haima Supergroup sedimentary rocks (fig. A.1).

As the older of the two formations, the Saiq Formation (10) or Khuff Formation (subsurface name) is well-exposed in the JAD and the SHD (e.g. Glennie *et al.* 1974; Villey *et al.* 1986*c*). The Saiq Formation's thickness varies from *c.* 450 m in the JAD (Saiq Plateau) to *c.* 1000 m at the eastern flank of the SHD (Le Métour *et al.* 1986*b*, 1992; Rabu *et al.* 1986). In the JAD, the Saiq Formation consists of two members. The lower member consists of a <25-m-thick unit of conglomerate, sandy limestone, siltstone, felsic volcaniclastic and volcanic rocks (Rabu *et al.* 1986), whereas the upper *c.* 425-m-thick member is composed mostly of fossil-rich (e.g. corals, foraminera, gastropods, echinoderms), shallow-marine dark limestone and dolostone towards the top (Rabu *et al.* 1986). In the JAD, the Saiq Formation has been studied as a Khuff outcrop analogue and has been divided into six sequences (Khuff sequences KS6 to KS1, from base to top; e.g. Haase and Aigner 2013). The Wuchiapingian Khuff Sequence KS4 has attracted much attention because it is a main reservoir of the world's largest-known gas field, located in Qatar and Iran (Haase and Aigner 2013). The same authors provided high-resolution sedimentological results of the lower KS4 sequence. The typical grainstone was deposited in an epeiric carbonate ramp with lateral facies changes (Haase and Aigner 2013).

In the eastern SHD, however, the lithology of the entire Saiq Formation is more diverse, including conglomerates, volcanic rocks, tuffite, yellow and black limestone, dolerite sills and mudstone (Le Métour *et al.* 1986*a*, *b*). The highly variable Saiq Formation of the SHD has been divided into six members on the 1:100 000-scale map (Le Métour *et al.* 1986*b*) that were correlated within the SHD (Chauvet *et al.* 2009, fig. 3). On the 1:250 000-scale map, the Saiq Formation has been divided into five informal members (Le Métour *et al.* 1992). In the JAD, the basal Saiq Formation has a Roadian (Kubergandian) age, whereas the top has uppermost Chansinghian (Dorashamian) age (Koehrer *et al.* 2012; fig. A.3). See below (Mahil Formation) for the age discussion of the top of the Saiq Formation and base of the Mahil Formation.

It is interesting to note that there is a marked difference in the distribution of Permian volcanic rocks with only limited occurrence in the JAD (Rabu *et al.* 1986; Blendinger *et al.* 1990) and abundant outcrops in the SHD (Le Métour *et al.* 1986*b*, 1992; Blendinger *et al.* 1990). There is also a marked difference in thickness between the two domes. These differences may be explained by the different distances of the domes to the rift axis (section 'Pangea rifting and formation of the Neo-Thethys Ocean and Hawasina Basin' in Scharf *et al.* 2021*c*, Chapter 5, this Memoir). The magmatic rocks within the Saiq Formation of the SHD show a characteristic high-titanium alkaline chemistry, typical of mantle-plume-related igneous rocks that were produced during the break-up of Pangea along the Arabian margin (Glennie *et al.* 1974; Searle *et al.* 1980; Maury *et al.* 2003; Chauvet *et al.* 2009).

At the northeastern margin of the SHD, the pre-Permian rocks and the Hajar Supergroup (Akhdar and Sahtan groups) consist of high-pressure/low-temperature metamorphic rocks with increasing metamorphism towards the east (e.g. Le Métour *et al.* 1986*b*). The petrology reveals eclogite- to high-pressure greenschist-facies conditions of the Permian mafic protoliths with *c.* 1.6–2.3 GPa at 550–580°C and *c.* 0.3–0.6 GPa at 280–315°C, respectively (Goffé *et al.* 1988; El-Shazly *et al.* 1990, 2001; Searle *et al.* 1994, 2004; El-Shazly 1995; Yamato *et al.* 2007; fig. A.1; section 'Intraoceanic thrusting, formation and obduction of the Semail Ophiolite' in Chapter 5). Carpholite-bearing rocks indicate >0.8–1.0 GPa at <330°C (Goffé *et al.* 1988; Agard *et al.* 2010; fig. A.1). Furthermore, the rocks of the eastern SHD including the Saiq Formation show intensive high-strain folding, including spectacular sheath folds (Searle and Alsop 2007; Cornish and Searle 2017; Fig. 2.3; section 'F4 folds' in Chapter 3) and thrusts.

The Saiq Formation is overlain by the Mahil Formation (11) or Jilh and Sudair formations (subsurface names). In the JAD, the thickness of the Mahil Formation measures 600 m and in the SHD 700–800 m. As for the underlying Saiq Formation, the thickness of the Mahil Formation is greater in the Saih Hatat area (see section 'Pangaea rifting and formation of the Neo-Tethys Ocean and Hawasina Basin' in Chapter 5). The Mahil Formation was first described by Glennie *et al.* (1974). The formation consists mainly of yellowish-greyish dolostone (Rabu *et al.* 1986). Detailed studies on the middle Mahil Formation have been carried out by Pöppelreiter *et al.* (2011).

The age for the base of this formation varies with author (Baud and Richoz 2013; table 1). This difference is related to two different definitions of the Saiq–Mahil Boundary in the JAD (SMB; Glennie *et al.* 1974; Baud and Richoz 2013); an 'upper SMB' and 'lower SMB' therefore exists.

The 'upper SMB' position is characterized by a breccia at the base of the Mahil Formation in Wadi Nakhl near Nakhl (Fig. 2.1) as well as 'thinly bedded, unfossiliferous sequence of burrowed limestone, interbedded with porcellanite, laminated siltstone and shale' (Glennie *et al.* 1974; Baud and Richoz 2013).

The 'lower SMB' position is defined by the sudden change in colour between the formations (Glennie *et al.* 1974). The Mahil Formation is lighter in colour than the dark Saiq Formation and is characterized by massive bedding (Glennie *et al.* 1974; Baud and Richoz 2013). The lower formation is rich in macrofossils, giving the Saiq Formation a dark colour, while the Mahil Formation is lacking in such fossils and is formed of light-grey-coloured rocks. The change in colour and disappearance of fossils is related to the Late Changhsingian extinction event (Baud *et al.* 2010), colloquially known as the 'Great Dying'.

The 75–100-m-thick section between the two SM boundaries has been assigned to the 'Saiq unit C' or 'lower Mahil member' (Baud and Richoz 2013). Based on the different definition of the SMB, the top Saiq or base Mahil has an age of uppermost Permian (e.g. Koehrer *et al.* 2012) or Early Triassic (Baud and Richoz 2013). We adopt the age for the SMB provided by Koehrer *et al.* 2012 (figs A.1 & A.2). Based on micropalaeontological studies, Koehrer *et al.* (2012) suggested uppermost Permian age for the basal part of the Mahil Formation.

Sahtan Group. The Sahtan Group was first introduced by Glennie *et al.* (1974). It is unconformably blanketing the Akhdar Group. Rousseau *et al.* (2005) conducted detailed work on the Sahtan Group, summarized in this section. The top of the Sahtan Group is marked by an angular

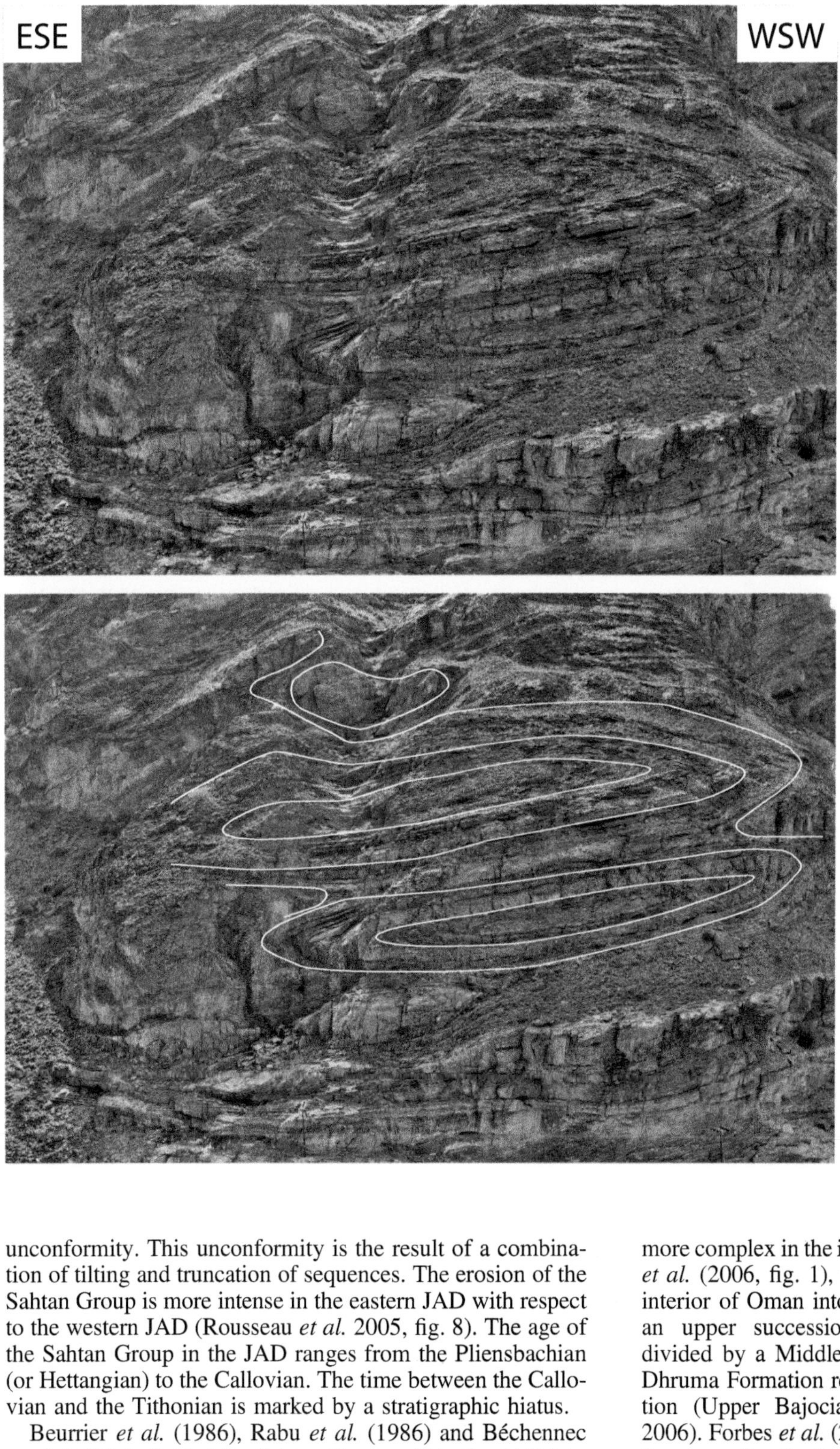

Fig. 2.3. Late Cretaceous sheath fold within the Saiq Formation (Wadi Mayh, eastern SHD; Fig. 2.1). Height of the cliff is *c.* 150 m. Courtesy of Christopher M. Bailey (see Appendix C for further information). See Cornish and Searle (2017) for a three-dimensional visualization of the sheath folds in Wadi Mayh.

unconformity. This unconformity is the result of a combination of tilting and truncation of sequences. The erosion of the Sahtan Group is more intense in the eastern JAD with respect to the western JAD (Rousseau *et al.* 2005, fig. 8). The age of the Sahtan Group in the JAD ranges from the Pliensbachian (or Hettangian) to the Callovian. The time between the Callovian and the Tithonian is marked by a stratigraphic hiatus.

Beurrier *et al.* (1986), Rabu *et al.* (1986) and Béchennec *et al.* (1992*a*) divided the Sahtan Group within the JAD into the Mafraq and Dhruma formations (12). The Mafraq Formation is older than the Dhruma Formation, and three additional formations within the Sahtan Group are known of (e.g. Rousseau *et al.* 2006; Forbes *et al.* 2010). These three additional formations are (from base to top): Tuwaiq Mountain Formation of the Upper Bathonian–Callovian; Hanifa Formation of the Kimmeridigian; and the Jubaila Formation of the Upper Kimmeridigian–Lower Tithonian (Forbes *et al.* 2010). These units are not present in the Oman Mountains due to erosion and non-deposition and/or an Upper Jurassic unconformity (Schlaich and Aigner 2017) and are only defined for the subsurface, and we do not include them in our compilation. Furthermore, the internal stratigraphy of the Sahtan Group is more complex in the interior of Oman. According to Rousseau *et al.* (2006, fig. 1), the Mafraq Formation is divided in the interior of Oman into a lower (Pliensbachian–Toarcian) and an upper succession (Upper Aalenian–Upper Bajocian) divided by a Middle–Upper Toarcian–Aalenian hiatus. The Dhruma Formation rests conformably on the Mafraq Formation (Upper Bajocian–Lower Callovian; Rousseau *et al.* 2006). Forbes *et al.* (2010) provided slightly different subsurface ages for the Lower Mafraq Formation (Late Pliensbachian–Toarcian), Upper Mafraq Formation (latest Toarcian to Late Bajocian) and Dhruma Formation (Late Bajocian to Bathonian–Callovian boundary). The base of the upper Mafraq Formation is characterized by a Late Toarcian unconformity (Ziegler 2001). Based on new stratigraphic information, Bendias and Aigner (2015) constrained an older age for the Lower Mafraq Formation in the western JAD. We will follow the ages for the Mafraq and Dhruma formations provided by Schlaich *et al.* (2021, fig. 1a), which are Hettangian–Late Pliensbachian for the Lower Mafraq Formation, Aalenian–Bajocian for the Upper Mafraq Formation and Bajocian to Bathonian–Callovian boundary for the Dhruma Formation (fig. A.2).

The Mafraq and Dhruma formations are grouped together in our map (fig. A.1), since both units are too thin for their separate depiction. The thickness of the entire group varies from 400 m in the western part of the JAD to 80 m at the eastern part of the JAD to 100–180 m in the SHD (Le Métour *et al.* 1986*b*). The upper limit of the Sahtan Group is marked by a stratigraphic hiatus (Forbes *et al.* 2010). The lower, up-to-150-m-thick Mafraq Formation is rust-coloured and consists of shallow-marine sandstone, micaeous siltstone and bioclastic clayey limestone (Rabu *et al.* 1986; Béchennec *et al.* 1992*a*). The Permo-Mesozoic formations of the Hajar Supergroup are largely carbonates, the main exception being the Mafraq Formation, representing a regional terrigenous interlude. Schlaich *et al.* (2021) provided detailed sedimentological information on the lowermost 40 m of the Lower Mafraq Formation in the western JAD. Clastic sand bodies are as thick as 12 m and are laterally correlatable for 4 km. The palaeoenvironment of the Lower Mafraq Formation includes shallow marine, tidal flat, coastal bar sands and estuarine multistory channel systems (Schlaich *et al.* 2021).

The Dhruma Formation comprises massive bluish shallow-marine limestone with low-energy facies dominated by lagoonal mud- and wackestone in the lower part and more high-energy facies, with peloidal and ooidal grainstone, towards the top (Schlaich and Aigner 2015). The Dhruma Formation of the JAD represents a low-energy carbonate platform facies, while at the SHD the facies reflects a palaeoenvironment of a higher-energy slope. This indicates that the shelf-slope break was between the JAD and SHD during the Mid to Late Jurassic (Coffield 1990). Detailed lithological and stratigraphic descriptions of the Sahtan Group have been published by Rousseau *et al.* (2005), Bendias and Aigner (2015), Bendias *et al.* (2015) and Schlaich and Aigner (2015, 2016, 2017). In the northeastern SHD the rocks of the Sahtan Group underwent high-pressure/low-temperature metamorphism (see 'Akhdar Group' above and sections on 'Upper–lower plate discontinuity or Hulw–As Sifah window' and 'Stage IV: Exhumation of high-pressure/low-temperature rocks' in Chapters 3 and 5, respectively).

Kahmah Group. The Kahmah Group unconformably overlies the Sahtan Group (Glennie *et al.* 1974). The Kahmah Group's age is Late Tithonian–Late Aptian (e.g. Hillgärtner *et al.* 2003; Rousseau *et al.* 2005; Forbes *et al.* 2010). In the subsurface and through detailed studies within the JAD, the Kahmah Group is divided into six formations (from base to top): Rayda, Salil, Habshan, Lekhwair, Kharaib and Shu'aiba (e.g. Hillgärtner *et al.* 2003; Forbes *et al.* 2010). However, in the JAD and at a map scale of 1:250 000, the Kahmah Group consists of only three formations: the lower two – Kahmah 1 and 2 (13) – and the upper Kahmah 3 (14) formations (compare Béchennec *et al.* 1992*a*). Synonyms for the Kahmah 1 and 2 formations are the older 'Rayda' (previously called 'Awabi Formation') and the younger 'Salil' (previous called 'Birkat Formation') formations. A synonym for the Kahmah 3 Formation is 'Shams Formation' (Rabu *et al.* 1990).

The Kahmah 3 Formation of the JAD represents the Habshan, Lekhwair, Kharaib and Shu'aiba formations (subsurface and surface studies of the JAD; van Buchem *et al.* 2002*a*; Hillgärtner *et al.* 2003). Béchennec *et al.* (1992*a*) did not divide the Kahmah 3 Formation into these four formations on the map at the scale of 1:250 000. We are following Béchennec *et al.* (1992*a*) and lump the four formations together into the Kahmah 3 Formation because their individual thickness in the JAD and SHD is too small to be properly depicted on the map at the scale of 1:250 000 (fig. A.1).

The entire Kahmah Group represents one large 'second-order' sequence, beginning with pelagic sedimentary rocks (Rayda Formation) and ending with shallow-marine platform carbonates of the Shu'aiba Formation (e.g. Hillgärtner *et al.* 2003). The top of this group is marked by a regional unconformity, caused by the emergence of the carbonate platform during the Late Aptian (Immenhauser *et al.* 1999). The Habshan, Lekhwair, Kharaib and Shu'aiba formations represent a succession of cyclic marine carbonates with very good reservoir properties for hydrocarbons (e.g. Hillgärtner *et al.* 2003; Forbes *et al.* 2010). Detailed sequence stratigraphic and biostratigraphic work of the entire Kahmah Group or parts of it have been carried out by, for example, van Buchem *et al.* (2002*a*), Hillgärtner *et al.* (2003), Schroeder *et al.* (2010), Razin and Grelaud (2011) and Celestino *et al.* (2017).

The Rayda Formation consists of 30 m of thick silicified, whitish, micritic limestone (Beurrier *et al.* 1986) with common belemnite rostra (Mattern and Scharf 2018). Celestino *et al.* (2017) subdivided the Rayda Formation in a lower unit consisting of reddish bioclastic wackstone/grainstone, and an upper unit of micritic limestone with chert nodules. The sediments were deposited during the Late Tithonian–Early Berriasian (Rousseau *et al.* 2005; Forbes *et al.* 2010) within an offshore to bathyal environment (Beurrier *et al.* 1986). Based on detailed stratigraphic and carbon isotope studies, Celestino *et al.* (2017) assigned a Late Valanginian age to the upper part of the Rayda Formation, which we are following in figure A.2.

The Salil Formation is up to 350 m thick, consists of alternating clayey-cherty limestone and marls, and has been dated in the subsurface as upper Berriasian–Valanginian (Forbes *et al.* 2010). Towards the top, the Salil Formation becomes more calcareous with grey and black micritic limestone (Beurrier *et al.* 1986). The palaeoenvironments of the formation are distal subtidal, bathyal, proximal to median subtidal (Beurrier *et al.* 1986).

The 450-m-thick Kahmah 3 Formation contains many different carbonate facies (Beurrier *et al.* 1986; Béchennec *et al.* 1992*a*). Most outcrops consist of rudists, bacinella and orbitolinids (see van Buchem *et al.* 2002*a*). Oolites are common in the Habshan Member while platform margin facies of the Shu'aiba and Kharaib members are common in the north. The ages of the four parts of the Kahmah 3 Formation are the following: Habshan, Valanginian–Hauterivian; Lekhwair, Hauterivian–Early Barremian; Kharaib, Early Barremian–earliest Aptian; and Shu'aiba, Early–Late Aptian (e.g. Hillgärtner *et al.* 2003; Forbes *et al.* 2010).

Wasia Group. The Wasia Group of Owen and Nasr (1958; Forbes *et al.* 2010) unconformably overlies the Kahmah Group (Glennie *et al.* 1974). It comprises two formations: the older Nahr Umr and the younger Natih formations. The Wasia Group represents one major transgressive regressive cycle with a major exposure surface at the top (Droste, H. 2019, pers. comm.).

The Nahr Umr Formation (15) has a thickness of 140–200 m in the JAD (Beurrier *et al.* 1986; Rabu *et al.* 1986) and is not present in the SHD. Black limestone beds of thickness 0.5–1 m alternate with 0.2–0.5-m-thick beds of orbitolinid-bearing marl with a significant clastic terrigenous fraction (Beurrier *et al.* 1986; Rabu *et al.* 1986). The Nahr Umr Formation has been dated as latest Aptian–Late Albian, based on their microfauna and isotope stratigraphy (Rabu *et al.* 1986; Immenhauser *et al.* 1999; Forbes *et al.* 2010). The depositional environment was a middle to proximal subtidal platform (Beurrier *et al.* 1986; Rabu *et al.* 1986) or an internal shelf (Rabu *et al.* 1990). The Nahr Umr Formation is an important seal for oil reservoirs in the subsurface of northern, central and southern Oman (Forbes *et al.* 2010).

The Natih Formation (16) is well-exposed at the southern flank of the JAD. The average thickness measures *c.* 200 m in the JAD and <400 m south of the JAD (Beurrier *et al.* 1986; Rabu *et al.* 1986; van Buchem *et al.* 2002b, fig. 9; Homewood *et al.* 2008, fig. 5), but decreases toward the NE of the JAD and is missing in the SHD. According to bio- and isotope stratigraphy, its age is Late Albian–Early Turonian (Simmons and Hart 1987; Smith *et al.* 1990; Philip *et al.* 1995; Homewood *et al.* 2008; Frijia, unpublished data). Piuz *et al.* (2014) and Wohlwend *et al.* (2016) suggested a Mid Turonian age for the upper Natih Formation in the Adam Foothills. Based on subsurface data, the formation has been subdivided into seven members (from bottom to top); these are referred to as 'g'–'a' in Hughes Clarke (1988) and 'G'–'A' in Forbes *et al.* (2010). The formation is characterized by limestone/marly limestone alternations in the lower part, suggesting a broad and flat, submerged to emergent benthic foraminifer-dominated ramp (g, f, d and c members), and by massive pure limestones of a rudist-dominated carbonate ramp, bordering a moderately deeper-water intra-shelf basin (fig. 5.1) with anoxic conditions, allowing deposition of organic-rich rocks (mainly e and b members). Within these 'source rocks' solid bitumen particles occur (Fink *et al.* 2016, who worked on the Natih a; Grobe *et al.* 2016, who worked on the Natih b). According to Forbes *et al.* (2010), the Natih Formation records the transitional sedimentation from a stable shallow-marine platform to a flexural basin. In the Rustaq area (western JAD; Fig. 2.1), Beurrier *et al.* (1986), Béchennec *et al.* (1986) and Rabu (1987; Béchennec *et al.* 1992*b*) introduced the 'Fitri Formation' at the top of the Natih Formation, a term now abandoned (Béchennec *et al.* 1992*b*). The former Fitri Formation correlated with the Natih a, b and c and was included into the Aruma Group by Rabu *et al.* (1990).

The excellently exposed carbonates of the JAD are a natural laboratory for fracture/vein analyses and fluid pathways during their formation (e.g. Holland *et al.* 2009*a*, *b*; Holland and Urai 2010; Arndt *et al.* 2014; Grobe 2017; Grobe *et al.* 2018, 2019). Hilgers *et al.* (2006), Gomez-Rivas *et al.* (2014) and Grobe *et al.* (2018) investigated fractures/veins from the Mesozoic carbonates of the JAD and revealed the geological history of the limestone from the Mesozoic until today, including the early diagenesis, obduction, exhumation, transpressional deformation and doming (see sections on 'Post-'Hercynian' deformation' and 'Post-obduction deformation' in Chapter 5).

Aruma Group. The Coniacian–Maastrichtian Aruma Group of Owen and Nasr (1958; Forbes *et al.* 2010) was defined in southern Iraq and Kuwait. It unconformably overlies the Wasia Group or older groups (Glennie *et al.* 1974; Hughes Clarke 1988). The Aruma Group represents the sediment fill of a flexural foreland basin in front of the advancing allochthonous units (e.g. Boote *et al.* 1990; Ravaut *et al.* 1997). According to an older opinion of Robertson (1987*a*), the forebulge was migrating continent-wards during the advancing allochthonous rocks (Fig. 2.4). However, according to Warburton *et al.* (1990), the forebulge was stationary at the southwestern side of the narrowing flexural foreland basin. The Aruma Group contains up to 3.5-km-thick sediments which have accumulated during emplacement of the Semail Ophiolite onto the Arabian Platform as well as post-obductional formations (e.g. Glennie *et al.* 1974; Searle 2007). Figure 2.5 schematically depicts a time chart of the evolution of the Aruma Group with respect to the Semail Ophiolite emplacement onto the Arabian Platform during the Late Cretaceous–earliest Paleocene.

Generally, the Aruma Group consists of the Coniacian–Lower Santonian clastic Muti Formation (17), the shale-rich Santonian–Campanian Fiqa Formation, the overlaying ophiolite clast-bearing Upper Campanian–Lower Maastrichtian Juweiza Formation, the fluvial clastic Al-Khod Formation and the calcareous Maastrichtian Simsima Formation (Glennie *et al.* 1974; Robertson 1987*a*, b; Skelton *et al.* 1990; Warburton *et al.* 1990; Glennie 2005; Searle 2007; Schlüter *et al.* 2008, fig. 1). The term 'Muti Formation' is used for outcrops, and 'Fiqa Formation' is applied for the subsurface. In

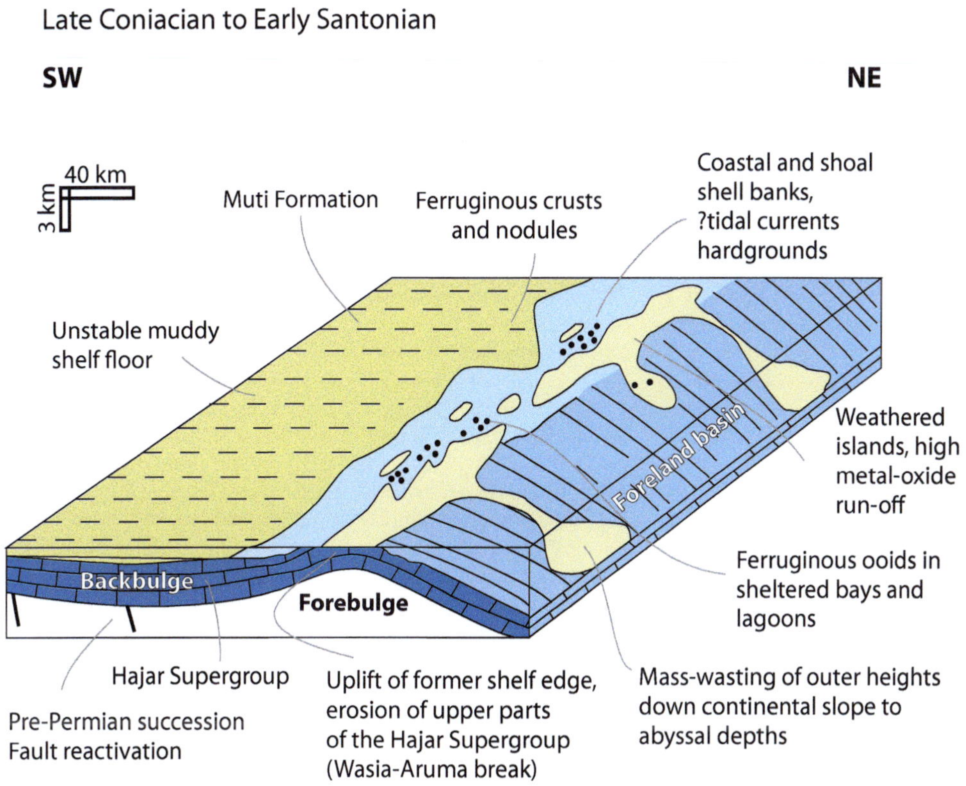

Fig. 2.4. Palaeogeographic sketch of the Arabian continental margin during deposition of the Muti Formation. Flexural platform uplift was followed by erosion and deposition of the Muti Formation in the foreland basin. The flexural bulge eventually submerged, resulting in a variety of high-energy palaeoenvironments. Redrawn after Robertson (1987*b*) with ages from Warburton *et al.* (1990).

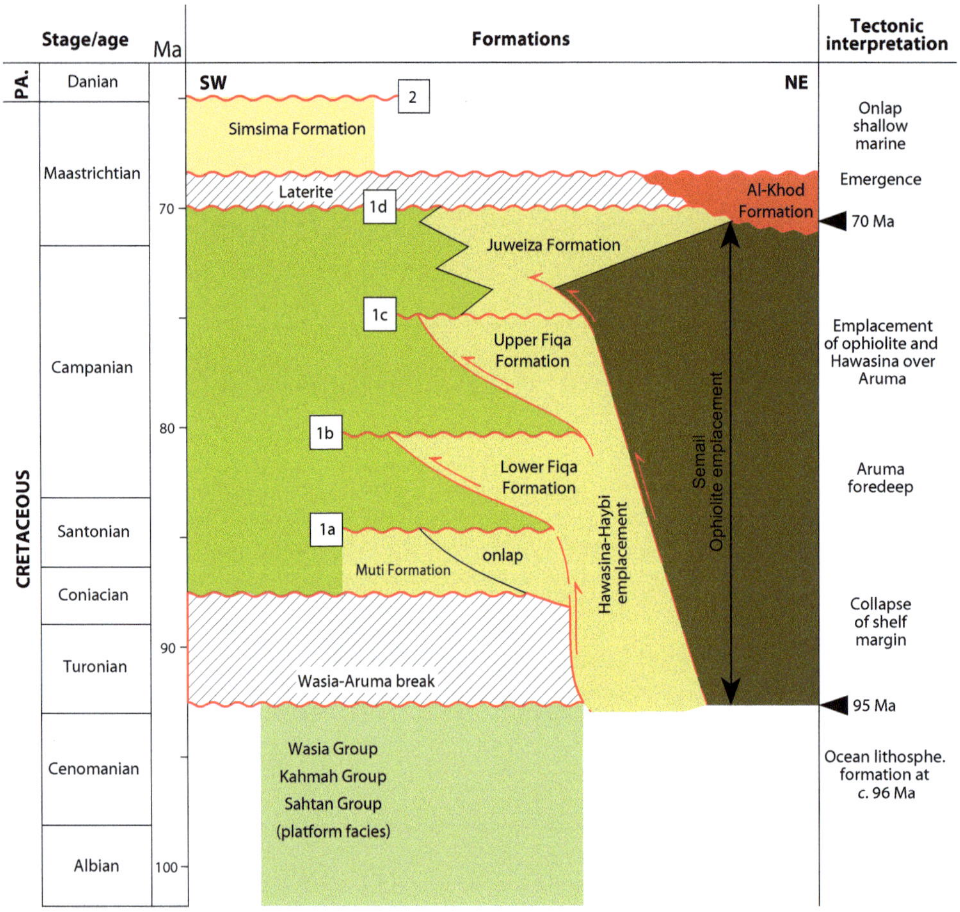

Fig. 2.5. Late Cretaceous–Early Paleozoic time chart of the Semail Ophiolite emplacement onto the Arabian platform and the associated Aruma Forland Basin in the Central Oman Mountains. Slightly modified after Searle (2007). Stages 1a, b indicate thrusts affecting the Fiqa and Juweiza formations during Semail Ophiolite obduction. Stage 2 is a gentle folding event, affecting the Simsima Formation.

publications relating to the subsurface close to the Oman Mountains, a Lower Fiqa unit is sometimes recognized which predates the emplacement of the nappes (see Boote *et al.* 1990). This Lower Fiqa unit would be the lateral equivalent of the Muti Formation. The Upper Fiqa section onlaps and post-dates nappe emplacement and is not exposed in the central part of the Oman Mountains (Fig. 2.5). In our map, we only use the term 'Muti Formation'.

The Al-Khod (Qahlah) and Simsima formations post-date thrusting of the allochthonous units. The Simsima Formation is as old as or slightly younger than the post-obductional Al-Khod (Qahlah) Formation (Hughes Clarke 1988; Glennie 2005). These authors therefore classified the Aruma Group as syn- (consisting of the Muti, Fiqa and Juwaiza formations) to post-obductional (consisting of the Simisma and Al-Khod formations).

The basal part of the Muti Formation has been dated as Early–Mid Turonian, based on microfauna (Béchennec *et al.* 1986; Beurrier *et al.* 1986; Rabu *et al.* 1986). Warburton *et al.* (1990) suggested a slightly older age for the post-tectonic infill (Fiqa Formation) of the foreland basin south of the JAD (Late Coniacian–Early Santonian; fig. A.2). We follow the ages provided by Warburton *et al.* (1990) throughout our text and in figure 2.5 and figures A.1 and A.2.

The lithology of the Muti Formation is mostly characterized by yellowish/beige siltstone and some reef limestone, and its depositional system alos includes proximal turbidities, derived from a subtidal carbonate platform (Fig. 2.4; Rabu *et al.* 1986, 1990). Ophiolite clasts within the Muti Formation have not been recorded.

The Muti Formation is exposed at the outermost margins of the JAD and SHD (fig. A.1). It overlies the Natih Formation at the western flank of the JAD, and overlies the Nahr Umr and Kahmah 1, 2 and 3 formations at the northeastern flank of the JAD. The gap between the Muti Formation and the underlying formations ('Wasia-Aruma break' of Glennie *et al.* 1974; Fig. 2.5) increases towards the SHD. At the northern SHD, the Muti Formation overlies the Mahil Formation and Sahtan Group, whereas at the southern margin of the SHD, the Muti Formation rests on top of the Kahmah 1, 2 and 3 formations. Robertson (1987*a*) correlated the stratigraphic gap with the uplift (as much as 600 m) and erosion of the Arabian Platform at the forebulge (see section on 'Intraoceanic thrusting, formation and obduction of the Semail Ophiolite' in Chapter 5).

Allochthonous

Hawasina nappes

The stratigraphy and nomenclature of the Hawasina nappes differs between authors (e.g. Robertson and Searle 1990; Blechschmidt *et al.* 2004; Searle 2007). The Bureau de Recherches Géologiques et Minières (BRGM) maps from the 1980s and early 1990s introduced a new stratigraphy and nomenclature for large parts of the Central and Southeastern Oman Mountains, which could lead to confusion when comparing them with existing nomenclature. We will largely use the nomenclature provided by the BRGM maps because their maps are the basis of figure A.1. The reader is referred to Robertson and Searle (1990) for further details on the different nomenclature provided in the literature.

The combined thickness of the Semail Ophiolite and the Hawasina nappes decreases from NE to SW in our study area (Aldega *et al.* 2017; Carminati *et al.* 2019). Based on clay-mineral geothermometers and ooid strain analyses,

Aldega *et al.* (2017) modelled a combined thickness of the allochthonous units with 4.5 km in the northern JAD area and 2 km further south in the foreland of the Oman Mountains.

The Hawasina nappes belong to an Upper Permian–Upper Cretaceous predominantly deep-marine succession with some rafts of isolated carbonate platform sediment units (Béchennec *et al.* 1990, table 1). The original position of the *c.* 450 km-wide Hawasina Basin and its sediments was the southwestern margin of the Neo-Tethys Ocean, that is, SW of the mid-ocean ridge and NE of the continental slope of the Arabian Platform (e.g. Glennie *et al.* 1974; Cooper 1987, 1988, 1989; Béchennec *et al.* 1990; Hanna 1995). The continental slope sediments at the transition from the Arabian margin to the Hawasina Basin are represented by the Sumeini Group, which is not exposed in our study area (e.g. Béchennec *et al.* 1990; Blechschmidt *et al.* 2004; Fig. 2.6).

Béchennec *et al.* (1992*a*) indicated that the Al-Aridh Group in the east-central and southeastern Oman Mountains is equivalent to the Sumeini Group. This differs from previous interpretations, for example those of Béchennec *et al.* (1990), who placed the Al-Aridh Group in a proximal position with respect to the Misfah micro-platform (Kawr Group). The interpretation of Béchennec *et al.* (1990) was based on the structural position of the Al-Aridh Group between the Kawr Group and the Hamrat Duru Group, and partly on their lithology (Béchennec *et al.* 1992*a*). The Al-Aridh Group contains pelagic deposits and a megabreccia of olistoliths derived from Upper Permian–Upper Triassic shallow-marine carbonates (see Béchennec *et al.* 1992*a*). The revised interpretation of Béchennec *et al.* (1992*a*) is partly based on lateral facies variations in beds between the continental slope and the proximal Hamrat Duru Group and on the idea that it is not necessary to have a fixed structural position of the Kawr nappe. Blechschmidt *et al.* (2004) and Wohlwend *et al.* (2017, fig. 2), however, placed the Al-Aridh Group between the Hamrat Duru Group and the Misfah Platform (Kawr Group), similar to the interpretation of Béchennec *et al.* (1990). Blechschmidt *et al.* (2004) suggested that the Hamrat Duru Group represents a base-of-slope to abyssal plain sedimentary facies, meaning that there is no need to consider the Al-Aridh Group as the slope facies of the Hawasina Basin. We follow the interpretation of Blechschmidt *et al.* (2004) and Wohlwend *et al.* (2017) in our text and place the Al-Aridh Group proximal to the Misfah Platform (Kawr Group; Fig. 2.6). A general correlation of the Hawasina nappes with the Sumeini sediments and the Arabian Platform is provided by Searle (2007, fig. 3).

The entire Hawasina Basin underwent two intervals of extension during the Late Paleozoic–Triassic (e.g. Béchennec *et al.* 1990; Hanna 1995; see section on 'Pangea rifting and formation of the Neo-Tethys Ocean and Hawasina Basin' in Chapter 5 for discussion of the time of rifting). A first phase of rifting and extension ensued during either the Late Carboniferous, Late Permian or Mid Permian, which later led to the creation of the Hamrat Duru Basin and the Baid Horst (e.g. Béchennec *et al.* 1990; see section on 'Pangea rifting and formation of the Neo-Tethys Ocean and Hawasina Basin' in Chapter 5). A second stage of rifting resulted in the formation of the Kawr Seamount or Horst with the adjacent Al-Aridh Trough and the Umar Basin (Fig. 2.6) during the Early–Mid Triassic (e.g. Béchennec *et al.* 1990). Both Hawasina subbasins were separated by a narrow chain of seamounts (Hanna 1995) or horsts (Béchennec *et al.* 1990) capped with reef and forereef carbonates, and were known as the 'Oman Exotics' (Searle and Graham 1982). The sediments overlying the volcanic edifices or horsts represent the Kawr Group. To the SW, the Kawr Zone was accompanied by the narrow Al-Aridh Trough (Béchennec *et al.* 1990; Blechschmidt *et al.* 2004; Fig. 2.6).

Two further extensional events affected the entire Hawasina Basin, one during the Tithonian–Berriasian and one during the Albian–Cenomanian (Béchennec *et al.* 1990). Blechschmidt *et al.* (2004) and Rousseau *et al.* (2005) postulated a Jurassic tectonic instable phase, resulting in uplift and erosion of the platform, block faulting and increased subsidence (section 'Arabian platform' in Chapter 5). The mid-Cretaceous extension was perhaps the result of slab pull of the subducting Arabian Plate (compare Tavani *et al.* 2021; see section on 'Stage I: Subduction initiation' in Chapter 5).

A detailed palinspastic reconstruction of the Hawasina Basin was suggested by Béchennec *et al.* (1988, 1990), Cooper (1989) and Blechschmidt *et al.* (2004; Fig. 2.6). The sedimentary rocks depicted in Figure 2.6 have been thrust towards the SW by the Semail Ophiolite. Searle (2007, fig. 3) grouped the different formations/units of the slope (Sumeini Group), the Hawasina Complex (Hamrat Duru

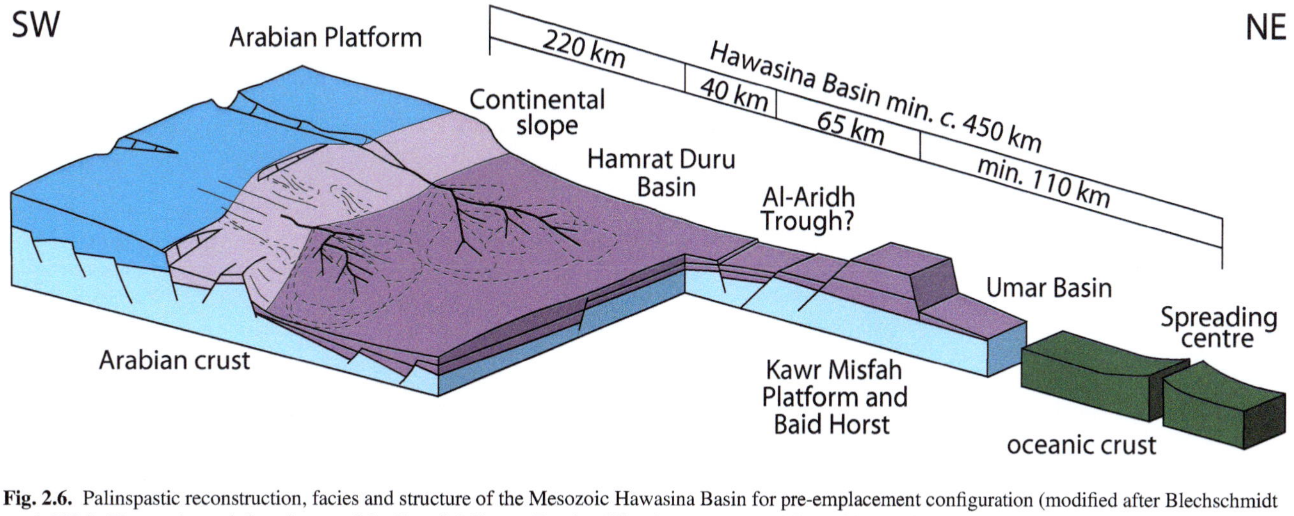

Fig. 2.6. Palinspastic reconstruction, facies and structure of the Mesozoic Hawasina Basin for pre-emplacement configuration (modified after Blechschmidt *et al.* 2004). The continental slope is part of the Sumeini Group. For simplification, we grouped the Baid Horst into the Kawr Group in our compilation (figs A.1 & A.2) following Rabu *et al.* (1986). The width of the Hawasina Basin and its different subunits are after Béchennec *et al.* (1990). Note that the different formations and/or groups of the Hawasina Basin form thrust sheets during the Late Cretaceous Semail Ophiolite thrusting (Searle 2007). Continental slope – Sumeini Group; Hamrat Duru Basin – Hawasina Complex; Al-Aridh Trough and Kawr Misfah Platform – Kawr-Mishaf Complex; Umar Basin – Haybi Complex distal ocean-trench. A general correlation of the Hawasina nappes with the Sumeini sediments and the Arabian Platform is provided by Searle (2007, fig. 3).

Group), the Kawr-Misfah and platform facies, the Haybi Complex distal ocean-trench (Umar Basin) and the Semail Ophiolite into different thrust sequences. These sequences (collectively known as the 'allochthonous') were thrust on top of the Arabian Platform ('autochthonous') during the Late Cretaceous (sections 'Allochthonous thrusting on autochthonous plates' and 'Intraoceanic thrusting, formation and obduction of the Semail Ophiolite' in Chapters 3 and 5, respectively). The thrust sheets have been intensely folded and faulted as well as dismembered into a multitude of small thrust units (section 'Allochthonous thrusting on autochthonous plates' in Chapter 3). On the map (fig. A.1) and in the following text, we subdivide the Hawasina nappes into four groups based on their palaeogeographic position within the Neo-Tethys Ocean, which is a simplification of the subdivision of Béchennec *et al.* (1990, 1992*a*; Fig. 2.6 & fig. A.2); for an alternative picture see Le Métour *et al.* (1986*b*).

Pillevuit *et al.* (1997, fig. 3) redefined parts of the Hawasina nappe systems. Some of the respective nappes are too small to be depicted on a geological map at a scale of 1:250 000; we therefore adopt the division of Béchennec *et al.* (1992*a*).

The ages of the Hawasina formations differ between authors (Robertson and Searle 1990, fig. 3; Pillevuit *et al.* 1997; Blechschmidt *et al.* 2004; Wohlwend *et al.* 2017). We follow the ages of Béchennec *et al.* (1990, 1992*a*), complemented by ages suggested by Pillevuit *et al.* (1997). We also use the ages of Blechschmidt *et al.* (2004, fig. 3) for the Sid'r and Nayid formations of the Hamrat Duru Group.

Thrust units of the Hamrat Duru Group originate from a proximal position in relation to Arabia (Fig. 2.6) at the SW margin of the Neo-Tethys Ocean. The Permo-Mesozoic formations of this group are classified into one thrust sheet (the Hawasina Complex of Searle 2007) that formed in the course of thrusting of the Semail Ophiolite (section 'Allochthonous thrusting on autochthonous plates' in Chapter 3). Sedimentological and isotopic investigations of the upper Jurassic–Cretaceous formations of the Hamrat Duru Group have been carried out by Cooper (1987, 1988, 1989, 1990), Blechschmidt *et al.* (2004) and Wohlwend *et al.* (2017). The detailed sedimentary evolution and palaeogeographical reconstruction of the Hamrat Duru Group has been summarized by Cooper (1988, 1989, 1990), Blechschmidt *et al.* (2004) and Phillips *et al.* (2013). The formations of the Hamrat Duru Group exhibit lateral facies changes (Béchennec *et al.* 1992*a*; Blechschmidt *et al.* 2004, figs 17, 18; Phillips *et al.* 2013, figs 4.15, 4.16, 4.17b, 4.18; Wohlwend *et al.* 2017, fig. 1b). According to Béchennec *et al.* (1992*a*), five formations in our study area are part of this group: the Al-Jil (18), Matbat (19), Guwayza (20), Sid'r (21) and Nayid (22) formations. Béchennec *et al.* (1992*a*) added two more formations into the Hamrat Duru Group: the Sabt and Wahrah formations. The Sabt Formation is not exposed in the JAD or SHD areas and the Wahrah Formation only occurs in a single outcrop at the very south of our study area (fig. A.1). We included the latter in the approximately coeval Sid'r Formation, following Phillips *et al.* (2013). Blechschmidt *et al.* (2004) suggested a detailed and new stratigraphic nomenclature of the Hamrat Duru Group, based on (bio-) stratigraphic results. The same authors also provided detailed information regarding the involved depositional systems and their respective lateral and along-strike facies variations. We follow the nomenclature of Béchennec *et al.* (1992*a*), because the geological map of figure A.1 is based on the compilation of Béchennec *et al.* (1992*a*). A correlation between different stratigraphic systems of the Hamrat Duru Group is shown in Figure 2.7.

The Al-Aridh Group (Béchennec *et al.* 1986) mainly comprises slope sediments that derived from the Kawr Zone (Fig. 2.1). This group is made up of the Sayfam (23), Buwaydah (24), and the Al-Ghafat and Musallah (25) formations (Glennie *et al.* 1974). We lumped together the latter two (figs A.1 & A.2) because the Al-Ghafat Formation was not described by Béchennec *et al.* (1992*a*). The Al-Aridh Group only occurs in the southwestern part of the JAD besides a few small outcrops of the Sayfam and Buwaydah formations in the Batinah Mélange (fig. A.1).

The sediments of the Kawr Group represent a topographic high (horst, seamount or a sliver of rifted Arabian continental margin; see Bernoulli *et al.* 1990; Fig. 2.6) separating the proximal Hamrat Duru Basin from the distal Umar Basin. The Baid (26), Misfah (27), Nadan (28) and Safil (29) formations are part of this group. We follow Rabu *et al.* (1986) in placing the Baid Formation into the Kawr Group. Béchennec *et al.* (1992*a*) assigned the Baid Formation into an extra group, palaeogeographically located between the Al-Aridh Group and the Hamrat Duru Basin (Fig. 2.6). The Kawr Group has been redefined and described in detail by Pillevuit (1993; Pillevuit *et al.* 1997). The Kawr Group partly forms impressive mountains and landmarks (e.g. Jabal Misfah/Jabal Kawr, SW of the JAD; Fig. 2.1). Some of the mountains are used as quarries, producing marble. The Kawr Group represents the Kawr-Misfah and platform facies thrust sequence, which originated during the Late Cretaceous (Searle 2007; section 'Allochthonous thrusting on autochthonous plates' in Chapter 3).

The Umar Basin was located NE of the 'Kawr High' (Béchennec *et al.* 1990, 1992*a*), and contains deep-sea sediments. The most distal formations within the Umar Group are the Sinni (30) and Aqil (31) formations The formations of this group are assigned to the Haybi Complex thrust sheet, which formed during the Late Cretaceous thrusting of the Semail Ophiolite (Searle 2007; section 'Allochthonous thrusting on autochthonous plates' in Chapter 3). The Haybi Complex was originally defined by Searle and Malpas (1980, 1982) and Searle (1985).

The term 'Oman Exotics' has two connotations: either widespread shallow-marine limestones of the Kawr Group, or numerous and comparatively small isolated limestone occurrences, resting on highly alkaline volcanic edifices of the Hamrat Duru or Umar groups (Searle *et al.* 1980; Searle and Graham 1982; Pillevuit *et al.* 1997).

The contact between the Hawasina nappes and the Semail Ophiolite is marked by a metamorphic sole (32). The metamorphic sole overprints the most distal rocks of the Hawasina nappes (Haybi Complex thrust sheet; Searle and Malpas 1980, 1982; Searle 1985).

Hamrat Duru Group. The stratigraphy of the Hamrat Duru Group is complex with different stratigraphic interpretations between authors. Figure 2.7 summarizes the different stratigraphic interpretations.

The Upper Permian–Middle Triassic Al-Jil Formation (18) or in part the Zulla Formation (see Cooper 1987, 1988, 1989, 1990; Blechschmidt *et al.* 2004, fig. 3) consist mainly of basal pillow lava, topped by shallow-marine limestone and radiolarian chert (Béchennec *et al.* 1992*a*). The carbonates contain reworked Late Permian biota (Béchennec *et al.* 1992*a*). The thickness of the Al-Jil Formation amounts to 80 m (Beurrier *et al.* 1986).

The Middle Triassic–Lower Jurassic (Béchennec *et al.* 1992*a*) Matbat Formation (19), defined by Béchennec (1987; Béchennec *et al.* 1992*a*), measures *c.* 140 m in thickness (Béchennec *et al.* 1992*a*). It comprises Triassic radiolarian chert, followed by Triassic–Lower Jurassic turbititic quartz-sandstone (Béchennec *et al.* 1992*a*).

The Bathonian–Kimmeridgian Guwayza Formation (20) consists of turbiditic oolitic calcarenites and conglomerates (Cooper 1987; Béchennec *et al.* 1992*a*). This formation

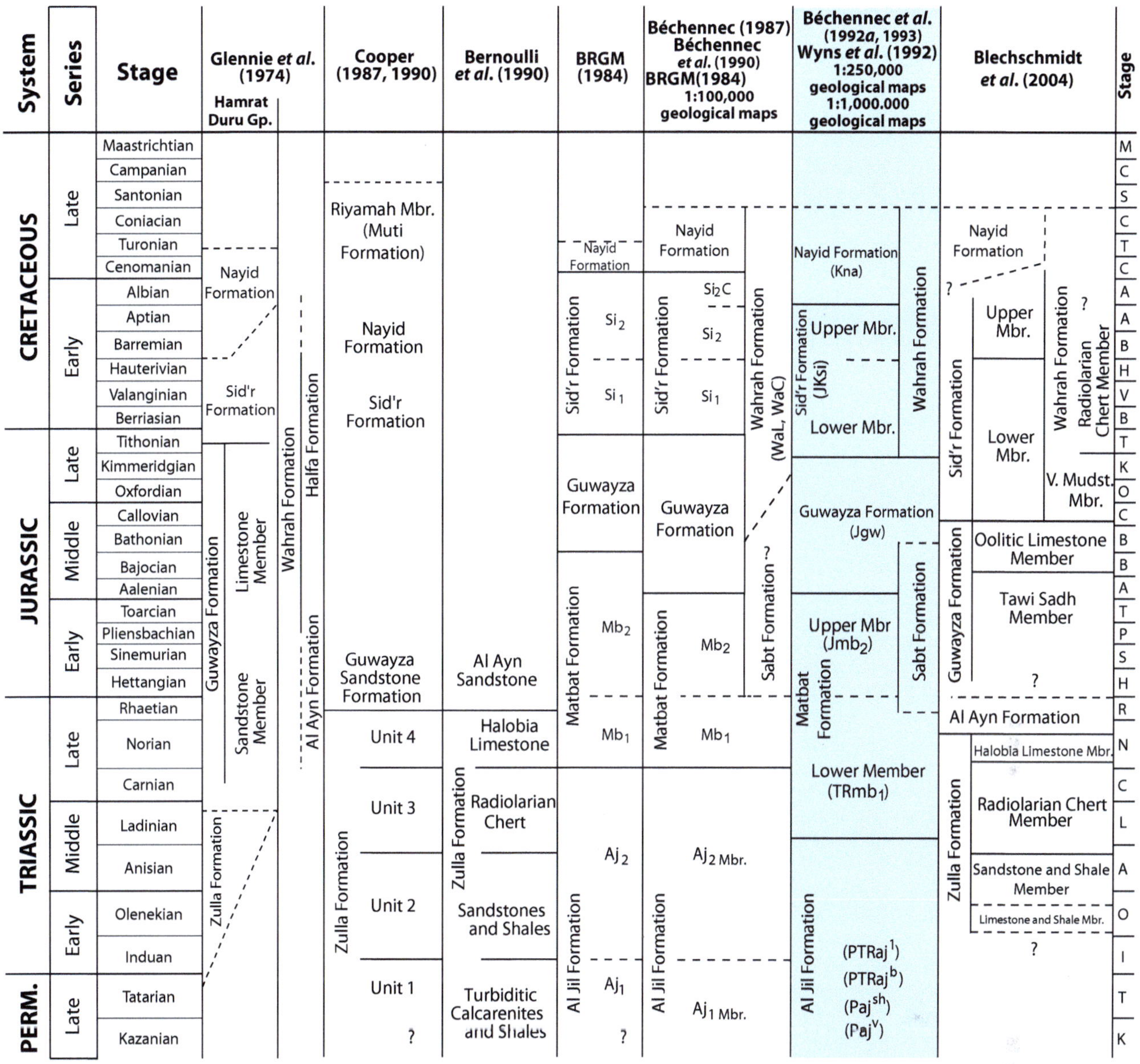

Fig. 2.7. Correlation of nomenclatures of the Hamrat Duru Group, slightly modified from Blechschmidt *et al.* (2004). The blue nomenclature column with some age modifications from Blechschmidt *et al.* (2004) is chosen for figures A.1 and A.2 (see text for further details).

represents a proximal depositional environment, indicated by frequent channels, massive beds of coarse-grained calcarenite and a significant variation in thickness from 0 to 220 m at the Semail Gap (Béchennec *et al.* 1992*a*) and 0 to 550 m in general (Cooper 1987, 1989, 1990, fig. 2). The oolitic member of the Guwayza Formation reveals similar depositional conditions to formations from the Sumeini Group in the Northern Oman Mountains.

The Upper Tithonian–Lower Cretaceous Sid'r Formation (21) contains radiolarian chert, micritic limestone and turbiditic calcarenite (Béchennec *et al.* 1992*a*), and is 120 m thick (Béchennec *et al.* 1992*a*). According to Wohlwend *et al.* (2017), the age of the lower part of this formation ranges between Oxfordian–Late Tithonian and Hauterivian. The upper part of the Sid'r Formation has an Aptian–Albian age (Wohlwend *et al.* 2017). We adopt the ages of Wohlwend *et al.* (2017).

The Nayid Formation (22) has been dated as Late Albian–Coniacian (Béchennec *et al.* 1992*a*). Wohlwend *et al.* (2017) indicated an Albian age for the base of the formation. The top of this formation has a Cenomanian–Coniacian to possibly Santonian age (Blechschmidt *et al.* 2004). We use the ages of Blechschmidt *et al.* (2004) and Wohlwend *et al.* (2017).

The 30-m-thick formation contains turbiditic pelletal calcarenite (Béchennec *et al.* 1986). In the southernmost part of our study area, the red shale of the Wahrah Formation (Glennie *et al.* 1974) is exposed in only one outcrop. The Wahrah Formation is considered as a distal facies of the Nayid and Sidr formations. The Wahrah Formation has a much larger age overlap with the Sidr Formation than the Nayid Formation. We therefore incorporated the Wahrah Formation, which represents a large portion of the Hawasina nappes, into the Sidr Formation south of our study area. The base of the Wahrah Formation has been dated as Oxfordian–Tithonian (Béchennec 1987; Béchennec *et al.* 1992*b*). Due to the absence of biota, the top of the formation has a poorly defined Coniacian or post-Coniacian age (Béchennec *et al.* 1992*b*). The Wahrah Formation represents a distal lithofacies within the Hamrat Duru Group (Robertson and Searle 1990).

Al-Aridh Group. The Sayfam Formation (23) of (Béchennec 1987; Béchennec *et al.* 1992*a*) contains a lower volcanic and an upper carbonate and/or radiolarite unit (Béchennec *et al.* 1992*a*). The upper member has a Mid Triassic–Early Norian age (Béchennec *et al.* 1992*b*). The thickness of the lower unit measures *c.* 350 m, and that of the upper unit 23 m (Béchennec *et al.* 1992*a*).

The Buwaydah Formation (24) of Béchennec (1987; Béchennec *et al.* 1992*b*) consists of thick-bedded, turbiditic calcarenite with an apparent thickness of 70 m south of the JAD and 80 m in the Batinah Coastal Plain (Béchennec *et al.* 1992*a*). The age of this formation is Early–Mid Jurassic (Béchennec *et al.* 1992*a*).

The Al-Ghafat and Musallah formations (25) are 50 m thick (combined thickness), consisting of radiolarian chert and calcirudite. The Al-Ghafat Formation is of Berriasian to probably Valanginian age (Béchennec *et al.* 1990, table 1; Pillevuit *et al.* 1997) while the Musallah Formation is of Valanginian–Santonian age (Béchennec *et al.* 1990, table 1). Pillevuit *et al.* (1997) indicated a Senonian age for the top of the Musallah Formation.

Kawr Group. Similar to the Hamrat Duru Group, the Kawr Group's stratigraphic interpretations differ between authors. We compiled the information in Figure 2.8 from Pillevuit *et al.* (1997, fig. 3).

The Upper Permian–Lower Triassic Baid Formation (26) (Robertson and Searle 1990, fig. 3) is characterized by shallow-marine, fusulinid-bearing carbonates of up to 400 m thickness (Béchennec *et al.* 1992*a*). The depositional environment was an inner shelf (Béchennec *et al.* 1992*a*). The age ranges from Late Permian to Early Triassic (Robertson and Searle 1990, fig. 3). Pillevuit *et al.* (1997) suggested a Wordian age for the formations's base and a Midian (Capitanian) to Dzulfian (Wuchiapingian) age for its top (fig. A.3). We follow the ages of Pillevuit *et al.* (1997).

The Triassic (Ladinian–Carnian) to Middle Jurassic (see Pillevuit *et al.* 1997; Robertson and Searle 1990, fig. 3) Misfah Formation (27) occurs at the SW flank of the JAD and forms the impressive cliffs of the Jabal Misfah (Béchennec *et al.* 1986, 1988, 1990, 1992*a*; Beurrier *et al.* 1986; Pillevuit *et al.* 1997; Fig. 2.1). According to Béchennec *et al.* (1992*a*), the formation is divided into a lower member of volcanic rocks (100 m thick) and an upper member of shallow-marine limestone and dolomite (800 m thick). The age of the lower volcanic member is not well-constrained, but benthic foraminifera in its upper part suggest a Ladinian–Carnian age (see Pillevuit *et al.* 1997). The upper member represents an inner shelf to intertidal-supratidal palaeoenvironment (Béchennec *et al.* 1992*a*). The thick limestone has been interpreted as time equivalent to the Mahil Formation.

Pillevuit *et al.* (1997) divided the Misfah Formation of Béchennec *et al.* (1992*a*) into three formations. The lower formation is the 80-m-thick volcanic Subayb Formation, followed by the Misfah Formation of 700–800 m thickness and the Fatah Formation of 1–6 m thickness. The latter has a late Early Bajocian age, which would represent the youngest ages of the Misfah Formation of Béchennec *et al.* (1992*a*); see Pillevuit *et al.* (1997, fig. 3) for a comparison of the different formations and their ages. We adopt the ages of Pillevuit *et al.* (1997) in our compilation (figs A.1 & A.2).

The Nadan Formation (28) mainly includes pale grey, silicified micritic limestone of thickness 25 m (Glennie *et al.* 1974; Beurrier *et al.* 1986, fig. 12). The age of this formation is Tithonian–Albian (Robertson and Searle 1990, fig. 3) and, according to Pillevuit *et al.* (1997), Late Oxfordian–Valanginian. Based on the Tithonian microfacies, Beurrier *et al.* (1986) deduced a subtidal to bathyal sedimentary environment. We use the ages provided by Pillevuit *et al.* (1997) (figs A.1 & A.2).

The Safil Formation (29) consists of calcirudite and pelagic clayey limestone, is 15 m thick (Beurrier *et al.* 1986, fig. 12) and is of Cenomanian–Santonian age (Robertson and Searle 1990, fig. 3). The depositional environment was bathyal (Beurrier *et al.* 1986). Pillevuit *et al.* (1997) suggested an Early Cenomanian–Mid Turonian age for this formation. We use the ages of Pillevuit *et al.* (1997) (figs A.1 & A.2).

Umar Group. Searle *et al.* (1980) as well as Searle and Malpas (1980) introduced the Haybi Complex as a highly deformed rock unit situated between the Hawasina Complex and the Semail Ophiolite. The Haybi Complex consists of sedimentary, metamorphic and igneous rocks. Searle (2007) defined the Haybi Complex as a distal ocean-trench facies rocks. The Haybi Complex represents one thrust sheet (Searle 2007). The authors of the BRGM maps do not use the term 'Haybi Complex'; the term 'Umar Group' was instead introduced to describe the distal Hawasina sediments, including volcanic rocks. Some of the units of the Haybi Complex have been assigned by the BRGM map authors to the other groups of the Hawasina nappes (Robertson and Searle 1990). This reorganization introduced controversy because the Haybi Complex is structurally very complicated, locally variable and largely comprises mélange (see Robertson and Searle 1990 for further details). Stratigraphically, we prefer the Umar Group nomenclature of the BRGM authors over the term 'Haybi Complex' because the geological map of figure A.1 is based on the BRGM maps.

The 150-m-thick Triassic Sinni Formation (30) mainly comprises pillow lava, overlain by radiolarian chert and boulders of shallow-marine carbonates (Béchennec *et al.* 1992*a*).

The Aqil Formation (31) rests conformably above the Sinni Formation. The thickness of this formation is *c.* 200 m (Béchennec *et al.* 1992*a*). The lithology includes in the lower part megabreccias and calcirudites with blocks and boulders of upper Permian–Late Triassic shallow-marine carbonates (Béchennec *et al.* 1992*a*). The upper part consists of

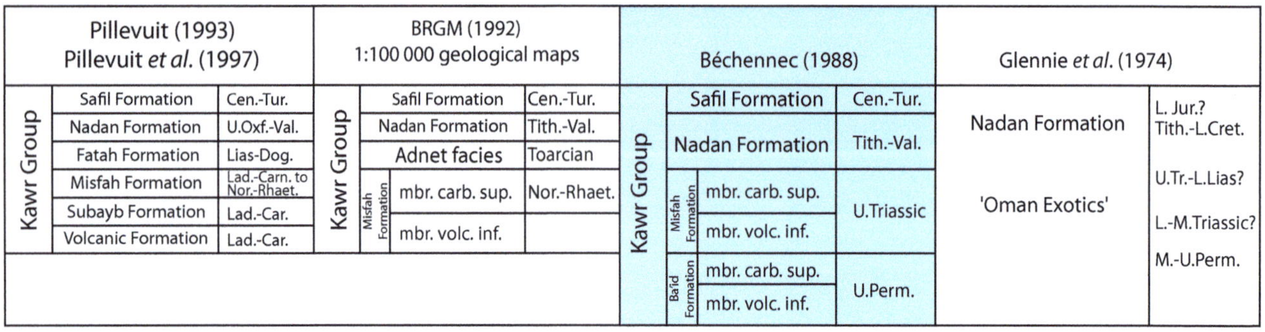

Fig. 2.8. Stratigraphic correlation of the Kawr Group members slightly modified from Pillevuit *et al.* (1997, fig. 3). The blue column with some age modifications from Pillevuit *et al.* (1997) has been selected for figures A.1 and A.2 (see text for further details).

radiolarian chert and micritic limestone. The base of this upper unit is Carnian–Norian in age (Beurrier *et al.* 1986), whereas the upper part of this formation has been dated as Hauterivian–Early Aptian (Béchennec 1987; Béchennec *et al.* 1992*a*).

Metamorphic sole. Obduction of the hot Semail Ophiolite caused the formation of a somewhat discontinuous metamorphic sole (32) among the upper sections of the uppermost Haybi Complex thrust sheet beneath the Semail Ophiolite (e.g. Searle 1985; Cooper 1989). The metamorphic sole is pronounced north of the JAD and along the southwestern SHD. However, the best exposure of the metamorphic sole in our study area is located in Wadi Tayin (southwestern margin of the SHD; Fig. 2.1). Occasionally, the metamorphic sole occurs directly above the Hawasina Complex due to Late Cretaceous–Cenozoic deformation (Béchennec *et al.* 1992*a*). Locally, up to three successive thrust events significantly increased the apparent thickness of the metamorphic sequence (Gnos 1998; Soret *et al.* 2017), doubling or tripling the average thickness of the metamorphic sole.

The metamorphic sole displays an inverse metamorphic gradient (e.g. Searle and Malpas 1980, 1982; Ghent and Stout 1981; Béchennec *et al.* 1992*a*; Searle and Cox 1999, 2002). The entire metamorphic sole has been affected by retrograde greenschist-facies metamorphism (Béchennec *et al.* 1992*a*). The metamorphic sole is subdivided into a lower and upper unit.

The lower unit consists of highly deformed metasediments (quartzite, marble and metachert), the protoliths of which are similar to the non-metamorphic cherts, distal volcanics and sediments seen in the unmetamorphosed Haybi Complex (e.g. Searle 1985).

The upper unit contains amphibolites with a primary composition of tholeiitic mid-ocean ridge basalts, which geochemically differ from the mafic rocks of the overlaying Semail Ophiolite (Searle and Malpas 1982; Béchennec *et al.* 1992*a*; Searle and Cox 2002). The protoliths of the metabasalts and/or amphibolites in the metamorphic sole are the Haybi volcanics (Searle and Malpas 1980, 1982). Besides the metamorphosed Haybi volcanics, metalimestones of the Oman Exotics are also present (Searle and Cox 2002). The rocks below the Semail Ophiolite were therefore suggested to be old, cold and Triassic–Jurassic in age (Searle and Cox 1999, 2002). The ocean floor below the Semail Ophiolite is older (*c.* 250–145 Ma) than the Semail Ophiolite (*c.* 96 Ma) and was exhumed from >35 km depth beneath the ultramafic rocks of the ophiolite (e.g. Cowan *et al.* 2014). They therefore seem to represent an upside-down sequence with basalts emplaced below ultramafic rocks (Hacker 1994; Hacker *et al.* 1996), and can be interpreted as relicts from a supra-subduction zone setting in which subduction occurred below a spreading centre (e.g. Cowan *et al.* 2014; Rioux *et al.* 2016). The formation of the discontinuous metamorphic sole took place during subduction infancy, and its evolution is associated with multiple stages of deformation and plate interface coupling (Soret *et al.* 2017). See section on 'Intraoceanic thrusting, formation and obduction of the Semail Ophiolite' in Chapter 5 for further discussion of the origin and thrusting of the metamorphic sole and the Semail Ophiolite.

The geochronological work of Rioux *et al.* (2016) produced U–Pb zircon ages of 94.815 ± 0.03 Ma (weighted mean of four zircon ages) from an amphibolite-facies diorite pod within the metamorphic sole of Wadi Tayin (Fig. 2.1). This age has been interpreted to indicate the age of peak metamorphism by the same authors. The ages provided by Rioux *et al.* (2016) partly overlap with ages provided by other researchers (see Table 2.2). Similar U–Pb zircon ages of 94.48 ± 0.24 Ma and 94.9 ± 0.38 to 94.69 ± 0.12 Ma on a trondhjemitic pod from Wadi Tayin were reported by Warren *et al.* (2005) and Rioux *et al.* (2013), respectively. The reader is referred to Cowan *et al.* (2014), Rioux *et al.* (2016) and Soret *et al.* (2017) for discussion of all previous studies, their uncertainties and re-interpretations. Recently, Guilmette *et al.* (2018) analysed three garnets from the metamorphic sole of Wadi Tayin and Sumeini using the garnet–whole-rock Lu–Hf technique. The obtained ages from Wadi Tayin are 104.1 ± 1.1 Ma and 103.2 ± 1.2 Ma, while the Sumeini sample has an age of 103.5 ± 1.6 Ma. The weighted mean of these ages is 103.7 ± 0.7 Ma. Based on microtextures and major- and trace-element distribution, Guilmette *et al.* (2018) interpreted that the garnets most likely formed during a single fast garnet growth event at sub- to supra-solidus conditions from *c.* 550°C and 0.8 GPa to 850°C and 1.1–1.3 GPa. The Albian garnet ages are critical for the interpretation of the subduction initiation (see section on 'Intraoceanic thrusting, formation and obduction of the Semail Ophiolite' in Chapter 5).

Furthermore, the same authors analysed five zircon populations and four titanite grains from Wadi Tayin using the $^{206}Pb/^{238}U$ method. The obtained average ages are 96.19 ± 0.14 Ma and 95.6 ± 0.27 Ma, respectively. The zircon ages are interpreted as crystallization ages while the titanite ages are cooling ages from 650 to 600°C (Guilmette *et al.* 2018).

Semail Ophiolite

The Semail Ophiolite in the Oman Mountains constitutes the best-preserved, least-deformed, best-exposed and most widely distributed ophiolitic sequence in the world. Here, a complete section of the oceanic lithosphere that has been thrust towards the SW can be observed. Many studies of the Semail Ophiolite have been conducted, for example, Reinhardt (1969), Allemann and Peters (1972), Glennie *et al.* (1974), Coleman and Hopson (1981), Lippard *et al.* (1986), Beurrier (1987), Nicolas *et al.* (1988*a*, *b*, 1996, 2000*a*), Goodenough *et al.* (2010), Kusano *et al.* (2012, 2014), MacLeod *et al.* (2013) and Rioux *et al.* (2013, 2016). The tectonic evolution of the ophiolite is under debate. The Semail Ophiolite either formed on a conventional, open-ocean, mid-ocean ridge-spreading centre or in a supra-subduction zone setting (see section on 'Intraoceanic thrusting, formation and obduction of the Semail Ophiolite' in Chapter 5 for further discussion). Spreading rates along the palaeo-ridge system are of the order of 5–12 cm a^{-1} (Hacker *et al.* 1996; Nicolas *et al.* 2000*a*) and therefore considered as a fast-spreading system (Pallister and Hopson 1981; Tilton *et al.* 1981; Nicolas *et al.* 2000*a*).

A schematic cross-section through the Semail Ophiolite with its mantle and crustal sequences, and the underlying metamorphic sole, is depicted in Figure 2.9. The ophiolite consists of an early magmatic (section on 'Early magmatic sequence' below; units 33–38) and a late magmatic suite (section on 'Late magmatic sequence' below; units 39–42). Rocks of the early magmatic sequence formed either on a normal mid-ocean ridge or above a subduction zone, whereas the rocks of the late magmatic sequence formed above a subduction zone (e.g. compare MacLeod *et al.* 2013; Goodenough *et al.* 2014; section on 'Intraoceanic thrusting, formation and obduction of the Semail Ophiolite' in Chapter 5). Weighted mean ages for rocks of the early and late magmatic sequences are 96.121 ± 0.051 to 95.504 ± 0.031 Ma and 95.447 ± 0.092 to 95.240 ± 0.025 Ma, respectively (high-precision single-grain U–Pb zircon technique; Rioux *et al.* 2016). The main

Table 2.2. *Compilation of geochronological data from the Semail Ophiolite and the metamorphic sole (Northern, Central and Southeastern Oman Mountains)*

	North	Central	Southeast
Early magmatic sequence	96.121 ± 0.051 to 95.504 ± 0.031 (U–Pb); Rioux *et al.* (2013, 2016)	96.03 ± 0.12 (U–Pb); Rioux *et al.* (2016)	96.46 ± 0.18 to 95.535 ± 0.063 (U–Pb); Rioux *et al.* (2016) 96.40 ± 0.17 to 95.50 ± 0.17 (U–Pb); Rioux *et al.* (2016)
Late magmatic sequence	95.74 ± 0.32 and 95.26 ± 0.31 (U–Pb); Goodenough *et al.* (2010) 96.4 ± 0.29 to 94.74 ± 0.54 (U–Pb); Styles *et al.* (2006) 93.3 ± 0.5 (hbl); Hacker *et al.* (1996) 89.1 ± 0.4 (wm) from garnet-bearing granite-intruding peridotites; Gnos and Peters (1993)	94.1 ± 0.5 (hbl) granitic dyke cutting harzburgite and cumulate layered gabbro; Hacker *et al.* (1996)	95.3 ± 0.2 (U–Pb); Warren *et al.* (2005) 94.8 ± 0.1 (U–Pb); Hacker *et al.* (1996) 96.3 ± 1.3 to 93.6 ± 0.5 (hbl); Hacker *et al.* (1996) (Weighted mean 94.1 ± 0.5 (hbl)); Hacker *et al.* (1996)
Metamorphic sole	103.5 ± 1.6 (Lu–Hf on garnet); Guilmette *et al.* (2018) (Weighted mean 96.169 ± 0.022 to 96.146 ± 0.035 (U–Pb)); Rioux *et al.* (2016) 95.69 ± 0.25 to 95.29 ± 0.21 (U–Pb); Styles *et al.* (2006) 95.7 ± 0.3 to 92.6 ± 0.6 (hbl); Hacker *et al.* (1996) (Weighted mean 94.9 ± 0.2 (hbl)); Hacker *et al.* (1996) (Weighted mean 92.7 ± 0.7 (wm)); Hacker *et al.* (1996) (Weighted mean 89.2 ± 0.4 (bt)); Hacker *et al.* (1996) 92.5 ± 4 to 85.5 ± 5 (bt); Allemann and Peters (1972) Searle *et al.* (1980), Gnos and Peters (1993)		104.1 ± 1.1 to 103.2 ± 1.2 (Lu–Hf on garnet); Guilmette *et al.* (2018) 96.19 ± 0.14 (U–Pb); Guilmette *et al.* (2018) 95.099 ± 0.055 to 94.777 ± 0.058 (U–Pb); Rioux *et al.* (2016) (Weighted mean 94.815 ± 0.030 (U–Pb); Rioux *et al.* (2016) 94.9 ± 0.38 to 94.69 ± 0.12 (U–Pb); Rioux *et al.* (2013) 94.48 ± 0.23 (U–Pb); Warren *et al.* (2005) 95.6 ± 0.27 (U–Pb on titanite); Guilmette *et al.* (2018) 94.9 ± 0.5 to 92.6 ± 0.6 (hbl); Hacker *et al.* (1996) (Weighted mean 93.5 ± 0.1 (hbl); Hacker *et al.* (1996) 93.6 ± 0.5 to 90.0 (wm); Hacker *et al.* (1996) (Weighted mean 92.4 ± 0.2 (wm); Hacker *et al.* (1996) 93.8 ± 0.9 to 90.4 ± 0.8 (ms); Gnos and Peters (1993)
Post metamorphic sole Formation intrusions	89.9 ± 0.4 (wm) stock intruding the metamorphic sole Gnos and Peters (1993)	93.7 ± 0.8 (hbl) diabase dyke cutting the metamorphic sole Hacker *et al.* (1996)	

Ages in Ma. Mineral systems provided in parantheses. U–Pb: U–Pb on zircon; hbl: K/Ar on hornblende; wm: K/Ar on white mica; ms: K/Ar on muscovite; bt: K/Ar on biotite.

part of the crust of the Semail Ophiolite therefore has an age of 96.12–95.50 Ma (Rioux *et al.* 2016).

The metamorphic sole in Wadi Tayin has an age of 94.815 ± 0.030 Ma (Rioux *et al.* 2016; see also section on 'Metamorphic sole' below). Rioux *et al.* (2016) reported U–Pb zircon ages of the metamorphic sole from the Northern Oman Mountains (Sumeini) of 96.169 ± 0.022 to 96.146 ± 0.035 Ma. Metamorphism on the sole took place just before or during the formation of the crust of the Semail Ophiolite. According to Rioux *et al.* (2016), this implies that the conventional mid-ocean ridge spreading centre model for the ophiolite origin is not valid as the sole would have to postdate ophiolite formation. Table 2.2 summarizes available geochronological data from the Semail Ophiolite and its metamorphic sole.

The uppermost part of the ophiolite contains Cenomanian–Turonian radiolarian cherts (Tippit *et al.* 1981) only in the Sohar area in the Northern Oman Mountains, immediately above the V1 lavas (Geotimes Unit), corroborating the radiometric dating of Tippit *et al.* (1981; fig. A.1).

As a result of the excellent exposure of the Semail Ophiolite, Nicolas *et al.* (1988*a*, 2000*a*), Ceuleneer *et al.* (1988) and Ceuleneer (1991) mapped several mantle diapirs up to the Moho with a size of *c.* 10 km (Fig. 2.10). Furthermore, Nicolas *et al.* (1988*b*, 1993) mapped internal structures of magma chambers, while Boudier and Nicolas (1995) determined ridge segments. Nicolas *et al.* (1996) constructed 32 cross-sections through the lower crust of the ophiolite. Figure 2.11 depicts 24 of those cross-sections, which are located in the vicinity of the JAD and SHD. Based on their detailed field work, these authors concluded that the gabbro section of the Semail Ophiolite is 0.5–1 km thinner than the average 6-km-thick section of fast-spreading ridges. Furthermore, the thickness of the gabbro varies by 1–2 km (Nicolas *et al.* 1996). Thinner gabbro sections are related to mantle diapirs. The Moho therefore shifts above a diapir towards the surface.

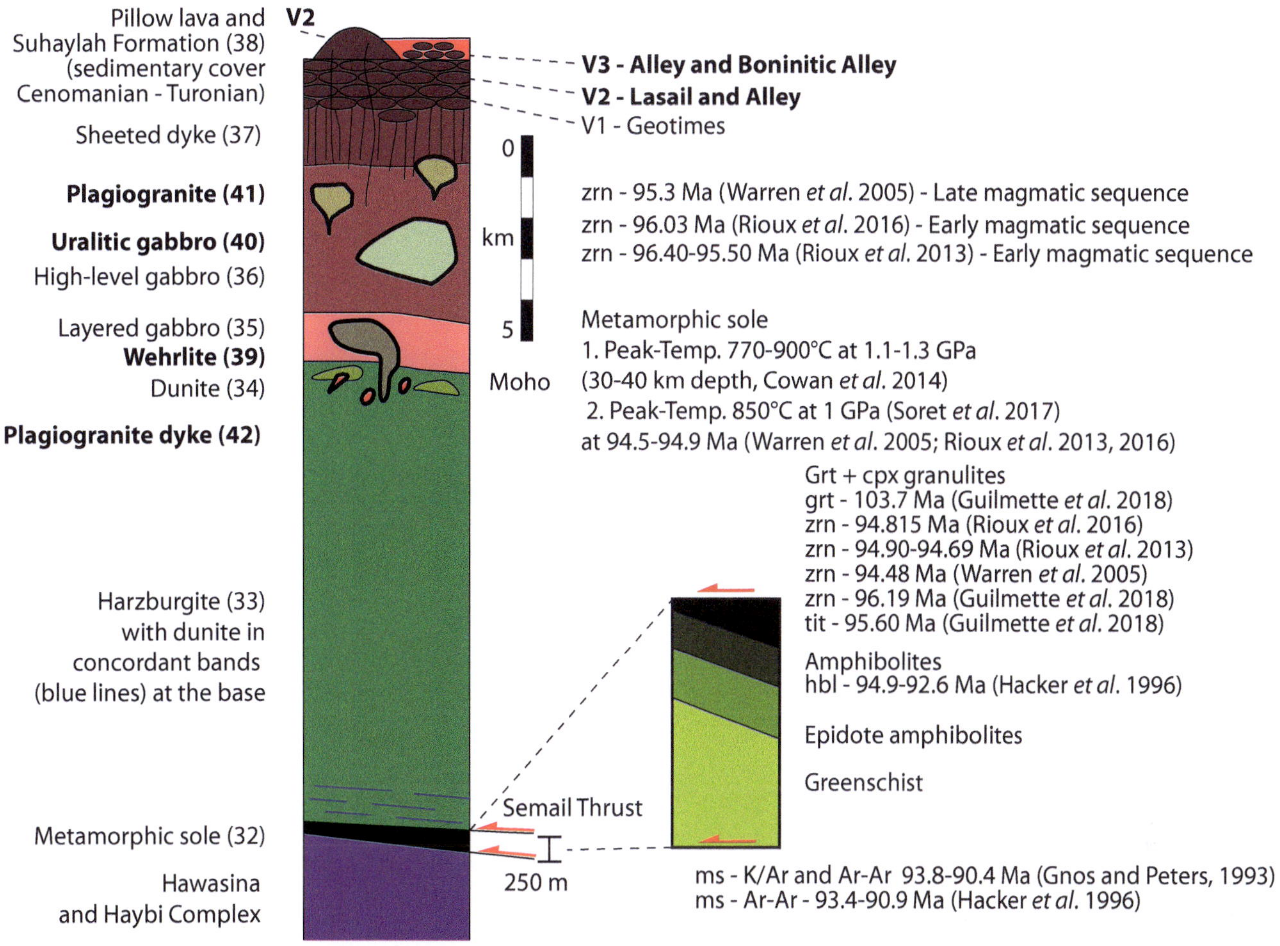

Fig. 2.9. Sketch of a complete section of the Semail Ophiolite, the subjacent metamorphic sole and Hawasina and Haybi Complex thrust sheets. Age data are compiled for the Southeastern Oman Mountains. The ages may differ slightly in the Northern Oman Mountains. More age data are listed in Table 2.2. Units, numbers in brackets and colours correlate with those in figures 4.1, A.1 and A.2. Modified after Hacker *et al.* (1996), Searle (2007) and Cowan *et al.* (2014). Units highlighted in regular and bold formatting belong to the early and late magmatic sequence, respectively. bt, K/Ar on biotite; hbl, K/Ar on hornblende; ms, K/Ar or Ar/Ar on muscovite; zrn, U–Pb on zircon; grt, garnet; cpx, clinopyroxene.

Early magmatic sequence. The lower part of the Semail Ophiolite mainly consists of harzburgite (33). It is a 5–7-km-thick mantle sequence, which also contains pyroxenite dykes, some dunite bodies and plagiogranites (trondhjemites).

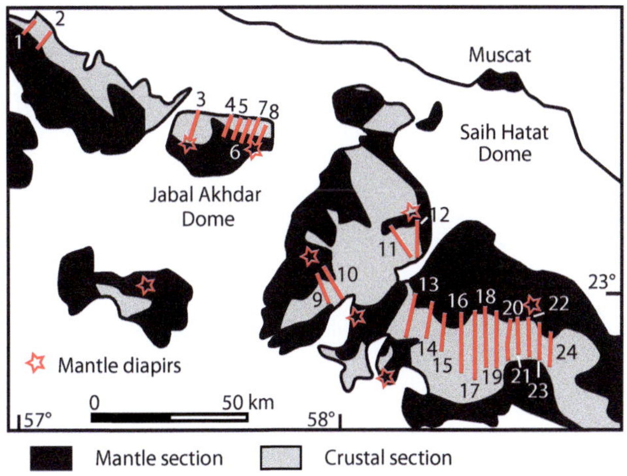

Fig. 2.10. Overview map of the Semail Ophiolite depicting the mantle and crustal rocks near the Jabal Akhdar and Saih Hatat domes, redrawn from Nicolas *et al.* (1996). The red lines with corresponding numbers are the traces for the cross-section of Figure 2.11.

Harzburgite typically forms dark, rolling hills and is locally highly serpentinized. The original base of the Harzburgite which formed during the Late Cretaceous is not exposed. Dunite (34) forms massive lensoid bodies in the upper part of the harzburgite, oriented parallell to the Moho with a long axis of up to *c.* 1000 m in width (Beurrier *et al.* 1986). Concordant bands of dunite are common at the base of the harzburgite (fig. A.1). Mineable podiform chromite deposits occur frequently within the dunite units, the transitional zone (TZ) between mantle and oceanic crust or at shear zones (Ceuleneer and Nicolas 1985; Schiano *et al.* 1997; Boudier and Al-Rajhi 2014; see section 'Metallogenic history' below). The thickness of the TZ between mantle and oceanic crust varies. The TZ measures several hundred metres in thickness above mantle diapirs, which can be recognized by steep foliations and lineations. Furthermore, magmatic and plastic flow is common (Nicolas *et al.* 1996). Away from the diapirs, the thickness of the TZ ranges from a few to tens of metres and is characterized by high-temperature plastic flow (Nicolas *et al.* 1996). Its composition is mainly dunitic with different amounts of plagioclase and clinopyroxene. It also contains pyroxenitic, gabbroic and chromitic segregations and xenolithic slabs of the overlying gabbros (Boudier and Nicolas 1995; Nicolas *et al.* 1996; Fig. 2.12).

The Moho is located at the base of the layered gabbro above the TZ (Nicolas *et al.* 1996), oriented parallel to the underlying mantle rocks and the overlying layered gabbro (Nicolas *et al.* 1996).

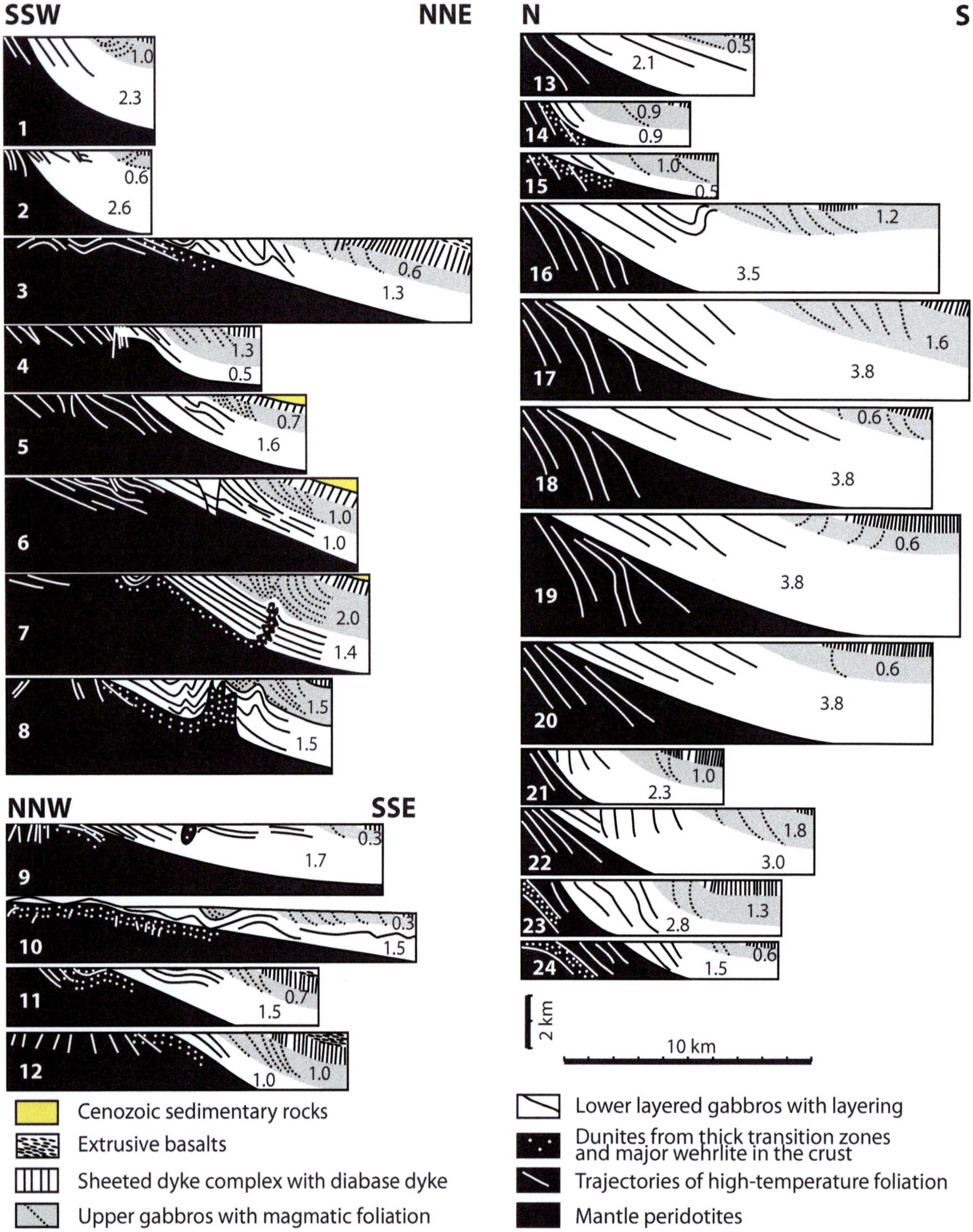

Fig. 2.11. Cross-section of the Semail Ophiolite redrawn from Nicolas *et al.* (1996). Traces are shown in Figure 2.10. Numbers to the right indicate the thickness of the respective section in kilometres.

Layered and planar-laminated gabbros (35) show a total thickness of 1–5 km, of which the layered gabbro encompasses 1–4 km. Planar-laminated gabbros, by contrast, may reach a maximum thickness of 900 m (Béchennec *et al.* 1992*a*). Variations in the mineral composition (mainly olivine, clinopyroxene and plagioclase) lead to centimetre- to decimetre-thick layers of leuco- and melagabbros (Béchennec *et al.* 1992*a*) that occasionally resemble sedimentary structures (Beurrier *et al.* 1986). Planar-laminated gabbros lack the compositional layering and derive their planar fabric from the alignment of plagioclase and pyroxene crystals. The contacts between the crustal units (35–38) are gradual

Fig. 2.12. The Moho in Wadi Al-Abyad, between Rustaq and Nakhl. The lower part of the outcrop consists of ultramafic rocks, and the upper part is represented by gabbro with layers of ultramafic rocks (not shown in Fig. 2.12).

and highly variable. The layered gabbros are overall parallel to the Moho. Layered gabbros may be tilted close to large-scale folds and wehrlite intrusions (Nicolas *et al.* 1996).

High-level gabbros (36) are isotropic and have a thickness of up to several hundred metres. The gabbros are fairly coarse-grained (millimetre- to centimetre-scale) and are dominated by plagioclase, green hornblende, clinopyroxene and Fe–Ti oxides. Epidote, actinolite and chlorite may have replaced the primary silicates during retrograde hydrothermal metasomatism (Béchennec *et al.* 1992*a*). The boundary between the dark harzburgite and the much lighter gabbro is commonly obvious, even from a distance and on satellite images. The gabbro units have been affected by large-scale folding in Wadi Tayin and the Central Oman Mountains. This might have resulted in overestimations of the thickness of the gabbro, previously thought to measure 3–7 km (Pallister and Hopson 1981; Lippard *et al.* 1986) but later remapped as 1.5–5 km (Fig. 2.11; Nicolas *et al.* 1996).

The thickness of the overlying sheeted dyke complex (37) amounts to aproximately 1500 m (Lippard *et al.* 1986), and it is characterized by subvertical dykes with an individual thickness of up to 3 m. The sheeted dyke complex consits of subparallel dolorite dykes (e.g. Rabu *et al.* 1986) composed of three different types of rock, as summarized by Béchennec *et al.* (1992*a*). The first type consists of plagioclase-rich doleritic microgabbros, clinopyroxene and Fe–Ti oxides that has been overprinted by hydrothermal minerals such as actinolite, chlorite, epidote, accessory quartz and leucoxene. The second type contains amphibole microgabbro and/or microdiorite with green hornblende and plagioclase, hydrothermally overprinted by a new mineral assemblage of actinolite, chlorite and albite. The third type consists of micro-quartz diorite with albitized plagioclase, green hornblende and some quartz; this rock type was also affected by a hydrothermal alteration.

Gabbro and basalt are found between the dykes, which consist of plagioclase-rich, doleritic micro-gabbros, amphibole micro-gabbros and micro-quartz-dolerite (Béchennec *et al.* 1992*a*). The gabbro and basalts within the sheeted dyke complex represent 30% of the total volume of the complex (e.g. Rabu *et al.* 1986). Most of the dykes display chilled margins and strike approximately NW–SE in the studied areas (Hacker *et al.* 1996; Nicolas *et al.* 2000*b*; fig. 1.1). The transition zone between the gabbros and the sheeted dykes complex amounts to <100 m (e.g. Pallister and Hopson 1981).

The uppermost pillow lavas (38) can reach a cumulative thickness of 800 m and consist of basaltic lava of mid-ocean ridge basalt (MORB) composition (Beurrier 1987; Béchennec *et al.* 1992*a*). The pillows display diameters of up to 80 cm and chilled margins (Béchennec *et al.* 1992*a*). At one location (*c.* 25 km NW of Rustaq), the pillows are covered by 5–10-m-thick chert of Santonian–Campanian age (Suhaylah Formation at the northwestern margin of fig. A.1; Beurrier *et al.* 1987; Béchennec *et al.* 1992*a*). More recent dating of the chert revealed a Cenomanian age (Cowan *et al.* 2014).

Following the classification of lavas by Alabaster *et al.* (1982), Godard *et al.* (2003) suggested a systematization scheme that defines the volcanostratigraphy of three major volcanic units for the basaltic suite of the ophiolite (from bottom to top): (1) Geotimes (V1); (2) Lasail and Alley (V2); and (3) Alley and Boninitic Alley (V3). The V1 unit represents the early magmatic sequence, resembling MORB.

The lowermost Geotimes unit (V1) covers the co-magmatic sheeted dyke complex. The thickness of the V1 unit measures *c.* 0.75 km. Reddish-brown oxidation colours are a typical feature, readily recognizable in the field. The basaltic lavas are aphyric and tabular, while the less common andesitic lavas show large pillows. Occasionally, they are accompanied by massive sheets, breccias and hyaloclastites. The Geotimes unit (V1) lavas and modern MORB have a similar geochemical signature (Fig. 2.13; Ernewein *et al.* 1988; Godard *et al.* 2006). However, compared with normal mid-ocean ridge basalts (N-MORB), Geotimes lavas show an anomalous depletion of high-field strength elements (HFSEs: rare earth elements, Zr, Hf, Th, U, Nb, Ta; Alabaster *et al.* 1982; Lippard *et al.* 1986; Godard *et al.* 2003; Gilgen *et al.* 2014).

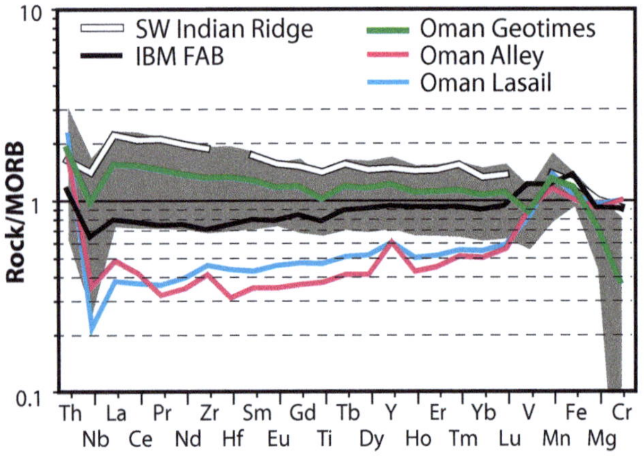

Fig. 2.13. Spider diagram normalized for the normal mid-oceanic ridge basalt (N-MORB) redrawn from MacLeod *et al.* (2013). Oman Geotimes (V1; *n* = 186–864, depending on element) axial lavas and dykes versus Oman Lasail (V2; *n* = 100) and Oman Alley (V3; *n* = 81) lavas and dykes. Thick V1 line represents the mean value while the shaded area is the ± 1 standard deviation. The SW Indian Ridge MORB and Izu-Bonin-Mariana (IBM) forearc basalts are from Coogan *et al.* (2004; *n* = 72) and Reagan *et al.* (2010 *N* = 23), respectively.

Late magmatic sequence. Plagioclase wehrlites and melagabbros (39) are associated with clinopyroxenite, dunite and enclaves of cumulate gabbro, representing crustal parts of the ophiolite and forming bodies of up to several kilometres in diameter that intruded the crust (Béchennec *et al.* 1992*a*). Associated with the wehrlites are sills and enclaves of cumulative layered gabbros.

According to Koga *et al.* (2001), several authors noticed that wehrlitic intrusions in Oman do not possess chilled margins and that they underwent ductile deformation together with their host rocks (Pallister and Hopson 1981; Benn *et al.* 1988; Juteau *et al.* 1988; Nicolas *et al.* 1988*a*; Reuber 1988). The intrusions must therefore have occurred while the gabbros were hot and close to the active spreading centre (Koga *et al.* 2001). Since wehrlite bodies have not been observed in the Oman mantle section, they must have originated from within the mantle transition zone (Koga *et al.* 2001).

Uralitic gabbros and gabbronorites (40) are rich in plagioclase, clinopyroxene, green hornblende, orthopyroxene and Fe–Ti oxides. Apparently, they intruded high-level gabbros, where intrusions of several kilometres in diameter have been found (Béchennec *et al.* 1992*a*). The uralitic gabbro resembles an anatectic migmatite due to partial melting of felsic components within a gabbronorite residue (Tsuchiya *et al.* 2013).

Unit 41 consists of large plagiogranite (see trondhjemite; Le Maître 1989) and quartz diorite bodies with a size of up to several kilometres across (Béchennec *et al.* 1992*a*), which intruded the Semail Ophiolite at a relatively late stage, partially cross-cutting the sheeted dyke complex (38). The unit mostly contains plagioclase as andesine and albite, quartz and minor green hornblende (Béchennec *et al.* 1992*a*).

Plagiogranitic dykes (42) of up to tens of metres in width mostly intruded the ultramafic units (especially harzburgite, unit 33). These rocks crystallized during 95.9–93.5 Ma (11 of 13 samples) and 97.9–97.0 Ma (2 of 13 samples; U–Pb zircon; Tilton *et al.* 1981). Rioux *et al.* (2013) dated a granite intrusion in the Central Oman Mountains (Haylayn Block) as 95.24 ± 0.43 Ma ($^{206}Pb/^{238}U$ zircon age). The composition of the granite intrusions ranges from diorite with amphibole, K-feldspar and biotite-bearing to granite-bearing muscovite and tourmaline (Béchennec *et al.* 1992*a*).

The late-stage magmatic sequence (including the V2 and V3 units) has been extensively studied to understand the tectonic and/or genetic 'environment' of the Semail Ophiolite. A mid-ocean ridge and a supra-subduction zone setting are the two discussed tectonic and/or genetic environments of the Semail Ophiolite (see section on 'Intraoceanic thrusting, formation and obduction of the Semail Ophiolite' in Chapter 5).

The V2 units have a typical geochemical subduction signature, indicating formation of the Semail Ophiolite above a subduction zone (e.g. Godard *et al.* 2003; MacLeod *et al.* 2013; Goodenough *et al.* 2014; Haase *et al.* 2015, 2016). The V3 unit is enriched with incompatible elements, post-dates intra-oceanic detachment of the Semail Ophiolite and pre-dates obduction onto the Arabian margin (Ernewein *et al.* 1988). The V3 volcanism is interpreted as an intraplate seamount-type magmatism (Ernewein *et al.* 1988) or as an off-axis volcanism (Haase *et al.* 2015).

The Lasail lavas (V2) comprise basaltic to andesitic pillow lavas and massive porphyritic units, and measure *c.* 0.75 km in thickness. With respect to N-MORB compositions, the Lasail unit has lower light rare earth elements (LREE) and incompatible element contents (Fig. 2.13). Refractory elements of the Lasail units are depleted when compared with the Geotimes unit (Godard *et al.* 2006). Alabaster *et al.* (1982) considered the Lasail lavas to be seamounts since they only occur in a few places, whereas Godard *et al.* (2003) suggested that unit V1 is the equivalent to the Lasail lavas (V2) and exists throughout the ophiolite. In contrast to Godard *et al.* (2003), Kusano *et al.* (2012) and Gilgen *et al.* (2014) explained the Lasail unit (V2) as a sub-unit of the Geotimes unit (V1). The V2 sequence is geochemically identical with island-arc lavas (Ishikawa *et al.* 2002), and has formed above a subducting slab (e.g. Haase *et al.* 2015).

The Alley and V3 sequence (Ishikawa *et al.* 2002; Gilgen *et al.* 2014) is also *c.* 0.75 km thick and comprises amygdaloidal, micro-porphyritic pillow basalts that generally cover the Lasail and Geotimes units. The basalts are often brecciated. Vacuoles are filled with zeolites and celadonite. Mineralogically and macroscopically, this clinopyroxene-phyric unit as outlined by Alabaster *et al.* (1982) is equivalent to the Boninitic Alley lavas. Geochemically, the Alley lavas are characterized by slightly lower chromium, higher REE and higher Zr contents than the underlying Lasail lavas.

Metallogenic history. The late-stage magmatic rock suite hosts sulfide ores. Lachize *et al.* (1996) and Lorand and Juteau (2000) also described an enrichment of platinum group elements (PGEs) within the ultramafic intrusions of Wadi Haymiliyah (NW of Rustaq; Fig. 2.1), which are quite rare in other locations of the Semail Ophiolite.

Within the Semail Ophiolite, several chromium and copper deposits occur as shown in Figure 2.14 (Michel 1993). At the regional scale, the chromite deposits and mineralizations show a roughly sub-parallell orientation to the northern coast of Oman (i.e. *c.* NW-striking), with two regions of major clustering (Michel 1993): (a) NW and near Sohar and (b) mainly east of the JAD and SW of the SHD (Fig. 2.14). Since chromite appears in podiform bodies, there are practically no sulfides and PGEs associated with the ores (Rollinson 2008). PGEs are commonly found in context with layered intrusions (e.g. Oberthür *et al.* 2003; Godel *et al.* 2007), while enrichments within podiform deposits are relatively rare (Prichard *et al.* 2008). Small uneconomic chromite placers in recent to subrecent wadi sediments and poorly consolidated sedimentary rocks derive from nearby chromite deposits. Chromite ore formation entails immiscible siliceous meltwater systems (Matveev and Ballhaus 2002) that cause a fractionation between silicates and oxides, concentrating chromite in particular positions within the melt and even occasionally producing chromite dykes within the Semail Ophiolite (Leblanc and Ceuleneer 1992; Zagrtdenov *et al.* 2018).

According to Partington (2009), the ophiolite contains a series of some 150 Cyprus-type volcanogenic massive sulfides (VMS) deposits in the NW of the country, which carry copper as the main commodity. Occasionally, they also show gold enrichment in some of the associated gossans as, for instance, in the Raqa and Bishara deposits near Yanqul (SW of the Hawasina Window; fig. 1.1), which have been mined for gold and silver. The deposits are concentrated at various stratigraphic levels of upper crustal units (pillow basalts: V1 Geotimes, V1–V2 Lasail, V2 Alley and boninitic Alley) and appear in distinct graben-like structures that reflect the respective spreading centres. However, a completely different geological setting controls the formation of many of the small- to medium-sized copper deposits east of the JAD, which occur as veins within mantle rocks (mainly harzburgites; Pracejus and Scharf 2020).

In both cases, the ore mineralogy is dominated by pyrite (FeS_2) and chalcopyrite ($CuFeS_2$), while bornite (Fe_5CuS_4), covellite (CuS) and a number of other Cu sulfides are less common. Sphalerite (ZnS) and galena (PbS) are generally scarce or absent (Pracejus and Scharf 2020). The primary mineralogy of the ultramafic-hosted ore minerals was deduced from sulfur-rich slags, associated with all the respective sites (Pracejus and Scharf 2020).

Many of the deposits mentioned above have been mined since ancient times in the land of 'Magan', 'Majan' or 'Makan' during the Bronze (third millennium BCE) and Iron ages (*c.* 800 BCE to AD 500) and reprocessed during the Islamic (ninth to tenth centuries) and Portuguese periods (sixteenth and seventeenth centuries; Goettler *et al.* 1976; Hauptmann and Weisgerber 1981). In early Mesopotamian documents, the translation of Magan is 'copper mountain' and Magan was located in northern Oman and the UAE (e.g. Belmonte and González-García 2014). Much of the copper from Oman was exported to the Sumerian city state Ur during the Bronze Age (Mesopotamia; documented on clay tablets, according to Goettler *et al.* 1976), where it was probably alloyed with tin to produce bronze (there are no tin deposits on the Arabian Peninsula). A large-scale export is reflected by the scarcity of copper and bronze artefacts in Oman (K. Douglas, pers. comm. 2018). Ancient copper mining and slag fields are listed in Table 2.3 after Goettler *et al.* (1976) and Weisgerber (1978).

Post-obduction units

Formations 43–52 described in the following accumulated after obduction of the Semail Ophiolite and date as latest Cretaceous and younger (Glennie *et al.* 1974). These formations are known as the 'Neo-Autochthonous' (de Wever *et al.* 1990; 'Neoautochthon' in Coffield 1990; figs A.1 & A.2). The oldest post-obduction sequences are characterized by the fluviatile and coastal clastics of the Al-Khod and Qahlah formations (e.g. compare with 'Al Khawd Formation' in Nolan *et al.* 1990; Abbasi *et al.* 2014; Pickford 2017), and the shallow-marine limestones of the Simsima Formation (Nolan *et al.* 1990; Skelton *et al.* 1990). These Cretaceous units are unconformably overlain by thick Cenozoic, mostly shallow-marine, successions ranging in age from Paleocene to Early Miocene and up to 1.2–3 km in thickness, in the Seeb-Muscat and Sur areas, respectively (Villey *et al.* 1986*b*; Wyns *et al.* 1992). The Cenozoic formations of predominantly platform carbonates accumulated around the Oman Mountains (palaeolatitude *c.* 10°N; Racey 1995; Fournier *et al.* 2006, fig. 2; fig. A.2). Upper Paleocene–Eocene platform limestones and marls of the Jafnayn, Rusayl and Seeb formations crop out extensively along the northern and eastern foot of the Southeastern Oman Mountains (White 1994; Racey 1995). These represent an almost continuous sedimentary sequence that is rich in grainstones with larger benthic foraminifera (LBF) and a diverse fauna including gastropods, red and green algae, echinoids, corals and bivalves, indicating normal marine conditions for most of this interval. Some of the limestone successions recorded intense bioturbation, which led to a nodular bedding style. The post-Eocene formations are represented by shallow-marine to terrestrial siliciclastic sediments.

The Al-Khod and Qahlah formations (43) of Glennie *et al.* (1974) belong to the Aruma Group. Unit 43 represents the Al-Khod Formation north of the JAD and SHD, and the Qahlah Formation SE of the SHD. The Al-Khod and Qahlah formations lie mainly nonconformably on the allochthonous units and are time-transgressive as sedimentation commenced and ceased at slightly different times during the Late Cretaceous (Nolan *et al.* 1990; Abbasi *et al.* 2014). The Al-Khod Formation has a highly variable thickness of up to 800–860 m (Pickford 2017). The Al-Khod Formation of the Seeb-Muscat area (Rusayl Embayment) contains alternating beds of mono- to polymictic conglomerate, sandstone and shale. The components of the conglomerate consist of radiolarite,

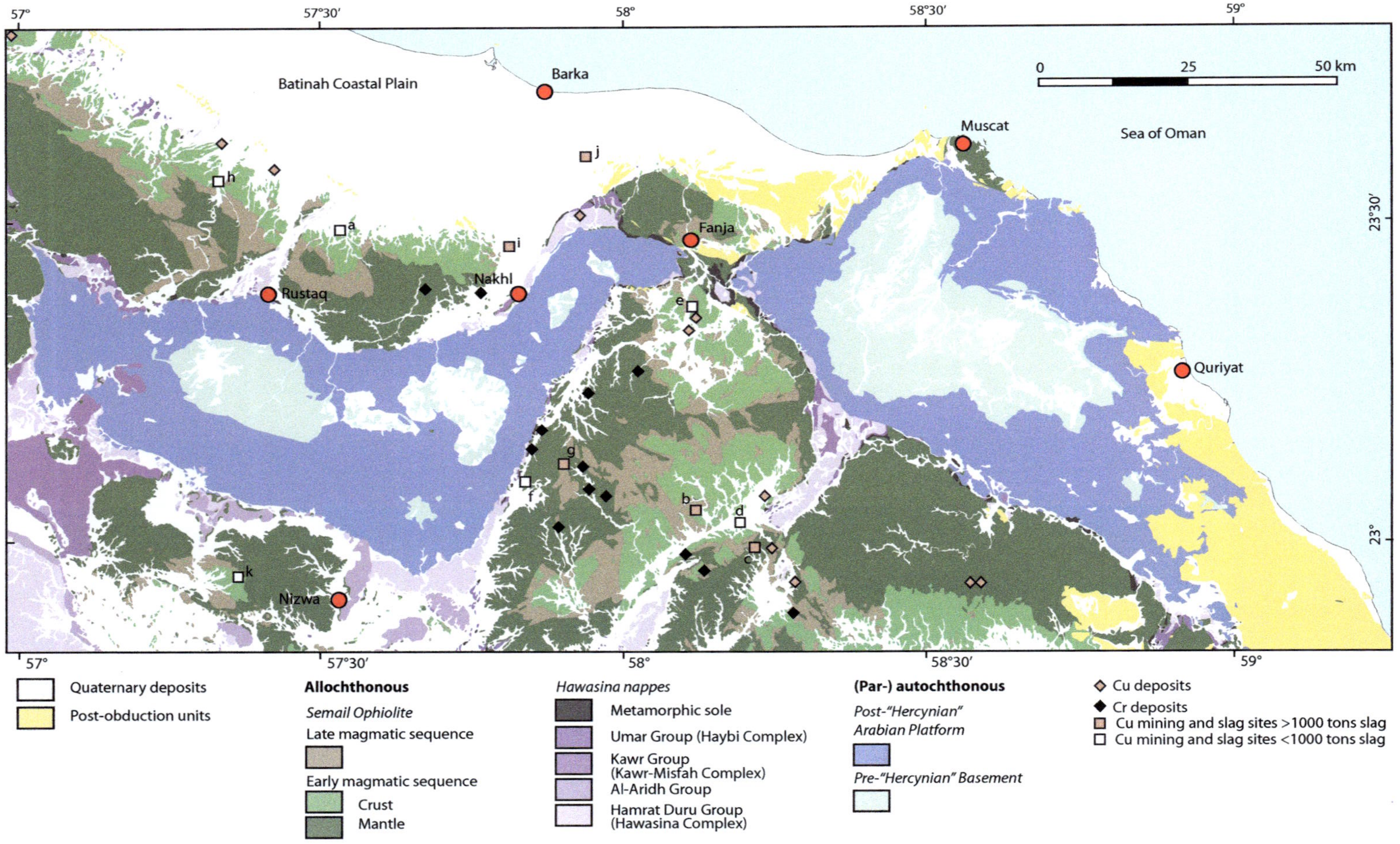

Fig. 2.14. Copper and chromium deposits after Michel (1993). Ancient copper mines and slag fields (a–k) after Goettler *et al.* (1976) and Weisgerber (1978), as listed in Table 2.3. See figure 4.1 for details on the tectonic nappes.

Table 2.3. *List of ancient copper mining and slag fields after Goettler* et al. *(1976) and Weisgerber (1978)*

Letter (Fig. 2.14)	Prospect no. of Weisgerber (1978)	Ancient mine sites	Estimated weight of slags (tons)
a	14	Al Muaeedin	500
b	17	Wadi Andam	1500
c	45	Musfa	3000
d	46	Mazra	200
e	47	Shell-Luzuc	100
f	48	Lashaya II	500
g	49	Maqsad	2000
h	102	Wadi Maidin	500
i	104	AFI	5000
j	108	Musayfiyah	5000
k	109	Bahla	25

limestone and ophiolite and range in size from granules to cobbles (Nolan *et al.* 1990; Abbasi *et al.* 2014). Usually, the clasts of the conglomerates come from progressively deeper stratigraphic and/or structural crustal levels, including ophiolite, carbonate and rock fragments from the SHD (Mann *et al.* 1990). Near the SHD the Al-Khod Formation displays such a normal unroofing sequence (*sensu* Colombo 1994) with ophiolite clasts dominating the basal layers. Carbonates derived from the Hajar Supergroup are common in the higher layers and pre-Permian-derived components are found in the highest parts (Mann *et al.* 1990; Nolan *et al.* 1990). This implies that the SHD was a major dome that had been eroded to the pre-Permian formations during the latest Cretaceous (Mann *et al.* 1990; Nolan *et al.* 1990). The vertical sequence of these clasts attests to the progressive uplift and erosion of the Oman Mountains during the Late Campanian–Early Maastrichtian. A several-metres-thick laterite horizon containing nickel may exist in places where the Al-Khod Formation rests on top of the Semail Ophiolite (Wyns *et al.* 1992; Al-Khirbash *et al.* 2014; Al-Khirbash 2015, 2021).

Informally, the Al-Khod Formation 10 km NNW of Fanja (Fig. 2.1) can be subdivided into three members (Pickford 2017). The basal member consists of predominantly freshwater facies with freshwater bivalves, gastropods and dinosaur remains, corresponding to braided river and alluvial fan deposits. The middle member yields coastal and/or brackish-water gastropods and shallow-water marine taxa. The upper member contains mainly shallow-marine molluscs, corresponding to a delta environment (Pickford 2017).

Invertebrate and vertebrate fossils suggest a Campanian–Maastrichtian age for this formation (Nolan *et al.* 1990). Dinosaur tracks, dinosaur remains and a crocodile have been found in this formation near Fanja (NE of the JAD: Fig. 2.1; Schulp *et al.* 2000, 2008; Buscalioni *et al.* 2004).

Mattern *et al.* (2018*b*) reported for the first time in Oman a higher-than-20-m dyke within the Al-Khod Formation The location of the dyke is *c.* 20 km to the WSW of Muscat in the hanging wall of the northern SHD. The shale dyke may be part of a left-lateral, negative flower structure which formed during the post-Eocene in the course of deformation of the Frontal Range Fault (Mattern and Scharf 2018; Mattern *et al.* 2018*b*; see sections on 'Main extensional fault systems' and 'Post-odcution deformation' in Chapters 3 and 5).

The Jafnayn Formation (44) of Nolan *et al.* (1990) or Umm Er Radhuma Formation (known elsewhere in the Arabian Plate; e.g. Wyns *et al.* 1992; Özcan *et al.* 2016) unconformably overlies the Al-Khod Formation (Nolan *et al.* 1990). At the type locality of Al-Khod Village (NE of the JAD), a fault contact exists between the Al-Khod Formation and the Jafnayn Formation (Scharf *et al.* 2016). The Jafnayn Formation reflects a regional transgression on the Arabian Peninsula during the Late Paleocene (Nolan *et al.* 1990).

The Jafnayn Formation has a Late Paleocene–Early Eocene age and an average thickness of 250–300 m N of the JAD and SHD (Béchennec *et al.* 1992*a*). Its thickness increases to 400–600 m SE of the SHD (Wyns *et al.* 1992). It is usually divided into two informal members, separated by a depositional earliest Eocene hiatus (Racey 1995; Haynes *et al.* 2010; Serra-Kiel *et al.* 2016). At the type locality the formation is >120 m thick (base not exposed). In the Fanja area, the formation consists of interbedded fossiliferous (LBF) marls, marly limestones and massive limestones, often displaying nodular bedding, interpreted as having formed in shallow, low-energy (lagoonal) shelf environments. An equivalent to the lower marly member of the Jafnayn Formation in the Fanja area might be the Selandian Shammar Formation of the subsurface (compare with Forbes *et al.* 2010).

In the lower part of the Jafnayn Formation, LBF and red algea are abundant (e.g. Mattern and Bernecker 2018). In some places, clinoforms can be found (e.g. Mattern and Bernecker 2018). The upper part of the formation consists of pure, massive limestone, deposited in shallow, often high-energy (open-marine shoal) environments (Nolan *et al.* 1990; Racey 1995; Özcan *et al.* 2016; Tomás *et al.* 2016). Increasing detrital quartz in the upper part of the formation might suggest a regressive phase or an increase of clastic input to the basin (Tomás *et al.* 2016).

A mafic dyke (basanite) has been reported at the base of the Jafnayn Formation *c.* 30 km to the SW of Muscat (fig. 4.1; Al-Harthy *et al.* 1991). This mafic rock unit contains abundant ultramafic xenoliths of a few centimetres in size. Geochemical and mineralogical descriptions of this mafic and ultramafic unit were provided by Gnos and Peters (2003) and Nasir *et al.* (2006). Nasir *et al.* (2006) identified a strong ocean-island basalt (OIB) geochemistry of the mafic rocks. Pressure estimates of the xenoliths indicate a depth of origin of 1.7–2.05 GPa (Nasir *et al.* 2006). The age of 42.7 ± 1 Ma (σ2 error) of this intrusion has been determined by Scharf *et al.* (2020*b*) using the $^{40}Ar/^{39}Ar$ whole rock stepwise heating technique.

The Abat Formation (45) of Roger *et al.* (1991) occurs SE of the SHD, is up to 700 m thick and consists of yellow, glauconitic marl, clayey or micritic grey to white limestone. The depositional setting may have been an open, intra-shelf slope or basin (Roger *et al.* 1991). The formation is of Late Paleocene–Early Eocene age (Late Thanetian–Early Ypresian; Ilerdian of Wyns *et al.* 1992; Pujalte *et al.* 2009; the Ilerdian occurred during the Early Ypresian). According to Serra-Kiel *et al.* (2016, fig. 2), the age of the Abat Formation is Mid Selandian–Ilerdian (Early Ypresian). Wyns *et al.* (1992, fig. 13) correlated the upper member of the Jafnayn Formation with the Abat Formation. Serra-Kiel *et al.* (2016, fig. 2) correlated the entire Jafnayn Formation with the Abat Formation. We use the correlation provided by Serra-Kiel *et al.* (2016) in our compilation (figs A.1 & A.2).

According to Nolan *et al.* (1990) and Racey (1995), the limestones of the Jafnayn Formation are overlain by the Lower–Middle Eocene deposits of the Rusayl Formation (46). Although the contact between the Rusayl Formation and the subjacent Jafnayn Formation appears conformable, it may represent a disconformity (Tomás *et al.* 2016).

The *c.* 150-m-thick Rusayl Formation consists of multicoloured marls with an up-to-20-m-thick limestone horizon in the middle of the formation (Villey *et al.* 1986*c*; Nolan *et al.* 1990; Keen and Racey 1991; Racey 1995). In some places, the thickness of the Rusayl Formation has been reduced to 45 m as a result of stretching along a local shear zone (Scharf *et al.*

2016). According to the fossil content, an Early Eocene age has been suggested (Villey *et al.* 1986*b*). The occurrence of the marly Rusayl Formation marks a change in the depositional environment, interpreted as including a lagoon, conglomeratic beach, peritidal facies, mangrove swamp and also a mudflow facies (Keen and Racey 1991; Dill *et al.* 2007).

At the southeastern shoulder of the SHD, the formation becomes more evaporitic and is referred to as the 'Rus Formation' (Powers *et al.* 1966; Wyns *et al.* 1992). The Rus Formation consists of grey, thin-bedded, dolomitic limestone and beige marl (Wyns *et al.* 1992). The formation has been dated as Early Eocene and is stratigraphically found between the Jafnayn and Seeb formations (Wyns *et al.* 1992). In parts, the Rus Formation is laterally replaced by the time-equivalent mudstones of the Abat Formation (Wyns *et al.* 1992). We include the Rus Formation in the Rusayl Formation because the more evaporitic Rus Formation is the stratigraphical equivalent of the Rusayl Formation (Keen and Racey 1991; Haynes *et al.* 2010).

A new marine transgression across central Oman during the Mid Eocene marks the onset of the deposition of the Seeb Formation (47) between the basal Middle Eocene and the lower Upper Eocene (Nolan *et al.* 1990; Jones and Racey 1994; Beavington-Penney *et al.* 2006). The Seeb Formation is known elsewhere on the Arabian Peninsula as the 'Dammam Formation' (Steineke *et al.* 1958; Nolan *et al.* 1990; Wyns *et al.* 1992; fig. A.2). The contact between the Seeb and Rusayl formations in the study area is sharp and conformable, while in the Batinah Coastal Plain, a disconformity is present between the Seeb Formation and the underlying Cretaceous formations (Béchennec *et al.* 1992*a*). The Seeb Formation is overlain by a regional unconformity that separates Eocene and Oligo-Miocene strata on the Arabian Peninsula (Nolan *et al.* 1990).

The Seeb Formation is made of brightly coloured shelf limestone (Nolan *et al.* 1990). The formation can be divided into five lithological units with an estimated thickness of 600 m (Hersi and Al-Harthy 2010). The base is characterized by yellowish-brown, sandy, bioclastic packstone to grainstone, containing large bioclasts and LBF-like Nummulites and Alveolina. Upwards, these limestones become nodular due to bioturbation. The top of the formation displays yellowish-brown, thickly to massively bedded, extensively burrowed, nodular, bioclastic wackestone, floatstone to rudstone with Nummulites and Alveolina (Hersi and Al-Harthy 2010).

Whitish bioclastic limestone of the Upper Eocene (Priabonian) to Oligocene Shama Formation (48) of Roger *et al.* (1991; Le Métour *et al.* 1992) rests disconformably on the Jafnayn Formation south of Quriyat where it has a maximum thickness of 40 m (Le Métour *et al.* 1992). In the Sur area, SE of the study area, the Shama Formation is at least 100–120 m thick. At this place, the Shama Formation overlies the Seeb Formation (Wyns *et al.* 1992).

In the study area, the Jafnayn and Seeb formations are unconformably overlain by either the Tahwah Formation of Roger *et al.* (1991; Le Métour *et al.* 1992) or the informal unit of Mu'askar Al Murtafa ('MAM') reef Formation (49) of Hersi and Al-Harthy (2010). The greater-than-100-m thick Tahwah Formation (near Quriyat) is interpreted as redeposited slope sediments of Late Eocene (Priabonian)–Early Miocene age (Roger *et al.* 1991; Béchennec *et al.* 1992*a*; Le Métour *et al.* 1992; Wyns *et al.* 1992). The depositional environment explains its highly different lithologies ranging from mostly green marl, sandstone and conglomerate to coral-rich limestone (Béchennec *et al.* 1992*a*). Based on foraminifera, Montenat and Blondeau (1977; Le Métour *et al.* 1992; Wyns *et al.* 1992) proposed a Late Eocene (Mid to Late Priabonian) to Early Miocene (Early Burdigalian) age of the Tahwah Formation, and a Priabonian–Chattian age has been considered by Fournier *et al.* (2006). We are following the ages of Wyns *et al.* (1992) (figs A.1 & A.2).

The islands of Jaza'ir Suwadi and Jaza'ir Daymaniyat, NW of Barka, are composed of the Sur Formation (Béchennec *et al.* 1992*b*; fig. A.1). The cliffs and slopes of these rugged islands expose interbedded sandy marl, yellow marly limestone and calcarenite with foraminifera (Béchennec *et al.* 1992*b*). Based on the latter, the rocks have been dated as Early Miocene (Aquitanian–Burdigalian) by Villey *et al.* (1986*b*) and Béchennec *et al.* (1992*a*). The formation also includes slump structures and some olistoliths, indicating synsedimentary deformation (Béchennec *et al.* 1992*b*). The type locality of the Sur Formation is in the Ja'alan area near Sur (Roger *et al.* 1991; Béchennec *et al.* 1992*b*; Fig. 2.1). We have assigned the Sur Formation to the Tahwah and MAM reefs formations for simplification.

The Barzaman Formation (50) was defined by Maizels (1988; Béchennec *et al.* 1992*a*) and reinterpreted by Béchennec *et al.* (1992*a*). It comprises white to reddish conglomerate, sandstone, shale and limestone. The age of the basal part of this formation is uncertain and has been assigned Mid Miocene south of the Oman Mountains (Béchennec *et al.* 1992*a*). However, the Barzaman Formation interfingers with the shallow-marine Dammam Formation and has no biostratigraphic age control south of the Oman Mountains, which makes its age somewhat difficult to understand (Rodgers and Gunatilaka 2002). The basal part of the Barzaman Formation could therefore have a Late Miocene age (see Rodgers and Gunatilaka 2002). The upper part of the Barzaman Formation is probably of Pliocene age (Béchennec *et al.* 1992*a*). In the study area, the thickness of the Barzaman Formation is *c.* 100 m (Béchennec *et al.* 1992*a*). Parts of the Barzaman Formation from the Northern Oman Mountains of the UAE have been interpreted as debris-flow deposits that formed during heavy, possibly seasonal, rainfall (Lacinska and Styles 2009). Macklin *et al.* (2012) and Lacinska *et al.* (2014) interpreted the diagenetically alterated rocks of the Barzaman Formation from the Northern Oman Mountains to have formed during changes in climate between humid and arid intervals. Furthermore, Lacinska *et al.* (2014) concluded that the Barzaman Formation in the UAE was deposited during increased erosion and uplift of the Oman Mountains during the Late Miocene.

Listwaenite (51) can be found near the NE flank of the JAD near Fanja and to the NW of the SHD (Fig. 2.15). At the same time, it occurs at or near the contact between the Semail Ophiolite and the Mesozoic shelf sediments. The listwaenite or 'Amqat' lithostratigraphic unit of Glennie *et al.* (1974) and Stanger (1985) is oriented sub-parallel to the basal thrust of the Semail Ophiolite and forms discontinuous planar features with a thickness of tens of metres. Within the listwaenite are numerous quartz and secondary calcite veins with a thickness of 0.2–3 cm (Nasir *et al.* 2007). The formation of the listwaenite is associated with serpentinization of the adjacent peridotite (e.g, Stanger 1985; Nasir *et al.* 2007; Falk and Kelemen 2015).

The age of the listwaenite is not precisely known, but its origin is either related to subduction of the Semail Ophiolite (Falk and Kelemen 2015) or Cenozoic extension (Stanger 1985; Wilde *et al.* 2002; Nasir *et al.* 2007; Bailey *et al.* 2019; Scharf *et al.* 2019, 2020*a*). A 97 ± 29-Ma Rb–Sr age of a fuchsite-bearing listwaenite sample east of Fanja (Fig. 2.1) implies a formation age of the listwaenite during thrusting of the Semail Ophiolite (Falk and Kelemen 2015). Furthermore, based on the geological map of Fanja (Villey *et al.* 1986*c*), the same authors wrote that the listwaenite is unconformably overlain by Upper Paleocene–Eocene conglomerates and that the geochemical signature of the listwaenite may be related to subduction. According

Fig. 2.15. Listwaenite (red rock) south of Fanja. The Semail Ophiolite marks the hanging wall. In the footwall is the metamorphic sole.

to Falk and Kelemen (2015), listwaenite formed at temperatures of 80–130°C within a subduction zone during obduction.

On the other hand, Scharf *et al.* (2019, 2020*a*) documented that carbonate veins, which emanate from the listwaenite, fill fissures in the post-obductional uppermost Cretaceous Al-Khod Formation near Fanja. Furthermore, two U–Pb ages of carbonate veins in the same listwaenite are 55 ± 4 Ma (Scharf *et al.* 2019, 2020*a*), suggesting a post-obductional age of the listwaenite (see section on 'Main extensional fault systems' in Chapter 3).

Listwaenite is the product of hydrothermal alteration of mafic and ultramafic rocks and mainly consists of quartz and carbonate as well as some other minerals (Buisson and Leblanc 1985, 1987; Bates and Jackson 1987; Leblanc 1991; Halls and Zhao 1995; Nasir *et al.* 2007). However, Nasir *et al.* (2007) stated, based on Kashkai and Allakverdiev (1965) and Ucurum (2000), that listwaenite cannot develop directly from the igneous protolith but must form in two successive stages: (1) autometasomatic serpentinization of ultramafic rocks; and (2) metasomatic overprinting of serpentinite. Such metasomatic rocks are either dominated by carbonate (Type I) or silicate-carbonate (Type II). Type I can be further subdivided into calcitic (Ia) and dolomitic (Ib) subtypes, depending on the predominant carbonate. Two major modes of listwaenite formation can also be distinguished: (1) according to the movement of hydrothermal fluids along thrust faults; and/or (2) within fine cracks and fissures of highly serpentinized ultramafic rocks.

Mineralogically, Oman's listwaenite belongs to the Type I rock associations (Wilde *et al.* 2002; Nasir *et al.* 2007). Besides carbonate, it also contains ankerite, siderite, magnesite, lizardite, chlorite, fuchsite, other phyllosilicates and goethite. The latter occurs along carbonate grain boundaries and gives the rock a brown colour. Rare relict textures of intercumulus chromite grains bear witness to the rock's ultramafic origin (Pracejus 2015). The carbonate-rich listwaenite of Type 1 generally carries only low and uneconomic concentrations of metals (Nasir *et al.* 2007).

The Quaternary deposits (52) include a wide variety of lithologies, including beach rock, beach sand, active sand dunes, ancient and recent slope colluvium, sabkha deposits, ancient and recent fluvial terraces, and alluvial fans. Common incisions of recent to subrecent wadi sediments of many metres (partly with overhangs) indicate recent and/or active uplift of the Oman Mountains (section 'Quaternary deformation' in Chapter 5).

Funding This research received no specific grant from any funding agency in the public, commercial, or not-for-profit sectors.

Author contributions **AS:** Conceptualization (lead), Writing original draft (lead), Visualization (lead), Supervision (lead), Investigation (lead); **FM:** Conceptualization (equal), Writing – review and editing (equal), Visualization (equal), Investigation (equal); **MAW:** Writing – review and editing (equal); **GF:** Writing – review and editing (equal); **DM:** Writing – review and editing (equal); **BP:** Writing – review and editing (equal); WB: Writing – review and editing (equal); **IC:** Writing – review and editing (equal).

Data availability statement Data sharing is not applicable to this article as no datasets were generated or analysed during the current study.

References

Abbasi, I.A., Hersi, O.S. and Al-Harthy, A. 2014. Late Cretaceous conglomerates of the Qalhah formation, north Oman. *Geological Society, London, Special Publications*, **392**, 325–341, https://doi.org/10.1144/SP392.17

Agard, P., Searle, M.P., Alsop, G. and Dubacq, B. 2010. Crustal stacking and expulsion tectonics during continental subduction: P-T deformation constraints from Oman. *Tectonics*, **29**, 1–19, https://doi.org/10.1029/2010TC002669

Alabaster, T., Pearce, J.A. and Malpas, J. 1982. The volcanic stratigraphy and petrogenesis of the Oman ophiolite complex. *Contribution to Mineralogy and Petrology*, **81**, 168–183, https://doi.org/10.1007/BF00371294

Al-Abri, F.G., Heward, A.P. and Abbasi, I.A. 2018. Regional 3D modelling of the Permo-Carboniferous Al Khlata Formation in the Rima area, eastern flank of the South Oman Salt Basin. *Journal of Petroleum Geology*, **41**, 29–46, https://doi.org/10.1111/jpg.12691

Al-Belushi, J.D., Glennie, K.W. and Williams, B.J.P. 1996. Permo-Carboniferous Al Khlata Formation, Oman: a new hypothesis of origin of its glaciation. *GeoArabia*, **1**, 389–404.

Aldega, L., Carminati, E., Scharf, A., Mattern, F. and Al-Wardi, M. 2017. Estimating original thickness and extent of the Semail Ophiolite in the eastern Oman Mountains by paleothermal indicators. *Marine and Petroleum Geology*, **84**, 18–33, https://doi.org/10.1016/j.marpet.geo.2017.03.024

Al-Harthy, M.S., Coleman, R.G., Hughes Clarke, M.W. and Hanna, S.S. 1991. Tertiary basaltic intrusions in the Central Oman Mountains. *In*: Peters, T., Nicolas, A. and Coleman, R.J. (eds) *Ophiolite Genesis and Evolution of the Oceanic Lithosphere*. Proceedings of the Ophiolite Conference, Muscat, Oman, Kluwer Academic Publishers, Dordrecht/Boston/London, **1990**, 475–682.

Al-Husseini, M.I. 2014. Ediacaran-Cambrian Middle East geologic time scale 2014, proposed correlation of Oman's Abu Mahara Supergroup and Saudi Arabia's Jibalah Group. *GeoArabia*, **19**, 107–132.

Ali, M.Y., Cooper, D.J.W., Searle, M.P. and Al-Lazki, A. 2014. Origin of gypsiferous intrusions in the Hawasina Window, Oman Mountains: implications from structural gravity investigations. *GeoArabia*, **19**, 17–48.

Al-Khirbash, S. 2015. Genesis and mineralogical classification of Ni-laterites, Oman Mountains. *Ore Geology Reviews*, **65**, 199–212, https://doi.org/10.1016/j.oregeorev.2014.09.022

Al-Khirbash, S. 2021. Mineralogical characterization of low-grade Nickel laterites from the North Oman Mountains: using Mineral Liberation analyses –Scanning Electrobe Microscopy-based automated quantitative mineralogy. *Ore Geology Reviews*, **120**, 103429, https://doi.org/10.1016/j.oregeorev.2020.103429

Al-Khirbash, S., Semhi, K., Richard, L., Nasir, S. and Al-Harthy, A. 2014. Rare earth element mobility during laterization of mafic rocks of the Oman ophiolite. *Arabian Journal of Geosciences*, **7**, 5443–5454, https://doi.org/10.1007/s12517-013-1189-6

Allemann, F. and Peters, T. 1972. The ophiolite-radiolarite belt of the North Oman Mountains. *Eclogae Geologicae Helvetica*, **65**, 657–697.

Allen, P.A. 2007. The Huqf Supergroup of Oman: basin development and context for Neoproterozoic glaciation. *Earth-Science Reviews*, **84**, 139–185, https://doi.org/10.1016/j.earscirev.2007.06.005

Allen, P.A. and Leather, J. 2006. Post-Marinoan marine siliciclastic sedimentation: the Masirah Bay Formation, Neoproterozoic Huqf Supergroup of Oman. *Precambrian Research*, **144**, 167–198, https://doi.org/10.1016/j.precamres.2005.10.006

Allen, P.A., Leather, J. and Brasier, M.D. 2004. The Neoproterozoic Fiq glaciation and its aftermath, Huqf Supergroup of Oman. *Basin Research*, **16**, 507–534, https://doi.org/10.1111/j.1365-2117.2004.00249.x

Allen, P.A., Leather, J. *et al.* 2011. Chapter 20 The Abu Mahara Group (Ghubrah and Fiq formations), Jabal Akhdar, Oman. *Geological Society, London, Memoirs*, **36**, 251–262, https://doi.org/10.1144/M36.20

Amthor, J.E., Grotzinger, J.P., Schröder, S., Bowring, S.A., Ramezani, J., Martin, M.W. and Matter, A. 2003. Extinction of Clodina and Namacalathus at the Precambrian-Cambrian boundary in Oman. *Geology*, **31**, 431–434, https://doi.org/10.1130/0091-7613(2003)031<0431:EOCANA>2.0.CO;2

Arndt, M., Virgo, S., Cox, S.F. and Urai, J.L. 2014. Changes in fluid pathways in a calcite vein mesh (Natih Formation, Oman Mountains): insights from stable isotopes. *Geofluids*, **14**, 391–418, https://doi.org/10.1111/gfl.12083

Bailey, C.M., Hurtado, C., Scharf, A. and Mattern, F. 2019. Structural controls on listwaenite genesis in the Semail Ophiolite, Northern Oman. EGU 2019–5794, Vienna, Austria.

Bates, R.L. and Jackson, J.A. 1987. *Glossary of Geology*. 3rd ed. American Geological Institute, Alexandria, VA, USA.

Baud, A. and Richoz, S. 2013. Permian-Triassic Transition and the Saiq/Mahil Boundary in the Oman Mountains: proposed correction for lithostratigraphic nomenclature. *GeoArabia*, **18**, 87–90.

Baud, A., Beauchamp, B., Henderson, C.M. and Richoz, S. 2010. The Permian-Triassic transition on the Saiq Plateau. *In*: Baud, A. and Bernecker, M. (eds) *IGCP 572 Field Guide Book 2: The Permian-Triassic transition in the Oman Mountains*. GUtech, Muscat, 34–47.

Bauer, W. 2019. Comment on: 'Alan P. Heward: further outcrops of diamictite in Wadi Amdeh: al Khlata, Mesozoic or Cenozoic Slope deposits?' *Al Hajar Magazine*, **26**, 32–33.

Beavington-Penney, S.J., Wright, V.P. and Racey, A. 2006. The Middle Eocene Seeb Formation of Oman: an investigation of acyclicity, stratigraphic completeness, and accumulation rates in shallow marine carbonate settings. *Journal of Sedimentary Research*, **76**, 1137–1161, https://doi.org/10.2110/jsr.2006.109

Béchennec, F. 1987. *Géologie des Nappes Hawasina dans les parties orientale et central des Montagnes d'Oman*. PhD thesis, Pierre and Marie Curie Université, Paris 6. Bureau de Recherches Géologiques et Minières, Document No. 127, Orléans, France, 427.

Béchennec, F. 1988. Géologie des nappes d'Hawasina dans les parties orientales et centrales des montagnes d'Oman. *Doc. BRGM Orléans*, **127**, 1–474.

Béchennec, F., Beurrier, M., Rabu, D. and Hutin, G. 1986. *Geological map of Bahla, sheet NF 40-07A, scale 1:100,000, with Explanatory Notes*. Directorate General of Minerals, Oman Ministry of Petroleum and Minerals.

Béchennec, F., Le Métour, J., Rabu, D., Villey, M. and Beurrier, M. 1988. The Hawasina Basin: a fragment of a starved passive continental margin, thrust over the Arabian Platform during obduction of the Sumail Nappe. *Tectonophysics*, **151**, 323–343, https://doi.org/10.1016/0040-1951(88)90251-X

Béchennec, F., Le Métour, J., Rabu, D., Bourdillon-de-Grissac, C., de Wever, P., Beurrier, M. and Villey, M. 1990. The Hawasina Nappes: stratigraphy, paleogeography and structural evolution of a fragment of the South-Tethyan passive continental margin. *Geological Society, London, Special Publications*, **49**, 213–223.

Béchennec, F., Roger, J., Le Métour, J. and Wyns, R. 1992*a*. *Geological map of Seeb, sheet NF 40-03, scale 1:250,000, with Explanatory Notes*. Directorate General of Minerals, Oman Ministry of Petroleum and Minerals.

Béchennec, F., Wyns, R., Roger, J., Le Métour, J. and Chevrel, S. 1992*b*. *Geological map of Nazwa, sheet NF 40-07, scale 1:250,000, with Explanatory Notes*. Directorate General of Minerals, Oman Ministry of Petroleum and Minerals.

Béchennec, F., Le Métour, J., Platel, J.P. and Roger, J. 1993. *Geological Map of the Sultanate of Oman, scale 1:1 000 000 with Explanatory Notes*. Directorate General of Minerals, Oman Ministry of Petroleum and Minerals, Muscat, Oman.

Belmonte, J.A. and González-García, A.C. 2014. On the orientation of early Bronze Age tombs in ancient Magan. *Mediterranean Archaeology and Archaeometry*, **14**, 233–246.

Bendias, D. and Aigner, T. 2015. Facies, sequence stratigraphy, reservoir and seal potential of the Mafraq Formation, Sultanate of Oman: an integrated outcrop analogue study. *GeoArabia*, **20**, 17–94.

Bendias, D., Aigner, T. and Schlaich, M. 2015. Reservoir and seal potential of the mixed carbonate-siliciclastic Mafraq Formation, Oman: an integrated outcrop analogue study. Fifth EAGE Arabian Plate Geology Workshop 2015.

Benn, K., Nicolas, A. and Reuber, I. 1988. Mantle-crust transition zone and origin of wehrlite magmas: evidence from the Oman ophiolite. *Tectonophysics*, **151**, 75–85, https://doi.org/10.1016/0040-1951(88)90241-7

Bernoulli, D., Weissert, H. and Blome, C.D. 1990. Evolution of the Triassic Hawasina Basin, Central Oman Mountains. *Geological Society, London, Special Publications*, **49**, 189–202, https://doi.org/10.1144/GSL.SP.1992.049.01.12

Beurrier, M. 1987. *Géologie des Montagnes d'Oman*. PhD thesis, Pierre and Marie Curie Université, Paris 6. Bureau de Recherches Géologiques et Minières, Document No. 128, Orléans, France, 412.

Beurrier, M., Béchennec, F., Rabu, D. and Hutin, G. 1986. *Geological map of Rustaq, sheet NF 40-03D, scale 1:100,000, with Explanatory notes*. Directorate General of Minerals, Oman Ministry of Petroleum and Minerals.

Beurrier, M., Bourdillon-Jeudy de Grissac, C., de Wever, P. and Lescuyer, J.L. 1987. Biostratigraphie des radiolarites associées aux volcanites ophiolitiques de la nappe de Samail (Sultanat d'Oman); conséquences tectogénétiques. Comptes Rendus des Séances de l'Academie des Sciences, Paris, 304, Série II, Mécanique, Physique, Cjimie, Sciences de l'univers. *Sciences de la Terre*, **304**, 907–910.

Blechschmidt, I., Dumitrica, P., Matter, A., Krystyn, L. and Peters, T. 2004. Stratigraphic architecture of the northern Oman continental margin – Mesozoic Hamrat Duru Group Hawasina Complex, Oman. *GeoArabia*, **9**, 81–132.

Blendinger, W., van Vliet, A.T. and Hughes Clarke, M.W. 1990. Updoming, rifting and continental margin development during the Late Paleozoic in northern Oman. *Geological Society, London, Special Publications*, **49**, 27–37, https://doi.org/10.1144/GSL.SP.1992.049.01.03

Boote, D.R.D., Mou, D. and Waite, R.I. 1990. Structural evolution of the Suneinah Foreland, Central Oman Mountains. *Geological Society, London, Special Publications*, **49**, 397–418, https://doi.org/10.1144/GSL.SP.1992.049.01.25

Boudier, F. and Al-Rajhi, A. 2014. Structural control on chromite deposits in ophiolites: the Oman case. *Geological Society, London, Special Publications*, **392**, 263–277, https://doi.org/10.1144/SP392.14

Boudier, F. and Nicolas, A. 1995. Nature of the Moho transition zone in the Oman ophiolite. *Journal of Petrology*, **36**, 777–796, https://doi.org/10.1093/petrology/36.3.777

Bowring, S.A., Grotzinger, J.P., Condon, D.J., Ramezani, J., Newall, M. and Allen, P.A. 2007. Geochronologic constraints of the chronostratigraphic framework of the Neoproterozoic Huqf Supergroup, Sultanate of Oman. *American Journal of Science*, **307**, 1097–1145, https://doi.org/10.2475/10.2007.01

Bowring, S.A., Grotzinger, J.P., Condon, D.J., Ramezani, J. and Newall, M.J. 2009. Reply to comment Oman chronostratigraphy. *American Journal of Sciences*, **309**, 91–96, https://doi.org/10.2475/01.2009.04

Brasier, M., McCarron, G., Tucker, R., Leather, J., Allen, P. and Shields, G. 2000. New U-Pb zircon dates for the Neoproterozoic Ghubrah glaciation and for the top of the Huqf Supergroup, Oman. *Geology*, **28**, 175–178, https://doi.org/10.1130/0091-7613(2000)28<175:NUZDFT>2.0.CO;2

BRGM (Bureau de Recherches Géologiques et Minières) 1984. Geological study of mineral occurrences of the central part of the northern Oman Mountains. Unpublished report, BRGM/OMAN, no. 2, 1–289.

Buisson, G. and Leblanc, M. 1985. Gold in carbonitized ultramafic rocks from ophiolite complexes. *Economic Geology*, **80**, 2026–2029, https://doi.org/10.2113/gsecongeo.80.7.2028

Buisson, G. and Leblanc, M. 1987. Gold in mantle peridotites from Upper Proterozoic ophiolites in Arabia, Mali, and Morocco. *Economic Geology*, **82**, 2091–2097, https://doi.org/10.2113/gsecongeo.82.8.2091

Burchette, T.P. and Wright, V.P. 1992. Carbonate ramp depositional systems. *Sedimentary Geology*, **79**, 3–57, https://doi.org/10.1016/0037-0738(92)90003-A

Burns, S.J. and Matter, A. 1993. Carbon isotopic record of the latest Proterozoic from Oman. *Eclogae Geologicae Helvetiae*, **86**, 595–607.

Buscalioni, A.D., Schulp, A.S., Jagt, J.W.M., Hanna, S.S. and Hartman, A.F. 2004. Late Cretaceous neosuchian crocodiles from the Sultanate of Oman. *Cretaceous Research*, **25**, 267–275, https://doi.org/10.1016/j.cretres.2003.12.004

Calver, C.R., Crowley, J.L., Wingate, M.T.D., Evans, D.A.D., Raub, T.D. and Schmitz, M.D. 2013. Globally synchronous Marinoan deglaciation indicated by U-Pb geochronology of the Cottons Breccia, Tasmania, Australia. *Geology*, **41**, 1127–1130, https://doi.org/10.1130/G34568.1

Carminati, E., Aldega, L., Smeraglia, L., Scharf, A. and Mattern, F. 2019. Obduction and collision tectonics in oman: constraints from structural and thermal analyses. *In*: Rossetti, F., Blanc, A.C., Riguzzi, F., Leroux, E., Pavalopoulos, K., Bellier, O. and Kapsimalis, V. (eds) *The Structural Geology Contribution to the Africa-Eurasia Geology: Basement and Reservoir Structure, Ore Mineralisation and Tectonic Modelling*, Proceedings of the 1st Springer Conference of the Arabian Journal of Geosciences (CAJG-1), Tunisia 2018, IEREK Interdisciplinary Series for Sustainable Development 319–321, https://doi.org/10.1007/978-3-030-01455-1_70

Celestino, R., Wohlwend, S., Reháková, D. and Weissert, H. 2017. Carbon isotope stratigraphy, biostratigraphy and sedimentology of the Upper Jurassic-Lower Cretaceous Rayda Formation, Central Oman Mountains. *Newsletters on Stratigraphy*, **50**, 91–109, https://doi.org/10.1127/nos/2016/0309.

Ceuleneer, G. 1991. Evidence for a paleo-spreading center in the Oman ophiolite: mantle structures in the Maqsad area. *In*: Peters, T., Nicolas, A. and Coleman, R.J. (eds) *Ophiolite Genesis and Evolution of the Oceanic Lithosphere*. Proceedings of the Ophiolite Conference Muscat, Oman, Kluwer

Academic Publishers, Dordrecht/Boston/London, **1990**, 147–173.

Ceuleneer, G. and Nicolas, A. 1985. Structures in podiform chromite from the Maqsad district (Sumail ophiolite, Oman). *Mineralium Deposita*, **20**, 177–185, https://doi.org/10.1007/BF00204562

Ceuleneer, G., Nicolas, A. and Boudier, F. 1988. Mantle flow patterns at an oceanic spreading centre: the Oman peridotites record. *Tectonophysics*, **151**, 1–26, https://doi.org/10.1016/0040-1951(88)90238-7

Chauvet, F., Dumont, T. and Basile, C. 2009. Structures and timing of Permian rifting in the central Oman Mountains (Saih Hatat). *Tectonophysics*, **475**, 563–574, https://doi.org/10.1016/j.tecto.2009.07.008

Coffield, D.Q. 1990. Structures associated with nappe emplacement and culmination collapse in the Central Oman Mountains. *Geological Society, London, Special Publications*, **49**, 447–458, https://doi.org/10.1144/GSL.SP.1992.049.01.28

Coleman, R.G. and Hopson, C.A. 1981. Introduction to the Oman Ophiolite Special Issue. *Journal of Geophysical Research*, **86**, 2495–2496, https://doi.org/10.1029/JB086iB04p02495

Colombo, F. 1994. Normal and reverse unroofing sequences in syntectonic conglomerates as evidence of progressive basinward deformation. *Geology*, **22**, 235–238, https://doi.org/10.1130/0091-7613(1994)022<0235:NARUSI>2.3.CO;2

Condon, D., Zhu, M., Bowring, S., Wang, W., Yang, A. and Jin, Y. 2005. U-Pb ages from the Neoproterozoic Doushantuo Formation, China. *Science*, **308**, 95–98, https://doi.org/10.1126/science.1107765

Coogan, L.A., Thompson, G.M., MacLeod, C.J., Dick, H.J.B., Edwards, S.J., Hosford Scheirer, A. and Barry, T.L. 2004. A combined basalt and peridotites perspective on 14 million years of melt generation at Atlantis Bank segment of the Southwest Indian Ridge: evidence for temporal changes in mantle dynamics? *Chemical Geology*, **207**, 13–30, https://doi.org/10.1016/j.chemgeo.2004.01.016

Cooper, D.J.W. 1987. Hamrat Duru Group: revised stratigraphy of a Mesozoic deep-water passive margin in the Oman Mountains. *Geological Magazine*, **124**, 157–164, https://doi.org/10.1017/S0016756800015971

Cooper, D.J.W. 1988. Structure and sequence of thrusting in deep-water sediments during ophiolite emplacement in the south-central Oman Mountains. *Journal of Structural Geology*, **10**, 437–485, https://doi.org/10.1016/0191-8141(88)90031-4

Cooper, D.J.W. 1989. A longitudinal carbonate fan from the Jurassic of the Oman Mountains: the Guweyza Limestone Formation of the Hamrat ad Duru. *Sedimentary Geology*, **61**, 253–275, https://doi.org/10.1016/0037-0738(89)90061-4

Cooper, D.J.W. 1990. Sedimentary evolution and paleogeographical reconstruction of the Mesozoic continental rise in Oman: evidence from the Hamrat Duru Group. *Geological Society, London, Special Publications*, **49**, 161–187.

Cooper, D.J.W., Searle, M.P. and Ali, M.Y. 2012. Structural evolution of Jabal Qumayrah: a salt-intruded culmination in the northern Oman Mountains. *GeoArabia*, **17**, 121–150.

Cooper, D.J.W., Ali, M.Y., Searle, M.P. and Al-Lazki, A.I. 2013. Salt intrusions in Jabal Qumayrah, northern Oman Mountains: implications from structural and gravity investigations. *GeoArabia*, **18**, 141–176.

Cooper, D.J.W., Ali, M.Y. and Searle, M.P. 2018. Origin and implications of a thrust-bound gypsiferous unit along the western edge of Jabal Sumeini, northern Oman Mountains. *Journal of Asian Earth Sciences*, **154**, 101–124, https://doi.org/10.1016/j.jseaes.2017.12.018

Cornish, S. and Searle, M. 2017. 3D geometry and kinematic evolution of the Wadi Mayh sheath fold, Oman, using detailed mapping from high-resolution photography. *Journal of Structural Geology*, **101**, 26–42, https://doi.org/10.1016/j.jsg.2017.06.009

Cowan, R.J., Searle, M.P. and Waters, D.J. 2014. Structure of the metamorphic sole to the Oman Ophiolite, Sumeini Window and Wadi Tayyin: implications for ophiolite obduction processes. *Geological Society, London, Special Publications*, **392**, 155–175, https://doi.org/10.1144/SP392.8

Cozzi, A. and Al-Siyabi, H.A. 2004. Sedimentology and play potential of the late Neoproterozoic Buah Carbonates of Oman. *GeoArabia*, **9**, 11–36.

Cozzi, A., Grotzinger, J.P. and Allen, P.M. 2004. Evolution of a terminal Neoproterozoic carbonate ramp system (Buah Formation, Sultanate of Oman): effects of basement paleotopography. *Geological Society of America Bulletin*, **116**, 1367–1384, https://doi.org/10.1130/B25387.1

Craddock, J.P., Ojakangas, R.W. *et al.* 2019. Detrital zircon provenance of Permo-Carboniferous glacial diamictites across Gondwana. *Earth Science Reviews*, **192**, 285–316, https://doi.org/10.1016/j.earscirev.2019.01.014

Csontos, L., Pocsai, T. *et al.* 2010. Structural evolution of the Hawasina Window, Oman Mountains. *GeoArabia*, **15**, 85–124.

de Wever, P., Bourdillon-de-Grissac, C.H. and Béchennec, F. 1990. Permian to Cretaceous radiolarian biostratigraphic data from the Hawasina Complex, Oman Mountains. *Geological Society, London, Special Publications*, **49**, 225–238, https://doi.org/10.1144/GSL.SP.1992.049.01.15

Dill, H.G., Wehner, H., Kus, J., Botz, R., Berner, Z., Stüben, D. and Al-Sayigh, A. 2007. The Eocene Rusayl Formation, Oman, carbonaceous rocks in calcareous shelf sediments: environment of deposition, alteration and hydrocarbon potential. *International Journal of Coal Geology*, **72**, 89–123. https://doi.org/10.1016/j.coal.2006.12.012

Droste, H. 2014. Petroleum geology of the Sultanate of Oman. *In*: Marlow, L., Kendall, C. and Yose, L. (eds) *Petroleum Systems of the Tethyan Region*. American Association of Petroleum Geologists, Memoirs, **106**, 713–755, https://doi.org/10.1036/12431870M1063039

El-Shazly, A.K. 1995. Petrology of Fe-Mg-carpholite-bearing metasediments from NE Oman. *Journal of Metamorphic Geology*, **13**, 379–396, https://doi.org/10.1111/j.1525-1314.1995.tb00227.x

El-Shazly, A.K., Coleman, R.G. and Liou, J.G. 1990. Eclogites and blueschists from Northeastern Oman: petrology and P-T evolution. *Journal of Petrology*, **31**, 629–666, https://doi.org/10.1093/petrology/31.3.629

El-Shazly, A.K., Broker, W., Hacker, B. and Calvert, A. 2001. Formation and exhumation of blueshists and eclogites from NE Oman: new perspectives from Rb-Sr and $^{40}Ar/^{39}Ar$ dating. *Journal of Metamorphic Geology*, **19**, 233–248, https://doi.org/10.1046/j.1525-1314.2001.00309.x

Ernewein, M., Pflumio, C. and Whitechurch, H. 1988. The death of an accretion zone as evidenced by the magmatic history of the Sumail ophiolite (Oman). *Tectonophysics*, **151**, 257–274, https://doi.org/10.1016/0040-1951(88)90248-X

Falk, E.S. and Kelemen, P.B. 2015. Geochemistry and petrology of listvenite in the Samail ophiolite, Sultanate of Oman: complete carbonation of peridotites during ophiolite emplacement. *Geochimica et Cosmochimica Acta*, **160**, 70–90, https://doi.org/10.1016/j.gca.2015.03.014

Fike, D.A., Grotzinger, J.P., Pratt, L.M. and Summons, R.E. 2006. Oxidation of the Ediacaran ocean. *Nature*, **444**, 744–747, https://doi.org/10.1038/nature05345

Fink, R., Virgo, S., Arndt, M., Visser, W., Littke, R. and Urai, J.L. 2016. Solid bitumen in calcite veins from the Natih Formation in the Oman Mountains: multiple phases of petroleum migration in a changing stress field. *International Journal of Coal Geology*, **157**, 39–51, https://doi.org/10.1016/j.coal.2015.07.012

Forbes, G.A., Jansen, H.S.M. and Schreurs, J. 2010. *Lexicon of Oman Subsurface Stratigraphy. Reference Guide to the Stratigraphy of Oman's Hydrocarbon Basins*. Gulf Petrolink, GeoArabia, Special Publication **5**.

Fournier, M., Lepvrier, C., Razin, P. and Jolivet, L. 2006. Late Cretaceous to Paleogene Post-obduction extension and subsequent Neogene compression in the Oman Mountains. *GeoArabia*, **11**, 17–40.

Ghent, E.D. and Stout, M.Z. 1981. Metamorphism at the base of the Samail ophiolite, southeastern Oman Mountains. *Journal of Geophysical Research*, **86**, 2557–2571, https://doi.org/10.1029/JB086iB04p02557

Gilgen, S.A., Diamond, L.W., Mercolli, I., Al-Tobi, K., Maidment, D.W., Close, R. and Al-Towaya, A. 2014. Volcanostratigraphic controls on the occurrence of massive sulfide deposits in the Semail ophiolite, Oman. *Economic Geology*, **109**, 1585–1610, https://doi.org/10.2113/econgeo.109.6.1585

Glennie, K.W. 2005. *The Geology of the Oman Mountains: An Outline of their Origin*. 2nd ed. Scientific Press, Buckinghamshire, 110.

Glennie, K.W., Boeuf, M.G.A., Hughes Clarke, M.W., Moody-Stuart, M., Pilaar, W. and Reinhardt, B.M. 1974. Geology of the Oman Mountains. Koninklijk Nederlands Geologisch en Mijnbouwkundig Genootschap. *Transactions*, **31**, 423.

Gnos, E. 1998. Peak metamorphic conditions of garnet amphibolites beneath the Semail ophiolite: implications for an inverted pressure gradient. *International Geological Review*, **40**, 281–304, https://doi.org/10.1080/00206819809465210

Gnos, E. and Peters, T. 1993. K-Ar ages of the metamorphic sole of the Semail ophiolite; implications for ophiolite cooling history. *Contributions to Mineralogy and Petrology*, **113**, 325–332, https://doi.org/10.1007/BF00286925

Gnos, R. and Peters, T. 2003. Mantle xenolith-bearing Maastrichtian to Tertiary alkaline magmatism in Oman. *Geochemistry, Geophysics, Geosystems*, **4**, https://doi.org/10.1029/2001GC000229

Godard, M., Dautria, J.-M. and Perrin, M. 2003. Geochemical variability of the Oman ophiolite lavas: relationship with spatial distribution and paleomagnetic directions. *Geochemistry Geophysics Geosystems*, **4**, 8609, https://doi.org/10.1029/2002GC000452

Godard, M., Bosch, D. and Einaudi, F. 2006. A MORB source for low-Ti magmatism in the Semail ophiolite. *Chemical Geology*, **234**, 58–78, https://doi.org/10.1016/j.chemgeo.2006.04.005

Godel, B., Barnes, S.-J. and Maier, W.D. 2007. Platinum-group elements in sulphide minerals, platinum-group minerals, and whole-rock of the Merensky Reef (Bushveld Complex, South Africa): implications for the formation of the reef. *Journal of Petrology*, **48**, 1569–1604, https://doi.org/10.1093/petrology/egm030

Goettler, G.W., Firth, N. and Huston, C.C. 1976. A preliminary discussion of ancient mining in the Sultanate of Oman. *The Journal of Oman Studies*, **2**, 43–55.

Goffé, B., Michard, A., Kienast, J.R. and le Mer, O. 1988. A case of obduction related high P, low T metamorphism in upper crustal nappes, Arabian continental margin, Oman: P T paths and kinematic interpretation. *Tectonophysics*, **151**, 363–386, https://doi.org/10.1016/0040-1951(88)90253-3

Gomez-Rivas, E., Bons, P.D. *et al.* 2014. The Jabal Akhdar Dome in the Oman Mountains: evolution of a dynamic fracture system. *American Journal of Science*, **314**, 1104–1139, https://doi.org/10.2475/07.2014.02

Goodenough, K.M., Styles, M.T. *et al.* 2010. Architecture of the Oman-UAE ophiolite: evidence for a multiple-phase magmatic history. *Arabian Journal of Geosciences*, **3**, 439–458, https://doi.org/10.1007/s12517-010-0177-3

Goodenough, K.M., Thomas, R.J., Styles, M.T., Schofield, D.I. and MacLeod, C.J. 2014. Records of ocean growth and destruction in the Oman-UAE ophiolite. *Elements*, **10**, 109–114, https://doi.org/10.2113/gselements.10.2.109

Gorin, G.E., Raacz, L.G. and Walter, M.R. 1982. Late Precambrian-Cambrian sediments of Huqf Group, Sultanate of Oman. *Bulletin of the American Association of Petroleum Geologists*, **66**, 2609–2627.

Gray, D.T., Gregory, R.T. and Miller, J.M. 2000. A new structural profile along the Muscat-Ibra transect, Oman: implications for emplacement of the Semail ophiolite. *Geological Society of America, Special Paper*, **349**, 513–523.

Grobe, A. 2017. *Evolution of Pressure, Temperature and Structures in the Footwall of Large-Scale Overthrusts – A Case Study from the Oman Mountains*. PhD thesis, RWTH Aachen, Germany, 225.

Grobe, A., Urai, J.L., Littke, R. and Lünsdorf, N.K. 2016. Hydrocarbon generation and migration under a large overthrust: the carbonate platform under the Semail Ophiolite, Jebel Akhdar, Oman. *International Journal of Coal Geology*, **168**, 3–19, https://doi.org/10.1016/j.coal.2016.02.007

Grobe, A., Virgo, S., von Hagke, C., Urai, J.L. and Littke, R. 2018. Multiphase structural evolution of a continental margin during obduction orogeny: insights from the Jebel Akhdar Dome, Oman Mountains. *Tectonics*, **37**, 888–913, https://doi.org/10.1002/2016TC004442

Grobe, A., von Hagke, C., Littke, R., Dunkl, I., Wübbeler, F., Muchez, P. and Urai, J.L. 2019. Tectono-thermal evolution of Oman's Mesozoic passive continental margin under the obducting Semail Ophiolite: a case study of Jebel Akhdar, Oman. *Solid Earth*, **10**, 149–175, https://doi.org/10.5194/se-10-149-2019

Grotzinger, J.P., Fike, D.A. and Fischer, W.W. 2011. Enigmatic origin of the largest-known carbon isotopic excursion in Earth's history. *Nature Geoscience*, **4**, 285–292, https://doi.org/10.1038/ngeo1138

Guilmette, C., Smit, M.A. *et al.* 2018. Forced subduction initiation recorded in the sole and crust of the Semail Ophiolite of Oman. *Nature Geoscience*, **11**, 688–695, https://doi.org/10.1038/s41561-018-0209-2

Haase, M.M-L. and Aigner, T. 2013. High-resolution anatomy of a grainstone package in Khuff Squence KS4, Oman Mountains, Sultanate of Oman. *GeoArabia*, **18**, 17–44.

Haase, K.M., Freund, S., Koepke, J., Hauff, F. and Erdmann, M. 2015. Melts of sediments in the mantle wedge of the Oman ophiolite. *Geology*, **43**, 275–278, https://doi.org/10.1130/G36354.1

Haase, K.M., Freund, S., Beier, C., Koepke, J., Erdmann, M. and Hauff, F. 2016. Constraints on the magmatic evolution of the oceanic crust from plagiogranite intrusions in the Oman ophiolite. *Contributions to Mineralogy and Petrology*, **171**, https://doi.org/10.1007/s00410-016-1261-9

Hacker, B.R. 1994. Rapid emplacement of young oceanic lithosphere: argon geochronology of the Oman ophiolite. *Science*, **265**, 1563–1565, https://doi.org/10.1126/science.265.5178.1563

Hacker, B.R., Mosenfelder, J.L. and Gnos, E. 1996. Rapid emplacement of the Oman ophiolite: thermal and geochronical constrains. *Tectonics*, **15**, 1230–1247, https://doi.org/10.1029/96TC01973

Halls, C. and Zhao, R. 1995. Listwaenite and related rocks: perspectives on terminology and mineralogy with reference to an occurrence at Creggan Baun, Co. Mayo, Republic of Ireland. *Mineralium Deposita*, **30**, 303–313, https://doi.org/10.1007/BF00196366

Hanna, S.S. 1990. The Alpine deformation of the central Oman Mountains. *Geological Society, London, Special Publications*, **49**, 341–359, https://doi.org/10.1144/GSL.SP.1992.049.01.21

Hanna, S.S. 1995. *Field guide to the Geology of Oman. Volume 1 Western Hajar Mountains and Musandam*. The Historical Association of Oman, Ruwi (Historical Association of Oman), 178.

Hartmann, B.H., Juhász-Bodnár, K., Ramseyer, K. and Matter, A. 2000. Polyphased quartz cementation and its sources: a case study from the Upper Palaeozoic Haushi Group sandstones, Sultanate of Oman. *In*: Worden, R.H. and Morad, S. (eds) *Quartz Cementation in Sandstones*. Blackwell Science, Oxford, International Association of Sedimentologists, Special Publication, **29**, 253–270.

Hauptmann, A. and Weisgerber, G. 1981. Third millennium BC copper production in Oman. *In*: *Revue d'Archéométrie, no. 1, 1981*. Actes du XXe Symposium International d'Archéométrie, Paris, 26–29 March 1980, Volume III, 131–138, https://doi.org/10.3406/arsci.1981.1138

Haynes, J.R., Recey, A. and Whittaker, J.E. 2010. A revision of the Early Palaeogene nummulitids (Foraminifera) from northern Oman, with implications for their classification. *In*: Whittaker, J.E. and Hart, M.B. (eds) *Micropalaeontology, Sedimentary Environments and Stratigraphy: A Tribute to Dennis Curry (1912–2001)*. Micropalaeontological Society, Special Publications, **4**, 29–89, https://doi.org/10.1144/TMS004.4

Hersi, O.S. and Al-Harthy, A. 2010. Lithofacies attributes of a transgressive carbonate system: the Middle Eocene Seeb Formation, Al Khoudh area, Muscat, Oman. *Sultan Qaboos University*

Journal for Science, **15**, 41–54, https://doi.org/10.24200/squjs.vol15iss0pp41-54
Heward, A.P. 1990. Salt removal and sedimentation in Southern Oman. *Geological Society, London, Special Publications*, **49**, 637–651, https://doi.org/10.1144/GSL.SP.1992.049.01.38
Heward, A.P. 2019. Further outcrops of diamictite in Wadi Amdeh: Al Khakata, Mesozoic or Cenozoic Slope deposits? *Al Hajar*, **35**, 14–17.
Heward, A.P. and Penney, R.A. 2014. Al Khlata glacial deposits in the Oman Mountains and their implications. *Geological Society, London, Special Publications*, **392**, 279–301, https://doi.org/10.1144/SP392.15
Heward, A.P., Booth, G.A., Fortey, R.A., Miller, C.G. and Sansom, I.J. 2018*a*. Darriwilian shallow-marine deposits from the Sultanate of Oman, a poorly known portion of the Arabian margin of Gondwana. *Geological Magazine*, **155**, 59–84, https://doi.org/10.1017/S0016756816000819
Heward, A.P., Miller, C.G. and Booth, G.A. 2018*b*. The Early Ordovician Middle Shale Member (Am3) of the Amdeh Formation and further evidence of conodont faunas from the Sultanate of Oman. *Geological Magazine*, 1–18, https://doi.org/10.1017/S0016756818000729
Hilgers, C., Kirschner, D.L., Breton, J.-P. and Urai, J. 2006. Fracture sealing and fluid overpressure in limestones of the Jabal Akhdar dome, Oman mountains. *Geofluids*, **6**, 168–184, https://doi.org/10.1111/j.1468-8123.2006.00141.x
Hillgärtner, H., van Buchem, F.S.P., Gaumet, F., Razin, P., Pittet, B., Grötsch, J. and Droste, H. 2003. The Barremian-Aptian evolution of the Eastern Arabian carbonate platform margin (northern Oman). *Journal of Sedimentary Research*, **73**, 756–773, https://doi.org/10.1306/030503730756
Hoffmann, K.H., Condin, D.J., Bowring, S.A. and Crowley, J.L. 2004. U-Pb zircon date from the Neoproterozoic Ghaub Formation, Namibia; constraints on Marinoan glaciation. *Geology*, **32**, 817–820, https://doi.org/10.1130/G20519.1
Holland, M. and Urai, J.L. 2010. Evolution of anastomosing crack-seal vein networks in limestones: insight from an exhumed high-pressure cell, Jabal Ahams, Oman Mountains. *Journal of Structural Geology*, **32**, 1279–1290, https://doi.org/10.1016/j.jsg.2009.04.011
Holland, M., Urai, J.L., Muchez, P. and Willemse, J.M. 2009*a*. Evolution of fractures in a highly dynamic thermal, hydraulic, and mechanical system – (I) Field observations in Mesozoic Carbonates, Jabal Shams, Oman Mountains. *GeoArabia*, **14**, 57–110.
Holland, M., Saxena, N. and Urai, J.L. 2009*b*. Evolution of fractures in a highly dynamic thermal, hydraulic, and mechanical system – (II) Remote sensing fracture analysis, Jabal Shams, Oman Mountains. *GeoArabia*, **14**, 163–194.
Homewood, P., Razin, P., Grelaud, C., Droste, H., Vahrenkamp, V., Mettraux, M. and Mattner, J. 2008. Outcrop sedimentology of the Natih Formation, northern Oman: a field guide to selected outcrops in the Adam Foothills and Al Jabal al Akhdar areas. *GeoArabia*, **13**, 39–120.
Hughes Clarke, M.W. 1988. Stratigraphy and rock unit nomenclature in the oil-producing area of Interior Oman. *Journal of Petroleum Geology*, **11**, 5–60, https://doi.org/10.1111/j.1747-5457.1988.tb00800.x
Immenhauser, A., Schlager, W. *et al.* 1999. Late Aptian to Late Albian sea-level fluctuations constrained by geochemical and biological evidence (Nahr Umr Formation, Oman). *Journal of Sedimentary Research*, **69**, 434–446, https://doi.org/10.2110/jsr.69.434
Ishikawa, T., Nagaishi, K. and Umino, S. 2002. Boninitic volcanism in the Oman ophiolite: implications for thermal condition during transition from spreading ridge to arc. *Geology*, **30**, 899–902, https://doi.org/10.1130/0091-7613(2002)030<0899:BVITOO>2.0.CO;2
Jones, R.W. and Racey, A. 1994. *Cenozoic Stratigraphy of the Arabian Peninsula and Gulf. Micropalaeontology and Hydrocarbon Exploration in the Middle East.* Chapman & Hall, London, 273–307.
Juteau, T., Ernewein, M., Reuber, I., Whitechurch, H. and Dahl, R. 1988. Duality of magmatism in the plutonic sequence of the Sumail nappe, Oman. *Tectonophysics*, **151**, 107–135, https://doi.org/10.1016/0040-1951(88)90243-0
Kapp, H. and Llewellyn, P.G. 1965. *The geology of the Central Oman Mountains.* Petroleum Development of Oman Ltd., Geological Group, Report **S00005-9**.
Kashkai, M.A. and Allakverdiev, Sh.I. 1965. *Listwaenites, their origin and classification (Izdat. Akad. Nauk Azerbaidzhanskoi SSSR, Baku, 1965).* US Geological Survey (translated by Vitaliano, D.B. for USGS in 1982).
Keen, M.C. and Racey, A. 1991. Lower Eocene ostracods from the Rusayl Shale Formation of Oman. *Journal of Micropalaeontology*, **10**, 227–233, https://doi.org/10.1144/jm.10.2.227
Kellerhals, P. and Matter, A. 2003. Facies analysis of a glaciomarine sequence, the Neoproterozoic Mirbat Sandstone Formation, Sultanate of Oman. *Eclogae Geologicae Helvetiae*, **96**, 49–70.
Kilner, B., Mac Niocaill, C. and Brasier, M. 2005. Low-latitude glaciation in the Neoproterozoic of Oman. *Geology*, **33**, 413–416, https://doi.org/10.1130/G21227.1
Koehrer, B., Aigner, T., Forke, H. and Pöppelreiter, M. 2012. Middle to Upper Khuff (Sequences KS1 to KS4) outcrop-equivalents in the Oman Mountains: grainstone architecture on a subregional scale. *GeoArabia*, **17**, 59–104.
Koga, K.T., Kelemen, P.B. and Shimizu, N. 2001. Petrogenesis of the crust-mantle transition zone and the origin of lower crustal wehrlite in the Oman ophiolite. *Geochemistry, Geophysics, Geosystems*, **2**, 2000GC000132.
Kusano, Y., Adachi, Y., Miyashita, S. and Umino, S. 2012. Lava accretion system around mid-ocean ridges: volcanic stratigraphy in the Wadi Fizh area, northern Oman ophiolite. *Geochemistry, Geophysics, Geosystems*, **13**, Q05012, 1-25, https://doi.org/10.1029/2011GC004006
Kusano, Y., Hayashi, M., Adachi, Y., Umino, S. and Miyashita, S. 2014. Evolution of volcanism and magmatism during initial arc stage: constraints on the tectonic setting of the Oman Ophiolite. *Geological Society, London, Special Publications*, **392**, 177–193, https://doi.org/10.1144/SP392.9
Lachize, M., Lonrand, J.P. and Juteau, T. 1996. Calc-alkaline differentiation trend in the plutonic sequence of the Wadi Haymiliyah section, Haylayn massif, Semail ophiolite, Oman. *Lithos*, **38**, 207–232, https://doi.org/10.1016/0024-4937(96)00009-6
Lacinska, A.M. and Styles, M.T. 2009. *The Barzaman Formation in the UAE – An Example of Intensive Alteration of Ultramafic and Mafic Rocks in the Near Surface Environment.* Mineralogia, Special Papers 35, Mineralogical Society of Poland.
Lacinska, A.M., Styles, M.T. and Farrant, A.R. 2014. Near-surface diagenesis of ophiolite-derived conglomerates of the Barzaman Formation, United Arab Emirates: a natural analogue for permanent CO_2 sequestration via mineral carbonation of ultramafic rocks. *Geological Society, London, Special Publications*, **392**, 343–360, https://doi.org/10.1144/SP392.18
Leather, J., Allen, P.A., Braiser, M.D. and Cozzi, A. 2002. A Neoproterozoic Snowball Earth under scrutiny: evidence from the Fiq glaciation of Oman. *Geology*, **30**, 891–894, https://doi.org/10.1130/0091-7613(2002)0302.0.CO.2
Leblanc, M. 1991. Platinum-group elements and gold in ophiolitic complexes: distribution and fractionation from mantle to oceanic floor. *In*: Peters, T., Nicolas, A. and Coleman, R.G. (eds) *Ophiolite Genesis and Evolution of the Oceanic Lithosphere.* Kluwer Academic Publishing, Dordrecht, 231–260.
Leblanc, M. and Ceuleneer, G. 1992. Chromite crystallization in a multicellular magma flow: evidence from a chromitite dike in the Oman ophiolite. *Lithos*, **27**, 231–257, https://doi.org/10.1016/0024-4937(91)90002-3
Lees, G.M. 1928. The geology and tectonics of Oman and parts of southeastern Arabia. *Quarterly Journal of the Geological Society of London*, **84**, 585–670, https://doi.org/10.1144/GSL.JGS.1928.084.01-04.24
Le Guerroué, E. 2006. *Sedimentology and Chemostratigraphy of the Ediacaran Shuram Formation, Nafun Group, Oman.* PhD thesis, ETH Zurich, Switzerland, Diss. ETH No. 16533.
Le Guerroué, E. 2010. Duration and synchroneity of the largest negative carbon isotope excursion on Earth: the Shuram/Wonoka

anomaly. *Comptes Rendus Geoscience*, **342**, 204–214, https://doi.org/10.1016/j.crte.2009.12.008

Le Guerroué, E. and Cozzi, A. 2009. Veracity of Neoproterozoic negative C-isotope values: the termination of the Shuram negative excursion. *Gondwana Research*, **17**, 653–661, https://doi.org/10.1016/j.gr.2009.11.002

Le Guerroué, E., Allen, P.A. and Cozzi, A. 2005. Two distinct glacial successions in the Neoproterozoic of Oman. *GeoArabia*, **10**, 17–34.

Le Guerroué, E., Allen, P.A. and Cozzi, A. 2006*a*. Chemostratigraphic and sedimentological framework of the largest negative carbon isotopic excursion in Earth history: the Neoproterozoic Shuram Formation (Nafun Group, Oman). *Precambrian Research*, **144**, 68–92, https://doi.org/10.1016/j.precamres.2006.01.007

Le Guerroué, E., Allen, P., Cozzi, A., Etienne, J.L. and Fanning, M. 2006*b*. 50 Myr recovery from the largest negative $\delta^{13}C$ excursion in the Ediacaran ocean. *Terra Nova*, **18**, 147–153, https://doi.org/10.1111/j.1365-3121.2006.00674.x

Le Guerroué, E., Rieu, R. and Cozzi, A. 2009. Comment Oman Chronostratigraphy. *American Journal of Science*, **309**, 85–90, https://doi.org/10.2475/01.2009.03

Le Maître, R.W. (ed.) 1989. *A Classification of Igneous Rocks and Glossary of Terms*. Blackwell Scientific Publications, Oxford.

Le Métour, J. 1988. *Géologie de l'Autochtone des Montagnes d'Oman: la fenêtre du Saih Hatat*. Documents du Bureau de Recherches Gèologiques et Minières, Orléans, **129**.

Le Métour, J., Villey, M. and de Gramont, X. 1986*a*. *Geological map of Masqat, sheet NF 40-4A, scale 1:100,000, Explanatory notes*. Directorate General of Minerals, Oman Ministry of Petroleum and Minerals.

Le Métour, J., Villey, M. and de Gramont, X. 1986*b*. *Geological map of Quryat, sheet NF 40-4D, scale 1:100,000, Explanatory notes*. Directorate General of Minerals, Oman Ministry of Petroleum and Minerals.

Le Métour, J.F., Rabu, D., Tegyey, M., Béchennec, F., Beurrier, M. and Villey, M. 1990. Subduction and obduction: two stages in the Eo-Alpine tectonometamorphic evolution of the Oman Mountains. *Geological Society, London, Special Publications*, **49**, 285–305, https://doi.org/10.1144/GSL.SP.1992.049.01.20

Le Métour, J.F., Béchennec, F., Roger, J. and Wyns, R. 1992. *Geological map of Muscat, sheet NF40-04, scale 1:250,000, with Explanatory notes*. Directorate General of Minerals, Oman Ministry of Petroleum and Minerals

Levell, B.K., Braakman, J.H. and Rutten, K.W. 1988. Oil-bearing sediments of Gondwana glaciation in Oman. *Bulletin of the American Association of Petroleum Geologists*, **72**, 775–796.

Lippard, S.J., Shelton, A.W. and Gass, I.G. 1986. The ophiolite of northern Oman. *Geological Society of London, Memoir*, **11**, 178.

Lorand, J.P. and Juteau, T. 2000. The Haymiliyah Sulphide Ores (Haylayn Massif, Oman Ophiolite): in-situ segregation of PGE-poor magmatic sulphides in a fossil oceanic magma chamber. *Marine Geophysical Researches*, **21**, 327–349, https://doi.org/10.1023/A:1004887103269

Lovelock, P.E.R., Potter, T.L., Walsworth-Bell, E.B. and Wiemer, W.M. 1981. Ordovician rocks in the Oman mountains: the Amdeh formation. *Geologie en Minjbouw*, **60**, 487–495.

Macklin, S., Ellison, R., Manning, J., Farrant, A. and Lorenti, L. 2012. Engineering geological characterization of the Barzaman Formation, with reference to coastal Dubai, UAE. *Bulletin of Engineering Geology and the Environment*, **71**, 1–19, https://doi.org/10.1007/s10064-011-0369-4

MacLeod, C.J., Lissenberg, L. and Bibby, L.E. 2013. 'Moist MORB' axial magmatism in the Oman ophiolite: the evidence against a mid-ocean ridge origin. *Geology*, **41**, 459, https://doi.org/10.1130/G33904.1

Maizels, J. 1988. Palaeochannels: Plio-Pleistocene raised channel systems of the western Sharqiyah. *In*: Dutton, R.W. (ed.) *The Scientific Results of the Royal Geographical Society's Oman Wahiba Project 1985-1987*. Journal of Oman studies Special Report, **3**, 95–112.

Mann, A. and Hanna, S.S. 1990. The tectonic evolution of pre-Permian rocks, Central and Southeastern Oman Mountains. *In*: Robertson, A.H.F., Searle, M.P. and Ries, A.C. (eds) *The Geology and Tectonics of the Oman Region*. Geological Society of London, Special Publications, **49**, 307–325, https://doi.org/10.1144/GSL.SP.1992.049.01.19

Mann, A., Hanna, S.S. and Nolan, S.C. 1990. The post-Campanian tectonic evolution of the Central Oman Mountains: Tertiary extension of the Eastern Arabian Margin. *Geological Society, London, Special Publications*, **49**, 285–305, https://doi.org/10.1144/GSL.SP.1992.049.01.33

Martin, J.R., Redfern, J. and Aitken, J.F. 2008. A regional overview of the late Paleozoic glaciation in Oman. *In*: Fielding, C.R., Frank, T.D. and Isbell, J.L. (eds) *Resolving the Late Paleozoic Ice Age in Time and space*. Geological Society of America, Special Paper, **441**, https://doi.org/10.1130/2008.2441(12)

Martin, J.R., Redfern, J. and Williams, B.J.P. 2012. Evidence for multiple ice centres during the late Palaeozoic ice age in Oman: outcrop sedimentology and provenance of the Late Carboniferous-Early Permian Al Khlata Formation. *Geological Society, London, Special Publications*, **368**, 229–256, https://doi.org/10.1144/SP368.18

Mattern, F. and Bernecker, M. 2018. A shallow marine clinoform system in limestones (Paleocene/Eocene Jafnayn Formation, Oman): geometry, microfacies, environment and processes. *Carbonates and Evaporites*, **34**, 101–113, https://doi.org/10.1007/s13146-018-0444-z

Mattern, F. and Scharf, A. 2018. Postobductional extension along and within the Frontal Range of the Eastern Oman Mountains. *Journal of Asian Earth Sciences*, **154**, 369–385, https://doi.org/10.1016/j.jseaes.2017.12.031

Mattern, F. and Scharf, A. 2019. Transition from the Hajir Formation to the Muaydin Formation: a facies change coinciding with extensional, syndepositional faulting (Ediacaran, Jabal Akhdar, Central Oman Mountains). *Journal of African Earth Sciences*, **152**, 237–244, https://doi.org/10.1016/j.jafrearsci.2019.02.016

Mattern, F., Pracejus, B. and Al Balushi, L. 2018*a*. Heavy mineral beach placers of the Ordovician Amdeh Formation (Member 4, Wadi Qahza, Saih Hatat, eastern Oman Mountains): where is the main source area? *Journal of African Earth Sciences*, **147**, 633–646, https://doi.org/10.1016/j.afrearsci.2018.06.031

Mattern, F., Scharf, A., Al-Sarmi, M., Pracejus, B., Al-Hinaai, A. and Al-Mamari, A. 2018*b*. Compaction history of Upper Cretaceous shale and related tectonic framework, Arabian Plate, Eastern Oman Mountains. *Arabian Journal of Geosciences*, **11**, 444, https://doi.org/10.1007/s12517-018-3781-2

Mattes, B.W. and Conway Morris, S. 1990. Carbonate/evaporate deposition in the Late Precambrian – Early Cambrian Ara Formation of Southern Oman. *Geological Society, London, Special Publications*, **49**, 617–636, https://doi.org/10.1144/GSL.SP.1992.049.01.37

Matveev, S. and Ballhaus, C. 2002. Role of water in the origin of podiform chromitite deposits. *Earth and Planetary Science Letters*, **203**, 235–243, https://doi.org/10.1016/S0012-821X(02)00860-9

Maury, R.C., Béchennec, F., Cotton, J., Caroff, M., Cordey, F. and Marcoux, J. 2003. Middle Permian plume-related magmatism of the Hawasina Nappes and the Arabian Platform: implications on the evolution of the Neotethyan margin in Oman. *Tectonics*, **22**, 12-11–12-13, https://doi.org/10.1029/2002TC001483

McCarron, G.M. 2000. *Sedimentology and stratigraphy of the Nafun Group: Huqf Supergroup, Oman*. PhD thesis, Trinity College, Dublin, Ireland, Oxford University.

Michel, J.C. 1993. *Mineral occurrence and metallogenic map of the Sultanate of Oman, scale 1:1,000,000, with Explanatory Notes*. Directorate General of Minerals, Oman Ministry of Petroleum and Minerals.

Miller, C.G., Heward, A., Mossoni, A. and Sansom, I.J. 2018. Two new early balognathid conodont genera from the Ordovician of Oman and comments on the early evolution of prioniodontid conodonts. *Journal of Systematic*

Palaeontology, **16**, 571–593, https://doi.org/10.1080/14772019.2017.1314985

Montenat, C. and Blondeau, A. 1977. Premier aperçu du Tertiaire d'Oman (Péninsule arabique orientale). *Bulletin de la Société Géologique de France*, **7**, 1285–1295, https://doi.org/10.2113/gssgfbull.S7-XIX.6.1285

Moraetis, D., Al Kindi, S.S. *et al.* 2020. Terrace agriculture in a mountainous arid environment – a study of soil quality and regolith provenance: Jabal Akhdar (Oman). *Geoderma*, **363**, 114152, https://doi.org/10.1016/j.geoderma.2019.114152

Mount, V.S., Crawford, R.I.S. and Bergman, S.C. 1998. Regional structural style of the Central and Southern Oman Mountains: Jabal Akhdar, Saih Hatat and the Northern Ghaba Basin. *GeoArabia*, **3**, 475–490.

Nasir, S., Al-Sayigh, A., Alharthy, A. and Al-Lazki, A. 2006. Geochemistry and petrology of Tertiary volcanic rocks and related ultramafic xenoliths from the central and eastern Oman Mountains. *Lithos*, **90**, 219–270, https://doi.org/10.1016/j.lithos.2006.03.002

Nasir, S., Al-Sayigh, A. *et al.* 2007. Mineralogical and geochemical characterization of listwaenite from the Semail Ophiolite, Oman. *Chemie der Erde*, **67**, 213–228, https://doi.org/10.1016/j.chemer.2005.01.003

Nicolas, A., Ceuleneer, G., Boudier, F. and Misseri, M. 1988*a*. Structural mapping in the Oman ophiolites: mantle diapirism along an oceanic ridge. *Tectonophysics*, **151**, 27–56, https://doi.org/10.1016/0040-1951(88)90239-9

Nicolas, A., Reuber, I. and Benn, K. 1988*b*. A new magma chamber model based on structural studies in the Oman ophiolite. *Tectonophysics*, **151**, 87–105, https://doi.org/10.1016/0040-1951(88)90242-9

Nicolas, A., Freydier, C., Godard, M. and Vauchez, A. 1993. Magma chambers at oceanic ridges. How large? *Geology*, **21**, 53–56, https://doi.org/10.1130/0091-7613(1993)021<0053:MCAORH>2.3.CO;2

Nicolas, A., Boudier, F. and Ildefonse, B. 1996. Variable crustal thickness in the Oman ophiolite: implications for oceanic crust. *Journal of Geophysical Research*, **101**, 17,941–17,950, https://doi.org/10.1029/96JB00195

Nicolas, A., Boudier, F., Ildefonse, B. and Ball, E. 2000*a*. Accretion of Oman and United Arab Emirates ophiolite – discussion of a new structural map. *Marine Geophysical Researches*, **21**, 147–179, https://doi.org/10.1023/A:1026769727917

Nicolas, A., Ildefonse, B., Boudier, F., Lenoir, X. and Ben Ismail, W. 2000*b*. Dike distribution in the Oman-United Arab Emirates ophiolite. *Marine Geophysical Researches*, **21**, 269–287, https://doi.org/10.1023/A:1026718026951

Nolan, S.C., Skelton, P.W., Clissold, B.P. and Smewing, J.D. 1990. Maastrichtian to Early Tertiary stratigraphy and paleogeography of the Central and Northern Oman Mountains. *Geological Society, London, Special Publications*, **49**, 307–325, https://doi.org/10.1144/GSL.SP.1992.049.01.31

Oberthür, T., Weiser, T.W. and Gast, L. 2003. Geochemistry and mineralogy of platinum-group elements at Hartley Platinum Mine, Zimbabwe. I. Primary distribution patterns in pristine ores of the Main Sulfide Zone of the Great Dyke. *Mineralium Deposita*, **38**, 327–343, https://doi.org/10.1007/s00126-002-0337-9

Osburn, M., Grotzinger, J. and Bergmann, K. 2014. Facies, stratigraphy, and evolution of a middle Ediacaran carbonate ramp: Khufai Formation, Sultanate of Oman. *American Association of Petrology Geologists Bulletin*, **98**, 1631–1667, https://doi.org/10.1306/07291312140

Osterloff, P., Al-Harthy, A. *et al.* 2004*a*. Depositional sequences of the Gharif and Khuff formations, subsurface interior Oman. *In*: Al-Husseini, M.I. (ed.) *Carboniferous, Permian and Early Triassic Arabian Stratigraphy*. GeoArabia Special Publication, Gulf PetroLink, Bahrain, **3**, 83–147.

Osterloff, P., Penney, R.A., Aitken, J., Clark, N. and Al-Husseini, M.I. 2004*b*. Depositional sequences of the Al Khlata Formation, subsurface Interior Oman. *In*: Al-Husseini, M.I. (ed.) *Carboniferous, Permian and Early Triassic Arabian Stratigraphy*. GeoArabia Special Publication, Gulf PetroLink, Bahrain, **3**, 61–81.

Owen, R.M.S. and Nasr, S.N. 1958. Stratigraphy of the Kuwait-Basra area. *In*: Weeks, L.G. (ed.) *Habitat of Oil*. American Association of Petroleum Geologist, Special Volume, 1252–1278.

Özcan, E., Abbasim, I.A., Drobne, K., Govindan, A., Jovane, L. and Boukhalfa, K. 2016. Early Eocene orthophragminids and alveolinids from the Jafnayn Formation, N Oman: significance of Nemkovella stockari Less & Özcan, 2007 in Tethys. *Geodinamica Acta*, **28**, 160–184, https://doi.org/10.1080/09853111.2015.1107437

Pallister, J.S. and Hopson, C.A. 1981. Samail ophiolite plutonic suite: field relations, phase variations, cryptic variation and layering, and a model of a spreading ridge magma chamber. *Journal of Geophysical Research*, **86**, 2593–2644, https://doi.org/10.1029/JB086iB04p02593

Partington, G. 2009. Developing models using GIS to assess geological and economic risk: an example from mineral exploration in Oman for VMS copper gold mineralisation. *Transport*, **2**, 1–85.

Penney, R.A., Al-Barram, I. and Stephenson, M.H. 2008. A high resolution palynozonation for the Al Khlata Formation (Pennsylvanian to Lower Permian), South Oman. *Palynology*, **32**, 213–229, https://doi.org/10.2113/gspalynol.32.1.213

Peters, J.M., Filbrandt, J.B., Grotzinger, J.P., Newall, M.J., Shuster, M.W. and Al-Siyabi, H.A. 2003. Surface-piercing salt domes of interior North Oman, and their significance for the Ara carbonate 'stringer' hydrocarbon play. *GeoArabia*, **8**, 231–270.

Philip, J., Borgomano, J. and Al-Maskiry, S. 1995. Cenomanian-Early Turonian carbonate platform of Northern Oman: stratigraphy and palaeo-environments. *Palaeogeography, Palaeoclimatology, Palaeoecology*, **119**, 77–92, https://doi.org/10.1016/0031-0182(95)00061-5

Phillips, E.R., Waters, C.N. and Ellison, R.A. 2013. The Jurassic-Cretaceous depositional and tectonic evolution of the southwestern margin of the Neotethys Ocean, Northern Oman and United Arab Emirates. *In*: Al Hosani, K., Roure, F., Ellison, R. and Lokier, S. (eds) *Lithosphere Dynamics and Sedimentary Basins: The Arabian Plate and Analogues*. Springer, Berlin, Heidelberg, 61–100, https://doi.org/10.1007/978-3-642-30609-9_4

Pickford, M. 2017. Late Cretaceous Lanistes (Mollusca, Gastropoda) from Al-Khodh, Oman. *Al Hajar*, **23**, 15–27.

Pillevuit, A. 1993. Les Blocs Exotiques du Sultanat d'Oman: evolution paléogéographique d'une marge passive flexural. *Mémoires de Géologie (Lausanne)*, **17**, 1–249.

Pillevuit, A., Marcoux, J., Stampfli, G. and Baud, A. 1997. The Oman Exotics: a key to the understanding of the Neotethyan geodynamic evolution. *Geodinamica Acta, Paris*, **10**, 209–238, https://doi.org/10.1080/09853111.1997.11105303

Piuz, A., Meister, C. and Vicedo, V. 2014. New Alveolinoidea (Foraminifera) from the Cenomanian of Oman. *Cretaceous Research*, **50**, 344–360, https://doi.org/10.1016/j.cretres.2014.04.009

Pöppelreiter, M.C., Schneider, C.J., Obermaier, M., Forke, H.C., Koehrer, B. and Aigner, T. 2011. Seal turns into reservoir: Sudair equivalents in outcrop, Al Jabal al-Akhdar, Sultanate of Oman. *GeoArabia*, **16**, 69–108.

Powers, R.W., Ramirez, L.F., Redmond, C.D. and Elberg, E.L. 1966. Geology of the Arabian Peninsula: Sedimentary geology of Saudi Arabia. *United States Geological Survey Professional Papers*, 560-D, p. 147.

Pracejus, B. 2015. *The Ore Minerals under the Microscope: An Optical Guide*. 2nd ed. Elsevier, Amsterdam.

Pracejus, B. and Scharf, A. 2020. Do Omani Cu-slags reflect the origin of hydrothermal fluids? International Conference on Ophiolites and the Oceanic Lithosphere: Results of the Oman Drilling Project and Related Research, 12–14h January 2020, Sultan Qaboos University, Muscat, Sultanate of Oman.

Prichard, H.M., Neary, C.R., Fisher, P.C. and O'Hara, M.J. 2008. PGE-rich podiform chromitites in the Al'Ays ophiolite complex, Saudi Arabia: an example of critical mantle melting to extract and concentrate PGE. *Economic Geology*, **103**, 1507–1529, https://doi.org/10.2113/gsecongeo.103.7.1507

Pujalte, V., Baceta, J.I. *et al.* 2009. Redefinition of the Ilerdian Stage (early Eocene). *Geologica Acta*, **7**, 177–194, https://doi.org/10.1344/105.000000268

Rabu, D. 1987. *Géologie de l'Autochtone des Montagnes d'Oman: la fenêtre du Jabal Akhdar. La semelle métamorphique de la Nappe ophiolitique de Samail dans les parties orientaler et central des Montagnes d'Oman: une revue.* PhD thesis, Pierre and Marie University, Paris 6, France, Bureau de Recherches Géologiques et Minières, Document No. 130, p. 582.

Rabu, D. 1988. *Géolgie de l'Autochtone des Montagnes d'Oman: la fenêtre du Jabal Akhdar. La semelle métamorphique de la Nappe ophiolitique de Samail dans les parties orientale et central des Montagnes d'Oman: une revue.* Documents du Bureau de Recherches Géologiques et Miniéres, Orléans **130**.

Rabu, D., Béchennec, F., Beurrier, M. and Hutin, G. 1986. *Geological map of Nakhl, sheet NF40-3E, scale: 1:100,000, with Explanatory Notes.* Directorate General of Minerals, Oman Ministry of Petroleum and Minerals.

Rabu, D., Le Métour, J., Béchennec, F., Beurrier, M., Villey, M. and Bourdillon-Jeudy De Grissac, C. 1990. Sedimentary aspects of the Eo-Alpine cycle on the northeast edge of the Arabian Platform (Oman Mountains). *Geological Society, London, Special Publications*, **49**, 49–68, https://doi.org/10.1144/GSL.SP.1992.049.01.05

Racey, A. 1995. Lithostratigraphy and larger foraminiferal (nummulitid) biostratigraphy of the Tertiary of northern Oman. *Micropaleontology*, **41**, 1–123, https://doi.org/10.2307/1485849

Ravaut, P., Bayer, R., Hassani, R., Rousset, D. and Yahya'ey, A.A. 1997. Structure and evolution of the northern Oman margin: gravity and seismic constraints over the Zagros-Makran-Oman collision zone. *Tectonophysics*, **279**, 253–280, https://doi.org/10.1016/S0040-1951(97)00125-X

Razin, P. and Grelaud, C. 2011. *Les systèmes carbonates crétacés de la plate-forme arabe.* Livret-Guide des Excursion du Groupe Français du Crétacé, <hal-01236365>.

Reagan, M.K., Ishizuka, O. *et al.* 2010. Fore-arc basalts and subduction initiation in the Izu-Bonin-Mariana system. *Geochemistry, Geophysics, Geosystems*, **11**, Q03X12, https://doi.org/10.1029/2009GC002871

Reinhardt, B.M. 1969. On the genesis and emplacement of ophiolites in the Oman Mountains geosyncline. *Schweizerische Mineralogische und Petrographische Mitteilungen*, **49**, 1–30.

Reuber, I. 1988. Complexity of the crustal sequence in the northern Oman ophiolite (Fizh and southern Aswad blocks): the effect of early slicing? *Tectonophysics*, **151**, 137–165, https://doi.org/10.1016/0040-1951(88)90244-2

Rickards, B.R., Booth, G.A., Paris, F. and Heward, A.P. 2010. Marine flooding events of the Early and Middle Ordovician of Oman and the United Arab Emirates and their graptolite, acritarch and chitinozoan associations. *GeoArabia*, **15**, 81–120.

Rieu, R., Allen, P.A., Cozzi, A., Kosler, J. and Bussy, F. 2007. A composite stratigraphy for the Neoproterozoic Huqf Supergroup of Oman: integrating new litho-, chemo- and chronostratigraphic data of the Mirbat area southern Oman. *Journal of the Geological Society, London*, **164**, 997–1009, https://doi.org/10.1144/0016-76492006-114

Rioux, M., Bowring, S., Kelemen, P., Gordon, S., Miller, R. and Dudás, F. 2013. Tectonic development of the Samail ophiolite: high-precision U-Pb zircon geochronology and Sm-Nd isotopic constraints on crustal growth and emplacement. *Journal of Geophysical Research: Solid Earth*, **118**, 2085–2101, https://doi.org/10.1002/jgrb.50139

Rioux, M., Garber, J., Bauer, A., Bowring, S., Searle, M.P., Kelemen, P. and Hacker, B. 2016. Synchronous formation of the metamorphic sole and igneous crust of the Semail ophiolite: new constraints on the tectonic evolution during ophiolite formation from high-precision U-Pb zircon geochronology. *Earth and Planetary Science Letters*, **451**, 185–195, https://doi.org/10.1016/j.epsl.2016.06.051

Robertson, A.H.F. 1987*a*. The transition from passive margin to Upper Cretaceous foreland basin related to ophiolite emplacement in the Oman Mountains. *Geological Society of America Bulletin*, **99**, 633–653, https://doi.org/10.1130/0016-7606(1987)99<633:TTFAPM>2.0.CO;2

Robertson, A.H.F. 1987*b*. Upper Cretaceous Muti Formation: transition of a Mesozoic carbonate platform to a foreland basin in the Oman Mountains. *Sedimentology*, **34**, 1123–1142, https://doi.org/10.1111/j.1365-3091.1987.tb00596.x

Robertson, A.H.F. and Searle, M.P. 1990. The northern Oman Tethyan continental margin: stratigraphy, structure, concepts and controversies. *Geological Society, London, Special Publications*, **49**, 3–25, https://doi.org/10.1144/GSL.SP.1992.049.01.02

Rodgers, D.W. and Gunatilaka, A. 2002. Bajada formation by monsoonal erosion of a subaerial forebulge, Sultanate of Oman. *Sedimentary Geology*, **154**, 127–146, https://doi.org/10.1016/S0037-0738(02)00126-4

Roger, J., Béchennec, F., Janjou, D., Le Métour, J., Wyns, R. and Beurrier, M. 1991. *Geological map of Ja'alan, sheet NF 40-08E, scale:100,000, with Explanatory Notes.* Directorate General of Minerals, Oman Ministry of Petroleum and Minerals.

Rollinson, H. 2008. The geochemistry of mantle chromitites from the northern part of the Oman ophiolite: inferred parental melt composition. *Contributions to Mineralogy and Petrology*, **156**, 273–288, https://doi.org/10.1007/s00410-008-0284-2

Rothman, D.H., Hayes, J.M. and Summons, R.E. 2003. Dynamics of the Neoproterozoic carbon cycle. *Proceedings of the National Academy of Sciences of the United States of America*, **100**, 8124-8129, https://doi.org/10.1073/pnas.0832439100

Rousseau, M., Dromart, G., Garcia, J-P., Atrops, F. and Guillocheau, F. 2005. Jurassic evolution of the Arabian carbonate platform edge in the central Oman Mountains. *Journal of the Geological Society, London*, **162**, 349–362, https://doi.org/10.1144/0016-764903-178

Rousseau, M., Dromart, G., Droste, H. and Homewood, P. 2006. Stratigraphic organization of the Jurassic sequence in Interior Oman, Arabian Peninsula. *GeoArabia*, **11**, 17–50.

Sansom, I.J., Miller, C.G., Heward, A., Davies, N.S., Booth, G.A., Fortey, R.A. and Paris, R.A. 2009. Ordovician fish from the Arabian Peninsula. *Palaeontology*, **52**, 337–342, https://doi.org/10.1111/j.1475-4983.2009.00846.x

Scharf, A., Mattern, F. and Al Sadi, S. 2016. Kinematics of post-obduction deformation of the Tertiary ridge at Al-Khod Village (Muscat Area, Oman). *Sultan Qaboos University Journal for Science*, **21**, 26–40, https://doi.org/10.24200/squjs.vol21iss1pp26-40

Scharf, A., Mattern, F., Bolhar, R., Bailey, C.M. and Ring, U. 2019. Early Cenozoic listwaenite formation at a major extensional fault zone of the Oman Mountains (Fanja area) – insights from structural analysis and U-Pb carbonate dating. AAPG Event, Structural styles in the Middle East, 9–11 December 2019, Muscat, Oman.

Scharf, A., Mattern, F., Bolhar, R., Bailey, C.M. and Ring, U. 2020*a*. U-Pb dating of postobductional carbonate veins in listwaenite of the Oman Mountains near Fanja. International Conference on Ophiolites and the Oceanic Lithosphere: Results of the Oman Drilling Project and Related Research, 12–14 January 2020, Sultan Qaboos University, Muscat, Sultanate of Oman.

Scharf, A., Sudo, M., Pracejus, B., Mattern, F., Callegari, I., Bauer, W. and Scharf, K. 2020*b*. Late Lutetian (Eocene) mafic intrusion into shallow marine platform deposits north of the Oman Mountains (Rusayl Embayment) and its tectonic significance. *Journal of African Earth Sciences*, **170**, 103941, https://doi.org/10.1016/j.jafrearsci.2020.103941

Scharf, A., Mattern, F., Al-Wardi, M., Frijia, G., Moraetis, D., Pracejus, B., Bauer, W. and Callegari, I. 2021*a*. Appendices to: The Geology and Tectonics of the Jabal Akhdar and Saih Hatat Domes, Oman Mountains. *Geological Society, London, Memoirs*, **54**, 113–116, https://doi.org/10.1144/M54.7

Scharf, A., Mattern, F., Al-Wardi, M., Frijia, G., Moraetis, D., Pracejus, B., Bauer, W. and Callegari, I. 2021*b*. Thrusts, extensional faults and fold patterns of the major units. *Geological Society, London, Memoirs*, **54**, 49–60, https://doi.org/10.1144/M54.3

Scharf, A., Mattern, F., Al-Wardi, M., Frijia, G., Moraetis, D., Pracejus, B., Bauer, W. and Callegari, I. 2021*c*. Tectonic evolution of the Oman Mountains. *Geological Society, London, Memoirs*, **54**, 67–104, https://doi.org/10.1144/M54.5

Schiano, P., Clocchiatti, R., Lorand, J.-P., Massare, D., Deloule, E. and Chaussidon, M. 1997. Primitive basaltic melts included in podiform chromites from the Oman Ophiolite. *Earth and Planetary Science Letters*, **146**, 489–497, https://doi.org/10.1016/S0012-821X(96)00254-3

Schlaich, M. and Aigner, T. 2015. The Geology of the Middle Jurassic Dhruma Formation in Outcrops of the Oman Mountains (Sultanate of Oman). 5th EAGE Arabian Plate Geology Workshop 2015.

Schlaich, M. and Aigner, T. 2016. Sequence Stratigraphy of the Middle Jurassic Dhruma Formation in the Oman Mountains (Sultanate of Oman). 78th EAGE Conference and Exhibition 2016.

Schlaich, M. and Aigner, T. 2017. Facies and integrated sequence stratigraphy of an epeiric carbonate ramp succession: Dhruma Formation, Sultanate of Oman. *The Depositional Record*, **3**, 92–132, https://doi.org/10.1002/dep2.28

Schlaich, M., Aigner, T. and Bendias, D. 2021. Facies distribution, depositional environment, and reservoir geometries of clastic sandbodies in the Lower Jurassic Mafraq Member (Oman Mountains, Sultanate of Oman). *In*: *Lower Triassic to Middle Jurassic Sequence of the Arabian Plate*, EAGE.

Schlüter, M., Steuber, T., Parente, M. and Mutterlose, J. 2008. Evolution of a Maastrichtian-Paleocene tropical shallow-water carbonate platform (Qalhat, NE Oman). *Facies*, **54**, 513–527, https://doi.org/10.1007/s10347-008-0150-8

Schroeder, R., van Buchem, F.S.P., Cherchi, A., Baghbani, D., Vincent, B., Immenhauser, A. and Grainer, B. 2010. Revised orbitolinid biostratigraphic zonation for the Barremian-Aptian of the eastern Arabian Plate and implications for regional stratigraphic correlations. *GeoArabia Special Publications*, **4**, 49–96.

Schröder, S., Schreiber, B.C., Amthor, J.E. and Matter, A. 2003. A depositional model for the terminal Neoproterozoic-Early Cambrian Ara Group evaporites in south Oman. *Sedimentology*, **50**, 879–898, https://doi.org/10.1046/j.1365-3091.2003.00587.x

Schulp, A.S., Hanna, S.S., Hartman, A.F. and Jagt, J.W.M. 2000. A late Cretaceous theropd caudal vertebra from the Sultanate of Oman. *Cretaceous Research*, **21**, 851–856, https://doi.org/10.1006/cres.2000.0235

Schulp, A.S., Al-Wosabi, M. and Stevens, N.J. 2008. First dinosaur tracks from the Arabian Peninsula. *PLoS One*, **3**, e2243, https://doi.org/10.1371/journal.pone.0002243

Searle, M.P. 1985. Sequence of thrusting and origin of culminations in the northern and central Oman Mountains. *Journal of Structural Geology*, **7**, 129–143, https://doi.org/10.1016/0191-8141(85)90127-0

Searle, M.P. 2007. Structural geometry, style and timing of deformation in the Hawasina Window, Al Jabal al Akhdar and Saih Hatat culminations, Oman Mountains. *GeoArabia*, **12**, 99–130.

Searle, M.P. and Alsop, G.I. 2007. Eye-to-eye with a mega-sheath fold: a case study from Wadi Mayh, northern Oman. *Geology*, **35**, 1043–1046. https://doi.org/10.1130/G23884A.1

Searle, M.P. and Cox, J.S. 1999. Tectonic setting, origin and obduction of the Oman ophiolite. *Geological Society of America Bulletin*, **111**, 104–122, https://doi.org/10.1130/0016-7606(1999)111<0104:TSOAOO>2.3.CO;2

Searle, M.P. and Cox, J.S. 2002. Subduction zone metamorphism during formation and emplacement of the Semail Ophiolite in the Oman Mountains. *Geological Magazine*, **139**, 241–255, https://doi.org/10.1017/S0016756802006532

Searle, M.P. and Graham, G.M. 1982. 'Oman Exotics' – oceanic carbonate build-ups associated with the early stages of continental rifting. *Geology*, **10**, 43–49, https://doi.org/10.1130/0091-7613(1982)10<43:OECBAW>2.0.CO;2

Searle, M.P. and Malpas, J. 1980. Structure and metamorphism of rocks beneath the Semail ophiolite of Oman and their significance in ophiolite obduction. *Earth and Environmental Science, Transactions of the Royal Society of Edinburgh*, **71**, 247–262, https://doi.org/10.1017/S0263593300013614

Searle, M.P. and Malpas, J. 1982. Petrochemistry and origin of subophiolitic metamorphic and related rocks in the Oman Mountains. *Journal of the Geological Society, London*, **139**, 235–248, https://doi.org/10.1144/gsjgs.139.3.0235

Searle, M.P., Lippard, S.J., Smewing, J.D. and Rex, D.C. 1980. Volcanic rocks beneath the Semail Ophiolite in the northern Oman Mountains and their tectonic significance in the Mesozoic evolution of Tethys. *Journal of Geological Society, London*, **137**, 589–604, https://doi.org/10.1144/gsjgs.137.5.0589

Searle, M.P., Waters, D.J., Martin, H.N. and Rex, D.C. 1994. Structure and metamorphism of blueschist – eclogite facies rocks from the northeastern Oman Mountains. *Journal of Geological Society, London*, **151**, 555–576, https://doi.org/10.1144/gsjgs.151.3.0555

Searle, M.P., Warren, C.J., Waters, D.J. and Parrish, R.R. 2004. Structural evolution, metamorphism and restoration of the Arabian continental margin, Saih Hatat region, Oman Mountains. *Journal of Structural Geology*, **26**, 451–473, https://doi.org/10.1016/j.jsg.2003.08.005

Serra-Kiel, J., Vicedo, V., Razin, P. and Grélaud, C. 2016. Selandian-Thanetian larger foraminifera from the lower Jafnayn Formation in the Sayq area (eastern Oman Mountains). *Geologica Acta*, **14**, 315–333, https://doi.org/10.1344/GeologicaActa2016.14.3.7

Simmons, M.D. and Hart, M.B. 1987. *The Biostratigraphy and Microfacies of the Early to mid-Cretaceous Carbonates of Wadi Mi'aidin, Central Oman Mountains. Micropalaeontology of Carbonate Environments*. Ellis Horwood, Chichester, 176–207.

Skelton, P.W., Nolan, S.C. and Scott, R.W. 1990. The Maastrichtian transgression onto the northwestern flank of the Proto-Oman Mountains: sequences of rudist-bearing beach to open shelf facies. *Geological Society, London, Special Publications*, **49**, 521–547, https://doi.org/10.1144/GSL.SP.1992.049.01.32

Smith, A,B., Simmons, M.D. and Racey, A. 1990. Cenomanian echinoids, larger foraminifera and calcareous algae from the Natih Formation, central Oman Mountains. *Cretaceous Research*, **11**, 26–69, https://doi.org/10.1016/S0195-6671(05)80041-2

Soret, M., Agard, P., Dubacq, B., Plunder, A. and Yamato, P. 2017. Petrological evidence for stepwise accretion of metamorphic soles during subduction infancy (Semail ophiolite, Oman and UAE). *Journal of Metamorphic Geology*, **35**, 1051–1080, https://doi.org/10.1111/jmg.12267

Stanger, G. 1985. Silicified serpentinite in the Semail nappe of Oman. *Lithos*, **18**, 13–22, https://doi.org/10.1016/0024-4937(85)90003-9

Steineke, M., Bramkamp, R.A. and Sander, N.J. 1958. Stratigraphic relations of Arabian Jurassic oil. *In*: Weeks, L.G. (ed.) *Habitat of Oil*. American Association of Petroleum Geologist, Special Volume, 1294–1329.

Stephenson, M.H., Al-Rawahi, A. and Casey, B. 2008. Correlation of the Al Khlata Formation in the Mukhaizna field, Oman, based on a new downhole, cuttings-based palynostratigraphy biozonation. *GeoArabia*, **13**, 15–34.

Stern, R.J., Avigad, D., Miller, N.R. and Beyth, M. 2006. Evidence for the Snowball Earth hypothesis in the Arabian-Nubian Shield and the East African Orogen. *Journal of African Earth Sciences*, **44**, 1–20, https://doi.org/10.1016/j.afrearsci.2005.10.003

Styles, M.T., Ellison, R.A. *et al.* 2006. *The Geology and Geophysics of the United Arab Emirates*, vol. **2**. Ministry of Energy, United Arab Emirates, Abu Dhabi.

Tavani, S., Corradetti, A., Sabbatino, M., Seers, T. and Mazzoli, S. 2021. Geological record of the transition from induced to self-sustained subduction in the Oman Mountains. *Journal of Geodynamics*, **133**, 101674, https://doi.org/10.1016/j.jog.2019.101674

Tilton, G.R., Hopson, C.A. and Wright, J.E. 1981. Uranium-lead isotopic ages of the Semail ophiolite, Oman, with applications to Tethyan oceanic ridge tectonics. *Journal of Geophysical Research*, **86**, 2763–2775, https://doi.org/10.1029/JB086iB04p02763

Tippit, P.R., Pessagno, E.A. and Smewing, J.D. 1981. The biostratigraphy of sediments in the volcanic unit of the Samail ophiolite. *Journal of Geophysical Research*, **86**, 2756–2762, https://doi.org/10.1029/JB086iB04p02756

Tomás, S., Frijia, G., Bömelburg, E., Zamagni, J., Perrin, C. and Mutti, M. 2016. Evidence for seagrass meadows and their response to paleoenvirnomental changes in the early Eocene (Jafnayn Formation, Wadi Bani Khalid, N Oman). *Sedimentary Geology*, **341**, 189–202, https://doi.org/10.1016/j.sedgeo.2016.05.016

Tsuchiya, N., Shibata, T. *et al.* 2013. Petrology of Lasail plutonic complex, northern Oman ophiolite, Oman: an example of arc-like magmatism associated with ophiolite detachment. *Lithos*, **156–159**, 120–138, https://doi.org/10.1016/j.lithos.2012.10.013

Ucurum, A. 2000. Listwaenites in Turkey: perspectives on formation and precious metal concentration with reference to occurrences in East-Central Anatolia. *Ofioliti*, **25**, 15–29.

van Buchem, F.S.P., Pittet, B. *et al.* 2002*a*. High-resolution sequence stratigraphic architecture of the Barremian-Aptian carbonate systems in northern Oman and the United Arab Emirates Kharaib and Shu'aiba formations. *GeoArabia*, **7**, 461–500.

van Buchem, F.S.P., Razin, P., Homewood, P.W., Oterdoom, W.H. and Philip, J. 2002*b*. Stratigraphic organization of carbonate ramps and organic-rich intrashelf basins: Natih Formation (middle Cretaceous) of northern Oman. *American Association of Petroleum Geologists Bulletin*, **86**, 21–53.

Villey, M., de Gramont, X. and Le Métour, J. 1986*b*. *Geological map of Seeb, sheet NF 40-3C, scale: 1:100,000, with Explanatory notes*. Directorate General of Minerals, Oman Ministry of Petroleum and Minerals.

Villey, M., Le Métour, J. and de Gramont, X. 1986*c*. *Geological map of Fanjah, sheet NF 40-3F, scale: 1:100,000, with Explanatory notes*. Directorate General of Minerals, Oman Ministry of Petroleum and Minerals.

Warburton, J., Burnhill, T.J., Graham, R.H. and Isaac, K.P. 1990. The evolution of the Oman Mountains foreland basin. *Geological Society, London, Special Publications*, **49**, 419–427, https://doi.org/10.1144/GSL.SP.1992.049.01.26

Warren, C.J., Parrish, R.R., Waters, D.J. and Searle, M.P. 2005. Dating the geologic history of Oman's Semail ophiolite: insight from U-Pb geochronology. *Contributions to Mineralogy and Petrology*, **150**, 403–422, https://doi.org/10.1007/s00410-005-0028-5

Weisgerber, G. , 1978. Evidence of Ancient Mining Sites in Oman: a Preliminary Report. *The Journal of Oman Studies*, **4**, 15–28.

White, M.R. 1994. Foraminiferal biozonation of the northern Oman Tertiary carbonate succession. *In*: Simmons, M.D. (ed.) *Micropalaeontology and Hydrocarbon Exploration in the Middle East*. Chapman & Hall, London, 309–339.

Wilde, A., Simpson, L. and Hanna, S. 2002. Preliminary study of Cenozoic hydrothermal alteration and platinum deposition in the Oman Ophiolite. *In*: Jessell, M.J. (ed.) *General Contributions, Journal of the Virtual Explorer*, **6**, 7–13.

Wohlwend, S., Hart, M. and Weissert, H. 2016. Chemostratigraphy of the Upper Albian to mid-Turonian Natih Formation (Oman) – how authigenic carbonate changes a global pattern. *The Depositional Record*, **2**, 97–117, https://doi.org/10.1002/dep2.15

Wohlwend, S., Celestino, R., Reháková, D., Huck, S. and Weissert, H. 2017. Late Jurassic to Cretaceous evolution of the eastern Tethyan Hawasina Basin (Oman Mountains). *Sedimentology*, **64**, 87–110, https://doi.org/10.1111/sed.12326

Wright, V.P., Ries, A.C. and Munn, S.G. 1990. Intraplatformal basin-fill deposits from the Infracambrian Huqf Group, east Central Oman. *Geological Society, London, Special Publications*, **49**, 601–616, https://doi.org/10.1144/GSL.SP.1992.049.01.36

Wyns, R., Béchennec, F., Le Métour, J. and Roger, J. 1992. *Geological map of Tiwi, sheet NF40-8B, scale 1:100,000, with Explanatory notes*. Directorate General of Minerals, Oman Ministry of Petroleum and Minerals.

Yamato, P., Agard, P., Goffé, B., De Andrade, V., Vidal, O. and Jolivet, L. 2007. New, high-precision P-T estimates for obduction, nappe stacking and exhumation processes. *Journal of Metamorphic Geology*, **25**, 657–682, https://doi.org/10.1111/j.1525-1314.2007.00722.x

Zagrtdenov, N.R., Ceuleneer, G., Rospabé, M., Borisova, A.Y., Toplis, M.J., Benoit, M. and Abily, B. 2018. Anatomy of a chromitite dyke in the mantle/crust transition zone of the Oman ophiolite. *Lithos*, **312–313**, 343–357, https://doi.org/10.1016/j.lithos.2018.05.012

Ziegler, M.A. 2001. Late Permian to Holocene paleofacies evolution of the Arabian Plate and its hydrocarbon occurrences. *GeoArabia*, **6**, 445–504.

Chapter 3

Thrusts, extensional faults and fold patterns of the major units

Andreas Scharf[1]*, Frank Mattern[1], Mohammed Al-Wardi[1], Gianluca Frijia[2], Daniel Moraetis[3], Bernhard Pracejus[1], Wilfried Bauer[4] and Ivan Callegari[4]

[1]Department of Earth Sciences, College of Science, Sultan Qaboos University, PO Box 36, PC 123, Al-Khod, Muscat, Sultanate of Oman

[2]Department of Physics and Earth Sciences, University of Ferrara, Via Saragat 1, 44122, Ferrara, Italy

[3]Department of Applied Physics and Astronomy, College of Sciences, University of Sharjah, PO Box 27272, Sharjah, United Arab Emirates

[4]Department of Applied Geosciences, German University of Technology GUtech, PO Box 1816, PC 130, Halban, Sultanate of Oman

*Correspondence: scharfa@squ.edu.om

Abstract: This chapter is concerned with the main faults and folds within the Southeastern Oman Mountains based on available literature. The main, best and most widely exposed thrusts are those related to the SW-directed late Cretaceous obduction of the allochthonous nappes onto the Arabian platform and margin. These thrusts are related to obduction of rocks, which had formed hundreds of kilometres offshore Oman. The thrusts were active from the Cenomanian to the Campanian. Obduction-related thrusts and folds are spectacularly exposed within the rocks of the Arabian platform in the eastern part of the Saih Hatat Dome, including large-scale recumbent cylindrical folds and sheath folds. At least six fold sets can be studied in the Southeastern Oman Mountains. At least two of them had formed prior to obduction and are exposed in the Pre-Permian formations of the Jabal Akhdar Dome. At least three fold sets formed in the course of obduction, while at least one fold set is post-obductional in age. Besides the compressional structures, the Oman Mountains expose major post-obductional extensional faults, mostly at the margins of the Jabal Akhdar and Saih Hatat domes. The throw of these faults amounts to a few to several kilometres. Finally, this chapter provides an overview of the enigmatic Batinah Mélange which consists of slivers of Hawasina rocks, resting (unusually) structurally above the Semail Ophiolite.

Main thrusts and thrust units

Allochthonous thrusting on autochthonous units

Several palaeogeographically different thrust units exist in the study area. Some of the Hawasina nappes are intensely deformed and a correlation or distinction is partly based on their structural position. This allows for some different interpretations of the positioning of different Hawasina nappes (Béchennec *et al.* 1990, 1992; see section 'Hawasine nappes' in Scharf *et al.* 2021*a*, Chapter 2, this Memoir). The structurally uppermost thrust units belong to the Cenomanian Semail Ophiolite. This unit was thrust onto the distal oceanic sedimentary rocks and mélanges of the Haybi Complex (Searle and Malpas 1980, 1982; Searle 1985), which was in turn thrust on top of the Kawr–Misfah and platform facies sheet complex and the Hawasina Complex, which represent the more proximal parts of the Permian–Cenomanian Tethys Basin (Searle 2007; marked by solid black line in figs 4.1 and A.1 in Scharf *et al.* 2021*b*, Chapter 7, this Memoir). The thrust sheets are characterized by several internal thrusts. These thrust sheets have been thrust onto the passive margin sediments of the Arabian Plate during the Late Cretaceous (thrust with a solid red line on figs 4.1 & A.1).

Parts of the Hawasina Basin were subducted to the NE during SW-directed obduction of the Semail Ophiolite. Some subducted sediments were sheared off their substratum and attached to the overriding Semail Ophiolite thrust sheets. The most distal basin sediments and volcanic rocks of the Haybi Complex were overridden first, separated by a thrust from the shallow-marine sediments of the Kawr–Misfah and platform facies sheet complex (Searle 2007). This nappe pile was then thrust over the sediments of the Hamrat Duru Basin or Hawasina Complex (Searle 2007). The aforementioned units were thrust over the Sumeini Group, which is not exposed in our study area. Finally, the stack of thrust sheets was emplaced above the Arabian continental margin (Béchennec *et al.* 1990), leading to the formation of the Aruma Foreland Basin, filled with the Aruma Group sediments (Robertson 1987), and also the 'Hawasina foreland fold-and-thrust belt' or 'Hamrat Duru Range' south of the JAD (Hanna 1990).

Upper–lower plate discontinuity or Hulw–As Sifah Window

The upper–lower plate discontinuity is a major fault zone in the northeastern Saih Hatat Dome (SHD) (Figs 3.1, 3.2 and fig. 4.2; e.g. Gregory *et al.* 1998; Miller *et al.* 1998, 2002; Gray *et al.* 2000; Searle 2007; Warren and Miller 2007; Yamato *et al.* 2007). The discontinuity is recognizable between the Hulw and As Sifah windows (e.g. Miller *et al.* 2002; Searle *et al.* 2004; Agard *et al.* 2010), which are separated by a major thrust zone (e.g. Searle *et al.* 2004; Agard *et al.* 2010). Gregory *et al.* (1998) interpreted a considerable jump in metamorphism along the upper–lower plate discontinuity. By contrast, Searle *et al.* (2004), Yamato *et al.* (2007) and Agard *et al.* (2010) pointed out that the metamorphic gradient across the upper–lower plate discontinuity is negligible. However, the strain is increasing from both sides towards the discontinuity, making this discontinuity a major shear zone (Miller *et al.* 2002; Searle *et al.* 2004) with a strong horizontal displacement (Agard *et al.* 2010). Miller *et al.* (2002) reported a heavily mineralized brittle fault at the discontinuity, truncating this major shear zone. Within the Hulw Window, Searle *et al.* (2004) showed that top-to-the-SW-directed thrust zones had formed within a subduction and/or exhumation channel of the NE-directed, down-going Arabian Plate during thrusting of the Semail Ophiolite. The SW-directed thrusting was later followed by NNE-directed normal faulting (e.g.

From: Scharf, A., Mattern, F., Al-Wardi, M., Frijia, G., Moraetis, D., Pracejus, B., Bauer, W. and Cellegari, I. 2021. *The Geology and Tectonics of the Jabal Akhdar and Saih Hatat Domes, Oman Mountains*. Geological Society, London, Memoirs, **54**, 49–60, https://doi.org/10.1144/M54.3

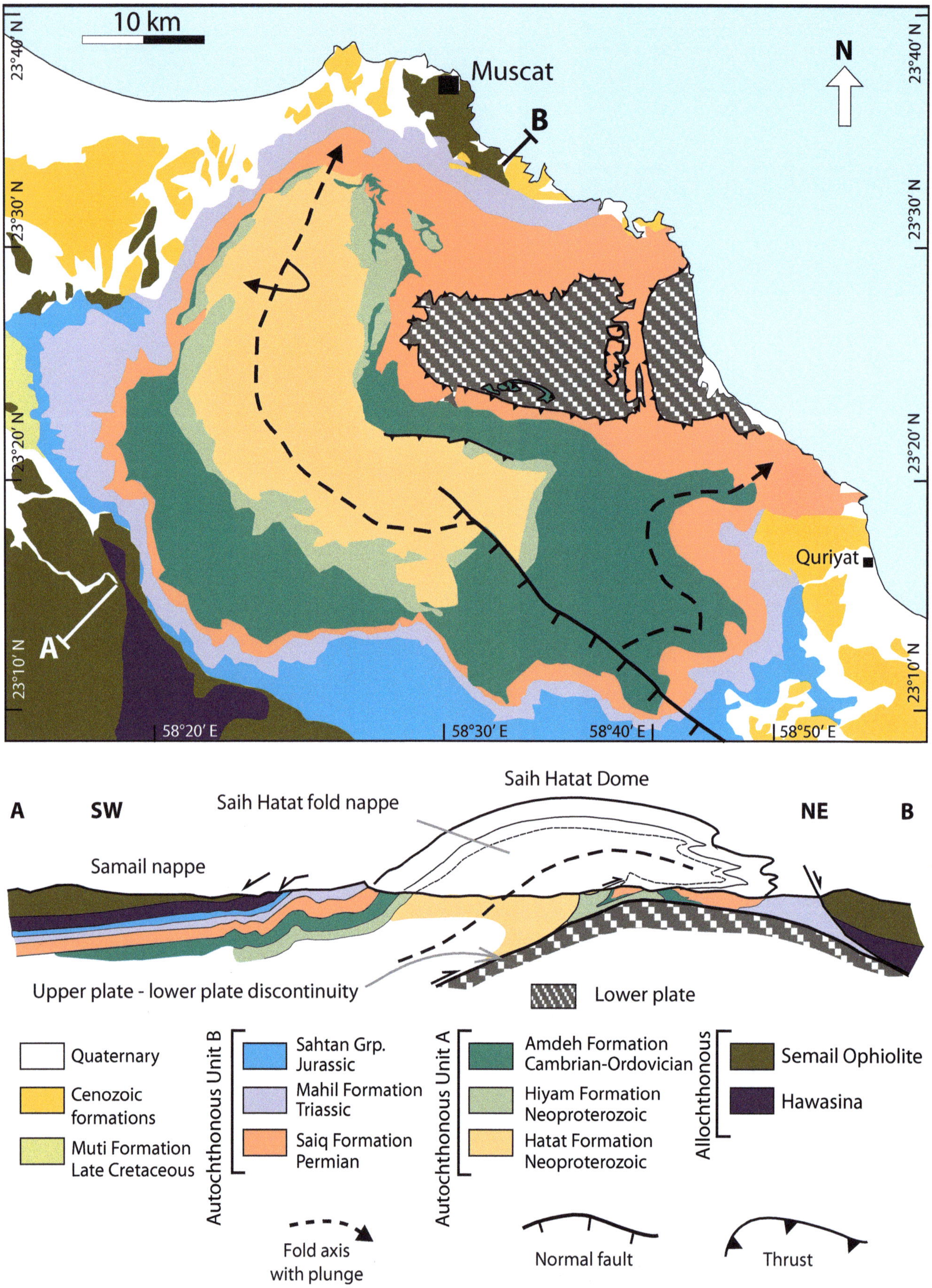

Fig. 3.1. Simplified geological map (upper) and cross-section (lower) of the SHD, depicting the 'upper–lower plate discontinuity'. Modified from Warren and Miller (2007) and Bauer *et al.* (2018). The cross-section depicts the recumbent Saih Hatat fold-nappe (Gray *et al.* 2005; 'F4 folds').

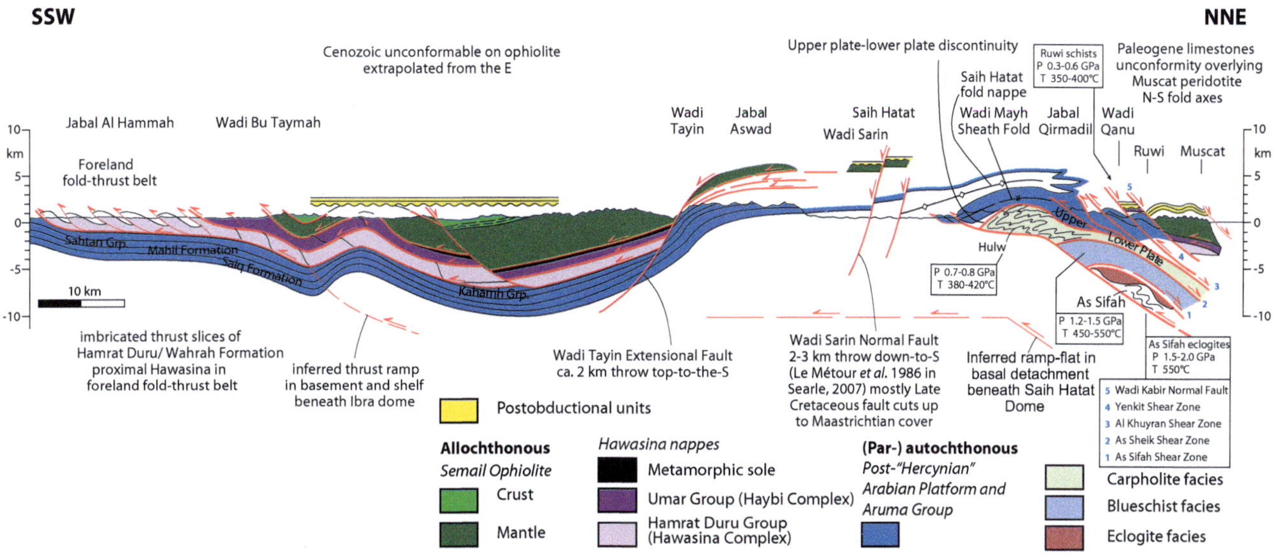

Fig. 3.2. Schematic cross-section through the Saih Hatat Dome and adjoining areas of Searle (2007; slightly modified). Cross-section strikes NNE–SSW with its northernmost point near Muscat. See Searle (2007, fig. 1) for exact position of the section and further discussion of the different units, localities and shear zones.

Agard *et al.* 2010). The rocks within or close to the upper–lower plate discontinuity are highly deformed and folded multiple times (e.g. Searle *et al.* 1994, 2004; Miller *et al.* 1998, 1999, 2002; Gray *et al.* 2004). An original >80-km-long segment of the distal Arabian Platform was reduced to 7 km length within the Hulw–As Sifah windows during the Late Cretaceous (Warren *et al.* 2005; Searle 2007). This indicates that the Arabian margin had extended considerably more to the NE before the Late Cretaceous deformation. The complex and multiple deformation acts of the eastern SHD during and after the Semail Ophiolite obduction are briefly described in Chapter 5 (see section on 'Intraoceanic thrusting, formation and obduction of the Semail Ophiolite'; fig. 5.8 in Scharf *et al.* 2021*c*, Chapter 5, this Memoir). The major discontinuity was formed during *c.* 76–74 Ma (Gray *et al.* 2005). The ages were constrained using the $^{40}Ar/^{39}Ar$ technique on phengites. The reader is referred to the comprehensive and detailed structural work of Miller *et al.* (1998, 1999, 2002), Searle *et al.* (2004), Gray *et al.* (2005), Warren and Miller (2007), Yamato *et al.* (2007) and Agard *et al.* (2010).

Main extensional fault systems

Post-obductional extensional fault systems, with a common throw of *c.* 5 km, are frame large parts of the northern and eastern flanks of the Jabal Akhdar Dome (JAD) and the northern and southwestern flanks of the SHD (Searle 2007; 'Frontal Range Fault' of Mattern and Scharf 2018; Scharf *et al.* 2019*b*; S1, S3, S4 and S5 of fig. 4.2). Field studies and geological map analyses show that large parts of the Hawasina nappes are missing at the surface along the respective faults (see also Searle *et al.* 2004; Al-Wardi and Butler 2007; Aldega *et al.* 2017; Mattern and Scharf 2018; Scharf *et al.* 2019*a*, *b*).

These extensional faults have two stages of activity. The first stage occurred during the Late Cretaceous until the Paleocene–Eocene boundary (compare with Grobe *et al.* 2019), while the second stage took place during the late Eocene–Miocene (compare with Fournier *et al.* 2006; Scharf *et al.* 2020*b*). The latter stage may be related to orogenic and/or gravitational collapse after the doming of the JAD and SHD (Mattern and Scharf 2018; Scharf *et al.* 2019*a*). Large listwaenite bodies are found along the extensional fault systems or its vicinity near Fanja (e.g. Bailey *et al.* 2019; Scharf *et al.* 2019*c*, 2020*a*), indicating that large amounts of hydrothermal fluids must have flowed through and/or circulated within the faults. U–Pb carbonate dating of two dolomite veins within such a listwaenite body near Fanja produced an age of 55 ± 4 Ma (Scharf *et al.* 2019*c*, 2020*a*). Furthermore, carbonate veins extending from the listwaenite body are cutting the latest Cretaceous Al-Khod Formation. These veins probably formed at a late stage of the listwaenite formation. Therefore, it is likely that the listwaenite have formed during the first extensional stage along the Frontal Range Fault, and are not related to subduction/convergence (Scharf *et al.* 2019*c*, 2020*a*).

Searle *et al.* (2004) reported that Paleocene–Eocene sediments of the Jafnayn Formation, resting on the Mahil Formation, are downthrown to the north by 450 m along the northern flank of the SHD (Wadi Kabir Fault). The age of some of the extensional faulting along the northern margins of the JAD and SHD is therefore post-early Eocene (Searle *et al.* 2004; Mattern and Scharf 2018). The Semail Ophiolite may have slid downwards driven by gravity in a radial fashion, away from the domes during the late Eocene–Miocene (Scharf *et al.* 2019*a*).

Post-Eocene extensional faults that are shallowly dipping away from the mountains towards the sea from the northern JAD along the Batinah Coastal Plain have been described by Mann *et al.* (1990; fig. 4.1). Post-Eocene normal faults are truncated by sinistral N-striking transfer faults with displacements of several hundred metres (Mann *et al.* 1990). The possible origin of this extension is described in the section on 'Post-obductional deformation' in Chapter 5 ('Tectonic Stage 2' of Fournier *et al.* 2006 and Mattern and Scharf 2018).

Another important element of this fault system is the *c.* 70-km-long, NNE-trending 'Semail Gap' or 'Semail Lineament' which marks the eastern margin of the JAD and separates the Central from the Southeastern Oman Mountains (Glennie *et al.* 1974; figs 2.1, 4.1, A.1 in Scharf *et al.* 2021*a*, *b*, *d*). At its southern end, the tectonic omission of the Hawasina nappes is gradually reduced.

Five different models for the development and deep structure of the Semail Gap have been proposed in the literature. (1) Hanna (1990) suggested that the Semail Gap is a culmination collapse structure which formed during doming of the JAD. (2) Searle (2007) pointed out that the Semail Gap is not a strike-slip fault since no lateral offset can be ascertained. He therefore interpreted the lineament as an S-shaped listric normal fault between the JAD and SHD. (3) Mount *et al.* (1998, fig. 3) suggested that the Semail Gap is underlain by a basement-rooted westward and steep-dipping blind reverse fault. (4) The Semail Gap may be underlain by a minimum 55-km-long, NW-dipping, lateral ramp with sinistral motion formed during the doming of the JAD in Eocene time (Hansman *et al.* 2017). (5) The present-day pattern of the Semail Gap reveals probable Eocene–Miocene deformation by juxtaposition of different lithologies on both sides (Fig. 3.3).

The eastern margin of the JAD may have undergone monoclinal bending with a fold axis parallel to the Semail Gap during the Eocene (Hansman *et al.* 2017). This possible monocline was subsequently deformed by a dextral transtensional fault, the Semail Gap Fault Zone (Scharf *et al.* 2019*a*; fig. 4.2). This interpretation is based on fault, map and satellite image analyses. Present-day seismicity along this structure is negligible (Scharf *et al.* 2019*a*). The Semail Gap probably formed in the course of orogenic and/or gravitational collapse of the JAD and SHD during the late Eocene–Miocene (Scharf *et al.* 2019*a*). The hanging-wall blocks of the two adjacent domes moved downwards during doming of the JAD and SHD in the ESE–SE and SSW–SW direction, respectively (Fig. 3.4). According to Scharf *et al.* (2019*a*), this movement ultimately resulted to dextral transtension along the eastern margin of the JAD (Semail Gap Fault Zone) and to sinistral transtensional motion along the southwestern margin of the SHD (Wadi Mansah Fault Zone, WMFZ; Fig. 3.4; see also Bailey *et al.* 2019).

The history of the lineament may be even older than late Eocene. The Proto-Semail Gap or Proto-Semail Gap Fault Zone probably formed during Cryogenian terrain accretion when the terrain boundaries were parallel to the Semail Gap

Fig. 3.3. Overview photograph from the Semail Gap, south of Semail (fig. 2.1), showing from left to right: limestone of the Arabian margin (Natih Formation), sediments of the Aruma Foreland Basin (Muti Formation), Hawasina nappes and the Semail Ophiolite.

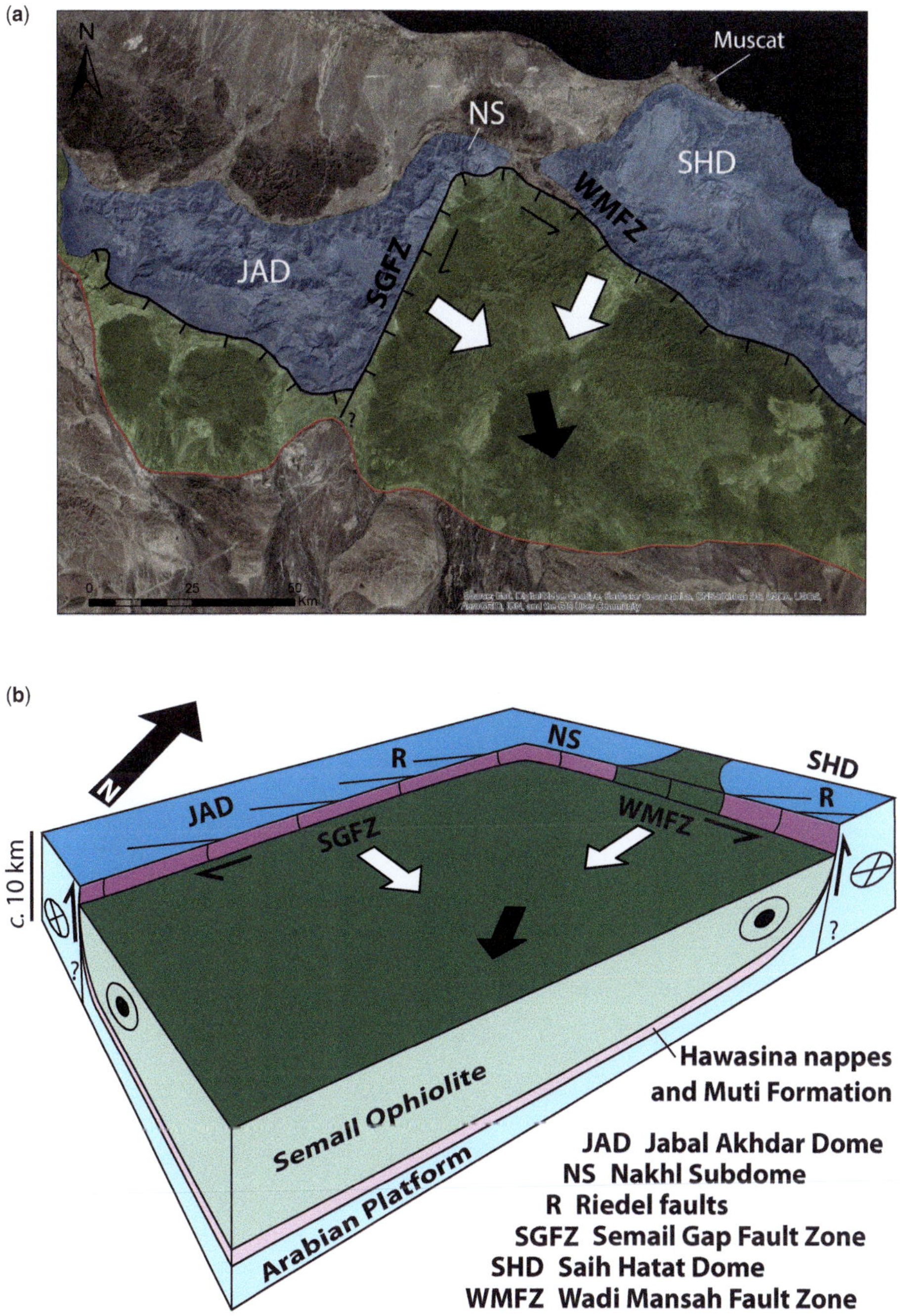

Fig. 3.4. (**a**) Schematic map depicting the sliding and/or downthrown direction of the hanging-wall segments (green) during exhumation and gravitational collapse of the JAD and SHD. The northern termination of the JAD is the Nakhl Subdome (NS). The resulting vector (black arrow) is towards the SSE. (**b**) Schematic block diagram during motion of the hanging-wall segments, resulting in detrital and sinistral transtensional motion along the Semail Gap and Wadi Mansah fault zones, respectively. The vertical extent of the two faults is unknown. Figure has been slightly modified after Scharf *et al.* (2019*a*).

(Romine *et al.* 2004). This structure was probably reactivated during the Neoproterozoic, explaining the different lithologies between the JAD and SHD. In particular, the occurrence of the Paleozoic Amdeh and Al Khlata formations and the absence of the Neoproterozoic Mu'aydin Formation in the SHD in contrast to the JAD are noteworthy. The NNE-trending Semail Gap may be an Early Paleozoic boundary of the Ghaba Salt Basin within the Oman Mountains (Mount *et al.* 1998; Rickards *et al.* 2010). The reader is referred to Scharf *et al.* (2019*a*, their supplementary material) for a detailed review of the pre-Cenozoic history of the Semail Gap.

Folds

Large-scale folds and the distribution of the main nappes are shown in figure 4.1. The weight of the Semail Ophiolite and the Hawasina nappes caused formation of a flexural bulge in the Arabian Plate (e.g. Le Métour *et al.* 1986; Warburton *et al.* 1990). In addition, six different sets of fold axes can be observed in the study area (F1 to F6). The F1 and F2 folds formed before the Late Carboniferous, while the F3 to F5 folds formed in the course of the Semail Ophiolite obduction and related processes. These processes include subduction and/or exhumation, subsequent doming and gravitational collapse of the JAD and SHD. The F3 to F5 folds may therefore have resulted from similar or overlapping deformation histories, which makes a clear distinction between and recognition of the F3 to F5 folds somewhat difficult. The last fold set (F6) is related to post-obductional tectonics.

F1 folds

Neoproterozoic–earliest Cambrian rocks of the study area were affected by the earliest Cambrian Cadomian Orogeny (Callegari *et al.* 2020). These authors presented the first structural evidence for the Cadomian Orogeny in Oman. The

Fig. 3.5. Village of Hajir (Wadi Hajir) in the central JAD. Permo-Mesozoic rocks in the background, unconformably overlying Neoproterozoic formations, which have been deformed during the Cadomian and/or Angudan Orogeny. E-oriented open anticline of the Hajir Formation is visible in the foreground.

following text is a summary of their results. The tight Cadomian folds with an amplitude of several to tens of metres are well-preserved in the Neoproterozoic Hajir Formation of the western JAD. They display subhorizontal fold axes and shallow-dipping fold axial planes. The direction of greatest stress during this folding event was NE–SW (in present coordinates). Associated with the folds are shallow-dipping thrusts that ductily deform the limestone of the Hajir Formation. The age of folding has been constrained to be between the deposition of the Fara Formation (*c.* 540 Ma; see section on 'Units below the Hercynian Unconformity' in Chapter 2) and the Angudan Orogeny (525 ± 5 Ma, Al-Husseini 2014). The vergence of the thrusts and the fold axial planes was towards the NE before the Angudan event (Fig. 3.6). The amplitude of the Cadomian folds is too small to be depicted on the tectonic map (fig. 4.1); the trace of the F1 folds is therefore presented in Figure 3.7 and schematically shown in Figure 3.6. No F1 folds have been mapped in the SHD.

F2 folds

The Cadomian F1 folds in the western JAD were refolded by tight kilometre-scale F2 folds (Figs 3.6 & 3.7; Callegari *et al.* 2020). The F2 folds have already been mapped by Glennie *et al.* (1974), Beurrier *et al.* (1986), Mann and Hanna (1990) and Béchennec *et al.* (1992). They were previously assigned to Late Paleozoic motion (e.g. Mann and Hanna 1990) and reinterpreted to be Anguadan in age (Callegari *et al.* 2020). These folds are accompanied by large NE-striking

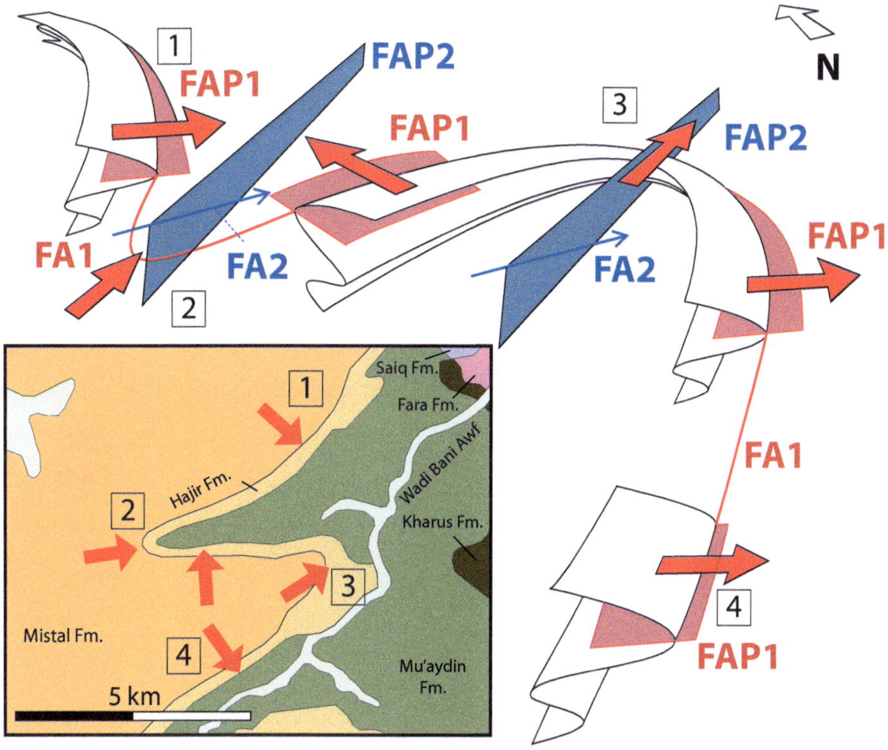

Fig. 3.6. Schematic depiction of the Cadomian folds (red), refolded by the Angudan folds (blue; modified after Callegari *et al.* 2020). The red arrows indicate the direction of the Cadomian vergence. FA – fold axis; FAP – fold axial plane. The map covers the area from the western JAD in the Sahtan Bowl between Wadi Sahtan to the west and Wadi Bani Awf towards the east (fig. 2.1 & Fig. 3.7; Béchennec *et al.* 1993).

thrusts (Mann and Hanna 1990). The direction of shortening was SE–NW, that is, at a high angle to that of the Cadomian event.

In the SHD and below the 'Hercynian' Unconformity, kilometre-scale gentle, upright folds exist. The fold axis is *c.* NNE-trending (Villey *et al.* 1986). Some exhumation occurred during the corresponding folding event and may explain the absence of the Hiyam Formation at the flanks (Villey *et al.* 1986). The age of this event is unknown, but we assign it to the Angudan event (F2) age because (1) folding pre-dates the 'Hercynian' Unconformity, (2) no such folds exist within the post-Angudan Cambro–Ordovician Amdeh Formation of the SHD (Mattern *et al.* 2020) and (3) because of the similar trend of the respective fold axes in the JAD and SHD. Within the northern SHD tight NNE-trending folds exist with a fold axial plane dipping moderately to shallowly towards the WNW. The orientation of the fold axes is parallel to that of the F2 folds. However, these tight folds affect units below and above the 'Hercynian' Unconformity. Thus, these tight folds are no F2 folds but are rather part of the F4 Saih Hatat fold nappe (see below in section F4 folds).

The Paleozoic Amdeh Formation of the SHD displays NW–SE-trending folds and no folds with a NE–SW orientation (Mattern *et al.* 2020). This indicates that a Hercynian deformation did not affect this area. Instead, these folds formed during the convergence between the Semail Ophilite and Arabia during the Late Cretaceous (Mattern *et al.* 2020).

F3 folds

The NNE-striking open F3 folds occur in the southwestern part of the JAD. They overprint the allochthonous and autochthonous nappes, and therefore must have formed during or after the arrival of the allochthonous rocks onto the Arabian Platform. Beurrier *et al.* (1986) mapped (from west to east) the Barut Anticline, Misfah Syncline and Wadi Nahur Anticline (fig. 4.1) in the southwestern JAD. Further F3 folds exist near Rustaq, Nizwa and SW of the SHD (fig. 4.1). They have subhorizontal fold axes, but display Hawasina nappes in their cores.

Even further south of the Oman Mountains, near the front of the Hawasina nappes, seismic data suggest that horizontal compression was oriented NW–SE during the Santonian–Campanian (possibly until the Early Paleogene; Filbrandt *et al.* 2006; Al-Kindy and Richard 2014). WNW-directed thrusting of the Masirah Ophiolite during the Maastrichtian–Early Paleocene may have formed the NNE–SSW-oriented F3 folds, especially near the coast around Masirah (Filbrandt *et al.* 2006; section on 'Arabian Platform' in Chapter 5).

F4 folds

Several NNE-trending and up to *c.* 15–25 km long sheath folds in the northern SHD have been described with typical eye-shaped cores that are bounded by double-vergence geometries (Searle *et al.* 2004; Searle and Alsop 2007; Agard *et al.* 2010; Cornish and Searle 2017; fig. 2.3). An ESE-striking section in Wadi Mayh (northern SHD; fig. 2.1) exposes a minimum of four stacked sheath folds within one mega-sheath fold in Permian rocks (Searle and Alsop 2007). Overall, the sheath folds show SSW-closing sheaths (Searle and Alsop 2007). The same authors reported that the eye-fold closures are bisected by NNE-trending mineral lineations. The envelope of the mega-sheath fold also consists of pre-Permian rocks. The rocks above the sheath fold have a planar detachment without internal complexities (Searle and Alsop 2007). The rocks above the mega-sheath fold are inverted, with Amdeh, Hiyam and Hatat formations from the bottom to the top. These formations bend around the southwestern closure of the mega-sheath fold (Searle and Alsop 2007). Agard *et al.* (2010) reported a consistently top-to-the-N, exhumation-related shear sense accompanying the sheath folds in Wadi Mayh. The strain increases from the bottom of the mega-sheath fold downwards to the upper–lower plate discontinuity (Searle and Alsop 2007; see above). The entire sequence of the mega-sheath fold and the inverted package is topped by the SW-directed thrust of the Semail Ophiolite (Searle and Alsop 2007). The same authors described the formation of these folds during the late-stage, high-strain deformation

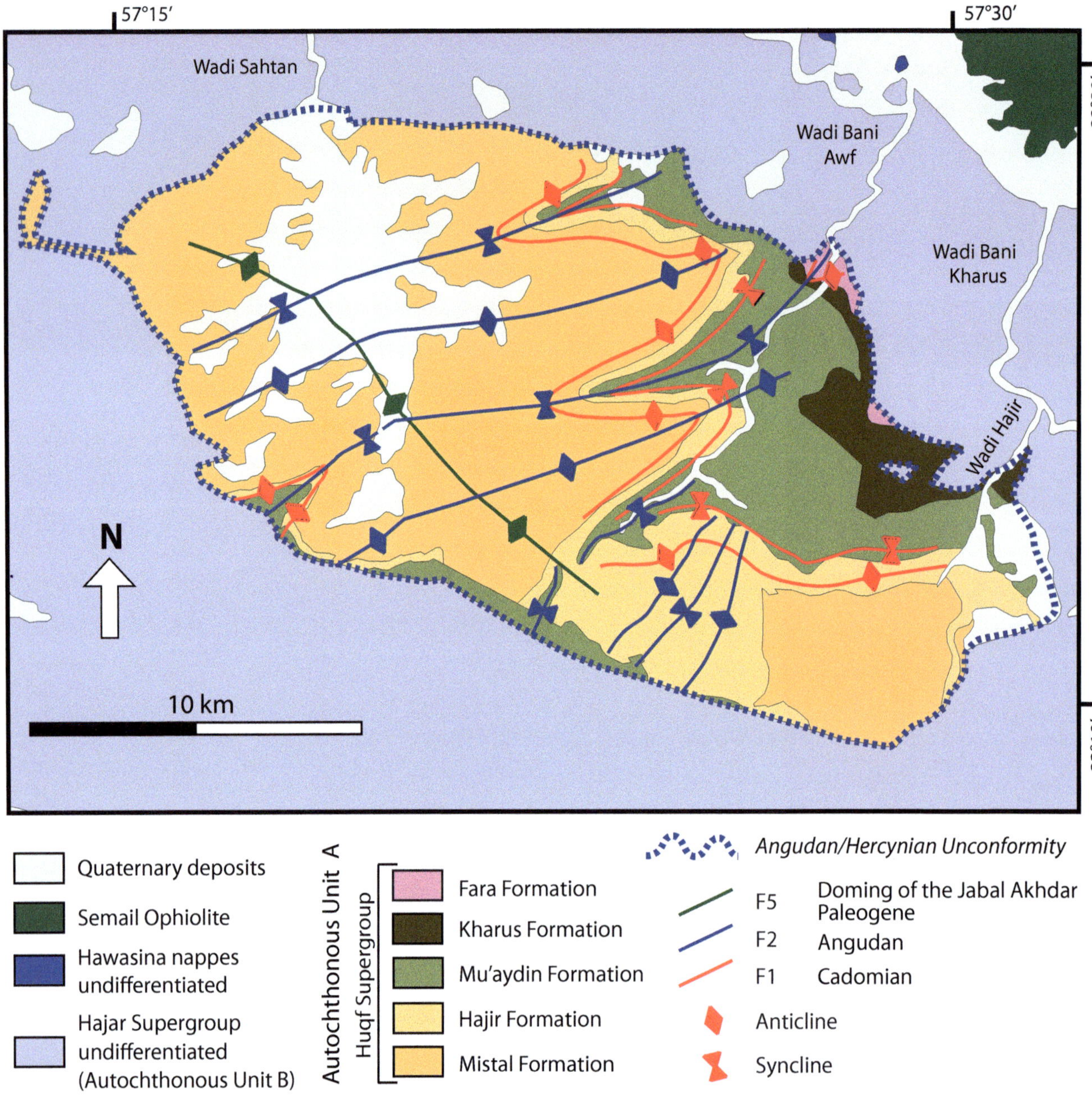

Fig. 3.7. Trace of the Cadomian (F1), Angudan (F2) and Jabal Akhdar Dome (F5) folds in the western JAD, modified from Béchennec *et al.* (1993) and Callegari *et al.* (2020).

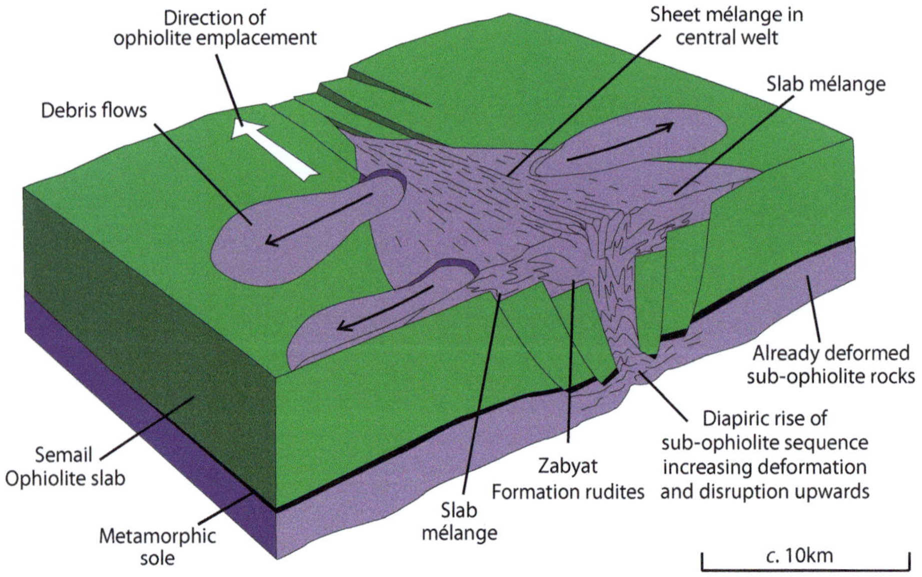

Fig. 3.8. Schematic sketch of the possible diapiric rise of the Batinah Mélange. Redrawn and modified after Robertson and Woodcock (1983).

within a NE-dipping exhumation channel during the obduction of the Semail Ophiolite onto the Arabian margin. The mega-sheath fold of Wadi Mayh is likely to have formed under carpholite-grade conditions at 0.8–1.0 GPa (Agard *et al.* 2010); they have an age of 76–70 Ma (Ar/Ar white micas; Late Campanian–Early Maastrichtian; Miller *et al.* 1999; Searle and Alsop 2007). Their formation and the Hulw–As Sifah Window ensued coevally (see 'Upper–lower plate discontinuity or Hulw–As Sifah Window' above; Agard *et al.* 2010). The sheath folds formed during post-peak metamorphic exhumation of the high-pressure rocks within a bottom-to-SSW-directed active footwall extrusion beneath a passive hanging-wall (Agard *et al.* 2010).

The Saih Hatat fold nappe formed during the late stage of margin deformation of the obduction (Gray *et al.* 2005; Figs 3.1, 3.2 & fig. 4.1). The Saih Hatat fold nappe is a recumbent fold (Figs 3.1 & 3.2). There is an increase in deformation towards the hinge and the lower limb of this large fold, where it is parallel with the upper–lower plate discontinuity (Gray *et al.* 2005). The age of the folding was dated as between *c.* 75.7 ± 0.6 and 70 ± 0.7 Ma, based on $^{40}Ar/^{39}Ar$ crystallization mica ages from fabrics defining the NE-trending lineation and axial surface foliation, respectively (Miller *et al.* 1998; Gray *et al.* 2005). Folding was in part coeval with shearing along the upper–lower plate discontinuity, active during *c.* 76–74 Ma (Gray *et al.* 2005; see section on 'Main thrusts and thrust faults' above).

F5 folds

The overall WNW- or NW-trending F5 fold axes (fig. 4.1) of the JAD and the SHD are parallel to the northern and southern margins of these domes. We attribute the development of the F5 folds to the overall convergence of Arabia and Eurasia once the Semail Ophiolite had started to override the Arabian Continent during the Turonian to Campanian–Early Maastrichtian. The large-scale F5 fold of the JAD underwent a significant developmental step during the Late Eocene–Early Miocene (Hansman *et al.* 2017, 2018; Corradetti *et al.* 2019; Grobe *et al.* 2019). At that time, the JAD experienced its main period of growth (4–6 km of uplift; Hansman *et al.* 2017), and both domes were uplifted (Saddiqi *et al.* 2006). The main phase of exhumation of the SHD may be dated as of latest Cretaceous, indicated by the presence of Amdeh Quartzite clasts within the Upper Cretaceous Al-Khod Formation (Hanna 1990; Nolan *et al.* 1990). The Late Eocene uplift of the SHD may amount to 1 km (Hansman *et al.* 2017) and post-Eocene surface uplift in the southern SHD amounts to 2 km or more, based on the presence of Middle Eocene shallow-marine sedimentary rocks at an elevation of *c.* 2000 m (Moraetis *et al.* 2018). The Late Eocene uplift pre-dates the onset of the Zagros collision phase by *c.* 10 Ma (Hansman *et al.* 2017). The proposed reason for the Late Eocene uplift could be a reduction of the subduction velocity within the Makran subduction zone (Hansman *et al.* 2017; section 'Post-obductional deformation' in Chapter 5).

F6 folds

A sixth set of fold axes can be found in the Cenozoic formations NW of the SHD near and within the Rusayl Embayment (figs 2.1 & 4.3; F6, purple lines in figs 4.1 & 4.3 and S4 of fig. 4.2). These open N-striking F6 folds (Villey *et al.* 1986; Béchennec *et al.* 1992) affected the Semail Ophiolite as well as Cenozoic sedimentary rocks (Mann *et al.* 1990). The same authors describe how the F6 folds tighten towards the south and open towards the north. The F6 folds may be related to the exhumation and the gravitational and/or culmination collapse of the domes (fig. 5.13; see Coffield 1990; Hanna 1990; Mann *et al.* 1990) or to unidentified E–W-aligned shortening. Tightening of the folds towards the south could be explained with the increasing confining pressure, related to gravitational movements of the Cenozoic sedimentary cover along the northern margin of the JAD and SHD towards the Rusayl Embayment NNW (fig. 5.13; Mann *et al.* 1990).

The Batinah Mélange

The geology differs in the northwestern part of the study area, along the Batinah Coastal Plain, as the Semail Ophiolite is overlain (not underlain) and disrupted by a mixture of slivers of Hawasina rocks and fragments of the Semail Ophiolite, including metamorphic rocks. This rock mixture is known as the 'Batinah Complex' or 'Batinah Mélange' (Woodcock and Robertson 1982). The Batinah Mélange (figs 4.1 & A.1) has been interpreted in several different ways. (1) It belongs to the post-obductional Late Cretaceous cover of the Semail Ophiolite (Lees 1928). (2) It is a part of the Hawasina nappes (Glennie *et al.* 1973, 1974), possibly a product of faulting (upturned or inverted element; Reinhardt 1969; Glennie *et al.* 1974). (3) It is a thrust sheet, unrelated to the Hawasina nappes (Carney and Welland 1974; Welland and Mitchell 1977; Woodcock and Robertson 1982). (4) It formed by a combination of complex faulting of the ophiolite and possible diapiric rise of mélange material in fault zones within the ophiolite, and subsequent spreading above the ophiolite (Woodcock and Robertson 1982). A slightly different interpretation was provided by Robertson and Woodcock (1983, Fig. 3.8). They suggested that during emplacement of the Semail Ophiolite a part of the ophiolite was split off. Along corresponding pathways, fragments of the underlying Hawasina nappes could have moved upwards through the ophiolite towards the top. At the surface, the mixture of Hawasina and ophiolite material was distributed above the Semail Ophiolite by surface processes, forming the Batinah Mélange.

Author contributions **AS:** Conceptualization (lead), Writing original draft (lead), Visualization (lead), Supervision (lead), Investigation (lead); **FM:** Conceptualization (equal), Writing – review and editing (equal), Visualization (equal), Investigation (equal); **MAW:** Writing – review and editing (equal); **GF:** Writing – review and editing (equal); **DM:** Writing – review and editing (equal); **BP:** Writing – review and editing (equal); **WB:** Writing – review and editing (equal); **IC:** Writing – review and editing (equal).

Funding This research received no specific grant from any funding agency in the public, commercial, or not-for-profit sectors.

Data availability Data sharing is not applicable to this article as no datasets were generated or analysed during the current study.

References

Agard, P., Searle, M.P., Alsop, G. and Dubacq, B. 2010. Crustal stacking and expulsion tectonics during continental subduction: P-T deformation constraints from Oman. *Tectonics*, **29**, 1–19, https://doi.org/10.1029/2010TC002669

Aldega, L., Carminati, E., Scharf, A., Mattern, F. and Al-Wardi, M. 2017. Estimating original thickness and extent of the Semail

Ophiolite in the eastern Oman Mountains by paleothermal indicators. *Marine and Petroleum Geology*, **84**, 18–33, https://doi.org/10.1016/j.marpet.geo.2017.03.024

Al-Husseini, M.I. 2014. Ediacaran-Cambrian Middle East geologic time scale 2014, proposed correlation of Oman's Abu Mahara Supergroup and Saudi Arabia's Jibalah Group. *GeoArabia*, **19**, 107–132.

Al-Kindy, M.H. and Richard, P.D. 2014. The main structural styles of the hydrocarbon reservoirs in Oman. *Geological Society, London, Special Publications*, **392**, 409–445, https://doi.org/10.1144/SP392.20

Al-Wardi, M. and Butler, R.W.H. 2007. Constrictional extensional tectonics in the northern Oman mountains, its role in culmination development and the exhumation of the subducted Arabian continental margin. *Geological Society, London, Special Publications*, **272**, 187–202, https://doi.org/10.1144/GSL.SP.2007.272.01.11

Bailey, C.M., Hurtado, C., Scharf, A. and Mattern, F. 2019. Structural controls on listwaenite genesis in the Semail Ophiolite, Northern Oman. EGU 2019–5794, Vienna, Austria.

Bauer, W., Callegari, I., Al Balushi, N., Al Busaidi, G., Al Barumi, M. and Al Shoukri, Y. 2018. Tectonic observations in the northern Saih Hatat, Sultanate of Oman. *Arabian Journal of Geosciences*, **11**, 94.

Béchennec, F., Le Métour, J., Rabu, D., Bourdillon-de-Grissac, C., de Wever, P., Beurrier, M. and Villey, M. 1990. The Hawasina Nappes: stratigraphy, paleogeography and structural evolution of a fragment of the South-Tethyan passive continental margin. *Geological Society, London, Special Publications*, **49**, 213–223, https://doi.org/10.1144/GSL.SP.1992.049.01.14

Béchennec, F., Roger, J., Le Métour, J. and Wyns, R. 1992. *Geological map of Seeb, sheet NF 40-03, scale 1:250,000, with Explanatory Notes*. Directorate General of Minerals, Oman Ministry of Petroleum and Minerals.

Béchennec, F., Le Métour, J., Platel, J.P. and Roger, J. 1993. *Geological map of the Sultanate of Oman, scale 1:1,000,000, with Explanatory Notes*. Directorate General of Minerals, Oman Ministry of Petroleum and Minerals.

Beurrier, M., Béchennec, F., Rabu, D. and Hutin, G. 1986. *Geological map of Rustaq, sheet NF 40-03D, scale 1:100,000, with Explanatory notes*. Directorate General of Minerals, Oman Ministry of Petroleum and Minerals.

Callegari, I., Scharf, A. *et al.* 2020. Gondwana accretion tectonics and implications for the geodynamic evolution of eastern Arabia: first structural evidence of the existence of the Cadomian Orogeny in Oman (Jabal Akhdar Dome, Central Oman Mountains). *Journal of Asian Earth Sciences*, **187**, 104070, https://doi.org/10.1016/j.jseaes.2019.104070

Carney, J.N. and Welland, M.J.P. 1974. Geology and mineral resources of the Oman Mountains. GB Institute of Geological Science Overseas Division Report, **27**, 504.

Coffield, D.Q. 1990. Structures associated with nappe emplacement and culmination collapse in the Central Oman Mountains. *Geological Society, London, Special Publications*, **49**, 447–458, https://doi.org/10.1144/GSL.SP.1992.049.01.28

Corradetti, A., Spina, V. *et al.* 2019. Late-stage tectonic evolution of the Al-Hajar Mountains, Oman: new constraints from Palaeogene sedimentary units and low-temperature thermochronometry. *Geological Magazine*, 1–14, https://doi.org/10.1017/S0016756819001250

Cornish, S. and Searle, M. 2017. 3D geometry and kinematic evolution of the Wadi Mayh sheath fold, Oman, using detailed mapping from high-resolution photography. *Journal of Structural Geology*, **101**, 26–42, https://doi.org/10.1016/j.jsg.2017.06.009

Filbrandt, J.B., Al-Dhahab, S. *et al.* 2006. Kinematic interpretation and structural evolution of North Oman, Block 6, since the Late Cretaceous and implications for timing of hydrocarbon migration into Cretaceous reservoirs. *GeoArabia*, **11**, 97–140.

Fournier, M., Lepvrier, C., Razin, P. and Jolivet, L. 2006. Late Cretaceous to Paleogene Post-obduction extension and subsequent Neogene compression in the Oman Mountains. *GeoArabia*, **11**, 17–40.

Glennie, K.W., Boeuf, M.G.A., Hughes Clarke, M.W., Moody-Stuart, M., Pilaar, W.F.H. and Reinhardt, B.M. 1973. Late Cretaceous nappes in Oman Mountains and their geological evolution. *American Association of Petroleum Geologists Bulletin*, **57**, 5–27.

Glennie, K.W., Boeuf, M.G.A., Hughes Clarke, M.W., Moody-Stuart, M., Pilaar, W. and Reinhardt, B.M. 1974. Geology of the Oman Mountains. Koninklijk Nederlands Geologisch en Mijnbouwkundig Genootschap. *Transactions*, **31**, 423.

Gray, D.R., Gregory, R.T. and Miller, J.M. 2000. A new structural profile along the Muscat-Ibra transect, Oman: implications for emplacement of the Semail ophiolite. *Geological Society of America, Special Paper*, **349**, 513–523.

Gray, D.R., Miller, J.M., Foster, D.A. and Gregory, R.T. 2004. Transition from subduction- to exhumation-related fabrics in glaucophane-bearing eclogites, Oman: evidence from relative fabric chronology and $^{40}Ar/^{39}Ar$ ages. *Tectonophysics*, **389**, 35-64, https://doi.org/10.1016/j.tecto.2004.06.016

Gray, D.R., Gregory, R.T., Armstrong, R.A., Richards, I.J. and Miller, J.McL. 2005. Age and stratigraphic relationship of structurally deepest level rocks, Oman Mountains: U/Pb SHRIMP evidence for Late Carboniferous Neotethys rifting. *The Journal of Geology*, **113**, 611–626, https://doi.org/10.1086/449325

Gregory, R.T., Gray, D.R. and Miller, J.M. 1998. Tectonics of the Arabian margin associated with the emplacement of the Oman margin along the Ibra transect: new evidence from northeast Saih Hatat. *Tectonics*, **17**, 657–670, https://doi.org/10.1029/98TC02206

Grobe, A., von Hagke, C., Littke, R., Dunkl, I., Wübbeler, F., Muchez, P. and Urai, J.L. 2019. Tectono-thermal evolution of Oman's Mesozoic passive continental margin under the obducting Semail Ophiolite: a case study of Jebel Akhdar, Oman. *Solid Earth*, **10**, 149–175, https://doi.org/10.5194/se-10-149-2019

Hanna, S.S. 1990. The Alpine deformation of the central Oman Mountains. *Geological Society, London, Special Publications*, **49**, 341–359, https://doi.org/10.1144/GSL.SP.1992.049.01.21

Hansman, R.J., Ring, U., Thomson, S.N., den Brock, B. and Stübner, K. 2017. Late Eocene uplift of the Al Hajar Mountains, Oman, supported by stratigraphic and low-temperature thermochronology. *Tectonics*, **36**, 3081–3109, https://doi.org/10.1002/2017TC004672

Hansman, R.J., Albert, R., Gerdes, A. and Ring, U. 2018. Absolute ages of multiple generations of brittle structures by U-Pb dating of calcite. *Geology*, **46**, 207–210, https://doi.org/10.1130/G39822.1

Lees, G.M. 1928. The geology and tectonics of Oman and parts of southeastern Arabia. *Quarterly Journal of the Geological Society of London*, **84**, 585–670, https://doi.org/10.1144/GSL.JGS.1928.084.01-04.24

Le Métour, J., Villey, M. and de Gramont, X. 1986. *Geological map of Quryat, sheet NF 40-4D, scale 1:100,000, Explanatory notes.* Directorate General of Minerals, Oman Ministry of Petroleum and Minerals.

Mann, A. and Hanna, S.S. 1990. The tectonic evolution of pre-Permian rocks, Central and Southeastern Oman Mountains. *Geological Society, London, Special Publications*, **49**, 307–325, https://doi.org/10.1144/GSL.SP.1992.049.01.19

Mann, A., Hanna, S.S. and Nolan, S.C. 1990. The post-Campanian tectonic evolution of the Central Oman Mountains: Tertiary extension of the Eastern Arabian Margin. *Geological Society, London, Special Publications*, **49**, 285–305, https://doi.org/10.1144/GSL.SP.1992.049.01.33

Mattern, F. and Scharf, A. 2018. Postobductional extension along and within the Frontal Range of the Eastern Oman Mountains. *Journal of Asian Earth Sciences*, **154**, 369–385, https://doi.org/10.1016/j.jseaes.2017.12.031

Mattern, F., Scharf, A. *et al.* 2020. Deformation of the Cambro-Ordovician Amdeh Formation (Members 1 and 2): characteristics, origins, and stratigraphic significance (Wadi Amdeh, Saih Hatat Dome, Oman Mountains). *Geosciences*, **10**, https://doi.org/10.3390/geosciences10020048

Miller, J.M., Gray, D.R. and Gregory, R.T. 1998. Exhumation of high-pressure rocks in northeastern Oman. *Geology*, **26**, 235–238, https://doi.org/10.1130/0091-7613(1998)026<0235:EOHPRI>2.3.CO;2

Miller, J.M., Foster, D.A., Gray, D.R. and Gregory, R.T. 1999. Geological and geochronological constraints on the exhumation of a high-pressure metamorphic terrane, Oman. *Geological Society, London, Special Publications*, **154**, 241–260, https://doi.org/10.1144/GSL.SP.1999.154.01.11

Miller, J.M., Gray, D.R. and Gregory, R.T. 2002. Geometry and significance of internal windows and regional isoclinal folds in northeast Saih Hatat, Sultanate of Oman. *Journal of Structural Geology*, **24**, 359–386, https://doi.org/10.1016/S0191-8141(01)00061-X

Moraetis, D., Mattern, F., Scharf, A., Frijia, G., Kusky, T.M., Yuan, Y. and Hussain, I.L. 2018. Neogene to Quaternary uplift history along the passive margin of the northeastern Arabian Peninsula eastern Hajar Mountains, Oman. *Quaternary Research*, **90**, 418–434, https://doi.org/10.1017/qua.2018.51

Mount, V.S., Crawford, R.I.S. and Bergman, S.C. 1998. Regional Structural style of the Central and Southern Oman Mountains: Jabal Akhdar, Saih Hatat and the Northern Ghaba Basin. *GeoArabia*, **3**, 475–490.

Nolan, S.C., Skelton, P.W., Clissold, B.P. and Smewing, J.D. 1990. Maastrichtian to Early Tertiary stratigraphy and paleogeography of the Central and Northern Oman Mountains. *Geological Society, London, Special Publications*, **49**, 307–325, https://doi.org/10.1144/GSL.SP.1992.049.01.31

Reinhardt, B.M. 1969. On the genesis and emplacement of ophiolites in the Oman Mountains geosyncline. *Schweizerische Mineralogische und Petrographische Mitteilungen*, **49**, 1–30.

Rickards, B.R., Booth, G.A., Paris, F. and Heward, A.P. 2010. Marine flooding events of the Early and Middle Ordovician of Oman and the United Arab Emirates and their graptolite, acritarch and chitinozoan associations. *GeoArabia*, **15**, 81–120.

Robertson, A.H.F. 1987. The transition from passive margin to Upper Cretaceous foreland basin related to ophiolite emplacement in the Oman Mountains. *Geological Society of America Bulletin*, **99**, 633–653, https://doi.org/10.1130/0016-7606(1987)99<633:TTFAPM>2.0.CO;2

Robertson, A.H.F. and Woodcock, N.H. 1983. Genesis of the Batinah mélange above the Semail ophiolite, Oman. *Journal of Structural Geology*, **5**, 1–17, https://doi.org/10.1016/0191-8141(83)90003-2

Romine, *et al.* 2004. *North Oman Haima-Huqf Tectonostratigraphic Study*. SRK Confidential Report to PDO, July 2004.

Saddiqi, O., Michard, A., Goffé, B., Poupeau, G. and Oberhänsli, R. 2006. Fission-track thermochronology of the Oman Mountains continental widows, and current problems of tectonic interpretation. *Bulletin de la Société Géologique de France*, **177**, 127–134, https://doi.org/10.2113/gssgfbull.177.3.127

Scharf, A., Mattern, F., Moraetis, D., Callegari, I. and Weidle, C. 2019*a*. Postobductional kinematic evolution and geomorphology of a major regional structure – the Semail Gap Fault Zone (Oman Mountains). *Tectonics*, **38**, 2756–2778, https://doi.org/10.1029/2019TC005588

Scharf, A., Mattern, F. and Pracejus, B. 2019*b*. Two new microscopic ductile kinematic indicators from the Oman Mountains. *Journal of Structural Geology*, **119**, 107–117, https://doi.org/10.1016/j.jsg.2018.12.006

Scharf, A., Mattern, F., Bolhar, R., Bailey, C.M. and Ring, U. 2019*c*. Early Cenozoic listwaenite formation at a major extensional fault zone of the Oman Mountains (Fanja area) – insights from structural analysis and U-Pb carbonate dating. AAPG Event, Structural styles in the Middle East, 9–11 December 2019, Muscat, Oman.

Scharf, A., Mattern, F., Bolhar, R., Bailey, C.M. and Ring, U. 2020*a*. U-Pb dating of postobductional carbonate veins in listwaenite of the Oman Mountains near Fanja. *International Conference on Ophiolites and the Oceanic Lithosphere: Results of the Oman Drilling Project and Related Research*, 12–14 January, 2020, Sultan Qaboos University, Muscat, Sultanate of Oman.

Scharf, A., Sudo, M., Pracejus, B., Mattern, F., Callegari, I., Bauer, W. and Scharf, K. 2020*b*. Late Lutetian (Eocene) mafic intrusion into shallow marine platform deposits north of the Oman Mountains (Rusayl Embayment) and its tectonic significance. *Journal of African Earth Sciences*, **170**, 103941, https://doi.org/10.1016/j.jafrearsci.2020.103941

Scharf, A., Mattern, F., Al-Wardi, M., Frijia, G., Moraetis, D., Pracejus, B., Bauer, W. and Callegari, I. 2021*a*. Tectonostratigraphy of the eastern part of the Oman Mountains. *Geological Society, London, Memoirs*, **54**, 11–47, https://doi.org/10.1144/M54.2

Scharf, A., Mattern, F., Al-Wardi, M., Frijia, G., Moraetis, D., Pracejus, B., Bauer, W. and Callegari, I. 2021*b*. Appendices to: The Geology and Tectonics of the Jabal Akhdar and Saih Hatat Domes, Oman Mountains. *Geological Society, London, Memoirs*, **54**, 113–116, https://doi.org/10.1144/M54.7

Scharf, A., Mattern, F., Al-Wardi, M., Frijia, G., Moraetis, D., Pracejus, B., Bauer, W. and Callegari, I. 2021*c*. Tectonic evolution of the Oman Mountains. *Geological Society, London, Memoirs*, **54**, 67–104, https://doi.org/10.1144/M54.5

Scharf, A., Mattern, F., Al-Wardi, M., Frijia, G., Moraetis, D., Pracejus, B., Bauer, W. and Callegari, I. 2021*d*. Large-scale structure of the study area. *Geological Society, London, Memoirs*, **54**, 61–66, https://doi.org/10.1144/M54.4

Searle, M.P. 1985. Sequence of thrusting and origin of culminations in the northern and central Oman Mountains. *Journal of Structural Geology*, **7**, 129–143, https://doi.org/10.1016/0191-8141(85)90127-0

Searle, M.P. 2007. Structural geometry, style and timing of deformation in the Hawasina Window, Al Jabal al Akhdar and Saih Hatat culminations, Oman Mountains. *GeoArabia*, **12**, 99–130.

Searle, M.P. and Alsop, G.I. 2007. Eye-to-eye with a mega-sheath fold: a case study from Wadi Mayh, northern Oman. *Geology*, **35**, 1043–1046. https://doi.org/10.1130/G23884A.1

Searle, M.P. and Malpas, J. 1980. Structure and metamorphism of rocks beneath the Semail ophiolite of Oman and their significance in ophiolite obduction. *Earth and Environmental Science Transactions of the Royal Society of Edinburgh*, **71**, 247–262, https://doi.org/10.1017/S0263593300013614

Searle, M.P. and Malpas, J. 1982. Petrochemistry and origin of sub-ophiolitic metamorphic and related rocks in the Oman Mountains. *Journal of the Geological Society, London*, **139**, 235–248, https://doi.org/10.1144/gsjgs.139.3.0235

Searle, M.P., Waters, D.J., Martin, H.N. and Rex, D.C. 1994. Structure and metamorphism of blueschist–eclogite facies rocks from the northeastern Oman Mountains. *Journal of Geological Society, London*, **151**, 555–576, https://doi.org/10.1144/gsjgs.151.3.0555

Searle, M.P., Warren, C.J., Waters, D.J. and Parrish, R.R. 2004. Structural evolution, metamorphism and restoration of the Arabian continental margin, Saih Hatat region, Oman Mountains. *Journal of Structural Geology*, **26**, 451–473, https://doi.org/10.1016/j.jsg.2003.08.005

Villey, M., de Gramont, X. and Le Métour, J. 1986. *Geological map of Seeb, sheet NF 40-3C, scale: 1:100,000, with Explanatory notes*. Directorate General of Minerals, Oman Ministry of Petroleum and Minerals.

Warburton, J., Burnhill, T.J., Graham, R.H. and Isaac, K.P. 1990. The evolution of the Oman Mountains foreland basin. *Geological Society, London, Special Publications*, **49**, 419–427, https://doi.org/10.1144/GSL.SP.1992.049.01.26

Warren, C.J. and Miller, J.M. 2007. Structural and stratigraphic controls on the origin and tectonic history of a subducted continental margin, Oman. *Journal of Structural Geology*, **29**, 541–558, https://doi.org/10.1016/j.jsg.2006.10.006

Warren, C.J., Parrish, R.R., Waters, D.J. and Searle, M.P. 2005. Dating the geologic history of Oman's Semail ophiolite: insight from U-Pb geochronology. *Contributions to Mineralogy and Petrology*, **150**, 403–422, https://doi.org/10.1007/s00410-005-0028-5

Welland, M.J.P. and Mitchell, A.H.G. 1977. Emplacement of the Oman ophiolite: a mechanism related to subduction and collision. *Bulletin Geological Society of America*, **88**, 1081–1088, https://doi.org/10.1130/0016-7606(1977)88<1081:EOTOOA>2.0.CO;2

Woodcock, N.H. and Robertson, A.H.F. 1982. Stratigraphy of the Mesozoic rocks above the Semail ophiolite, Oman. *Geological Magazine*, **119**, 67–76, https://doi.org/10.1017/S0016756800025668

Yamato, P., Agard, P., Goffé, B., De Andrade, V., Vidal, O. and Jolivet, L. 2007. New, high-precision P-T estimates for obduction, nappe stacking and exhumation processes. *Journal of Metamorphic Geology*, **25**, 657–682, https://doi.org/10.1111/j.1525-1314.2007.00722.x

Chapter 4

Large-scale structure of the study area

Andreas Scharf[1]*, Frank Mattern[1], Mohammed Al-Wardi[1], Gianluca Frijia[2], Daniel Moraetis[3], Bernhard Pracejus[1], Wilfried Bauer[4] and Ivan Callegari[4]

[1]Department of Earth Sciences, College of Science, Sultan Qaboos University, PO Box 36, PC 123, Al-Khod, Muscat, Sultanate of Oman

[2]Department of Physics and Earth Sciences, University of Ferrara, Via Saragat 1, 44122, Ferrara, Italy

[3]Department of Applied Physics and Astronomy, College of Sciences, University of Sharjah, PO Box 27272, Sharjah, United Arab Emirates

[4]Department of Applied Geosciences, German University of Technology GUtech, PO Box 1816, PC 130, Halban, Sultanate of Oman

*Correspondence: scharfa@squ.edu.om

Abstract: The Southeastern Oman Mountains are dominated by two major culminations: the Jabal Akhdar and Saih Hatat domes, surrounded by allochthonous and/or neo-autochthonous rocks. In the cores of both domes, folded autochthonous and par-autochthonous pre-Permian meta-sedimentary rocks are exposed, subjacent to the 'Hercynian' Unconformity. Above the unconformity are Permo–Mesozoic shelfal sedimentary rocks, characterized by carbonates. These sedimentary rocks were openly folded. The open folds are large-scale elongate structures that define the shapes of both domes. The main elongation direction is NW–SE. Doming is syn- to post-obductional. Most margins of the domes are marked by major post-obductional, extensional faults. Reactivated basement faults along the eastern margin of the Jabal Akhdar Dome may be responsible for the straight NNE-striking eastern margin which is perpendicular to the main elongation direction of the domes. The deep structure of both domes is poorly known. However, the Moho depth below the centre of the Jabal Akhdar Dome is at 50 km. We present a geological map of both domes, depicting the main faults and folds, and schematic cross-sections, parallel and perpendicular to the Oman Mountains.

Five cross-sections (S1–S5) illustrate the overall geometry of the study area (Figs 4.1 and 4.2), omitting local-scale deformation. The thickness of individual nappes was taken from the geological maps of Oman (1:100 000 scale). The traces of the sections (Fig. 4.1) are either perpendicular or parallel to the main structures of the trend of the Oman Mountains. The sections of Figure 4.2 are only generalized and simplified, showing the overall structure at the scale of Figure 4.1. For more detailed and larger cross-sections, see Glennie *et al.* (1974), Manghnani and Coleman (1981), Michard *et al.* (1984), Beurrier *et al.* (1986), de Gramont *et al.* (1986), Le Métour *et al.* (1986*a*, *b*, 1992), Rabu *et al.* (1986), Villey *et al.* (1986*b*), Robertson (1987), Béchennec *et al.* (1990, 1992), Cawood *et al.* (1990), Coffield (1990), Hanna (1990), Mount *et al.* (1998), Al-Lazki *et al.* (2002), Searle (2007), Cowan *et al.* (2014) and Hansman *et al.* (2017). The rocks of the northeastern Saih Hatat Dome (SHD) display complex and polyphase folding under high-pressure/low-temperature conditions which cannot be depicted in detail in S3 and S4 (Fig. 4.2) due to the scale of the structures (see Le Métour *et al.* 1990; Gregory *et al.* 1998; Jolivet *et al.* 1998; Searle and Cox 1999; Miller *et al.* 2002; Searle *et al.* 2004; Gray *et al.* 2005*a*, *b*; Saddiqi *et al.* 2006; Searle 2007; Agard *et al.* 2010; Agard and Vitale-Brovarone 2013).

According to Mount *et al.* (1998), the Semail Gap is cored by a W-dipping blind reverse fault beneath the Semail Gap. We interpret the present Semail Gap in S5 as part of a monoclinal fold which formed during the Eocene (Hansman *et al.* 2017; Scharf *et al.* 2019). The fold axis of this monocline is parallel to the Semail Gap. Within the basement rocks below the monocline, a blind fault may exist. The dip of the blind fault is unknown but certainly acted as a transfer fault during doming. The monocline was truncated by a dextral transtensional fault zone, the Semail Gap Fault Zone (SGFZ), with a vertical and horizontal displacement of *c.* 5.5 km and *c.* 2 km, respectively (Scharf *et al.* 2019). It is a structural element of the post-Eocene extensional regime (section on 'Main extensional fault systems' in Chapter 3). The vertical extent of the SGFZ is not known, but it certainly affected the Hajar Supergroup (Scharf *et al.* 2019). Similar configurations of extensional faults may be present along large segments of the Jabal Akhdar Dome (JAD) and SHD margins (see section on 'Allochthonous thrusting on autochthonous units' in Chapter 3).

Figure 4.3 depicts the cross-section A–A' (shown in Fig. 4.1) that strikes WNW–ESE across the Frontal Range Fault of the SHD. This figure depicts several open to gentle F6 folds in the hanging-wall of the Frontal Range Fault that could not be shown in the cross-section of Figure 4.2 because of scale.

The present-day crustal thickness measures *c.* 47 and 48–51 km at the JAD (compare with fig. 11a, b of Ravaut *et al.* 1997 and Al-Lazki *et al.* 2002, respectively), and 42–39 km below the Batinah Coastal Plain (Al-Lazki *et al.* 2002). N-striking gravity profiles across the SHD were analysed by Manghnani and Coleman (1981). The same authors identified the Semail Gap as a gravity anomaly and estimated the depth of the Moho at the SHD and to its south to be 42 and *c.* 35 km, respectively. A major conclusion of Manghnani and Coleman (1981) is that the Semail Ophiolite is not rooted in the oceanic lithosphere offshore NE of Oman. Weidle *et al.* (2020) set up a temporary broadband seismic network in the Oman Mountains with 40 instruments for continuous passive seismic registrations for 27 months. This network was completed by 18 permanent stations. Based on their dataset, Weidle *et al.* (2020) depicted a Moho depth of 40–45 km NW of the Semail Gap (i.e. in the JAD area), while it is only 20 km to the east of the Semail Gap (i.e. in the SHD area).

Lithospheric mantle Pn-velocity tomography studies of northern Oman and the adjacent areas have been conducted by Al-Lazki *et al.* (2014). They show a very low Pn-tomography velocity anomaly below northern Oman, the Sea of Oman and the western Makran subduction zone (Fig. 4.4). The western border of the anomaly coincides with the Dibba Zone in

From: Scharf, A., Mattern, F., Al-Wardi, M., Frijia, G., Moraetis, D., Pracejus, B., Bauer, W. and Cellegari, I. 2021. *The Geology and Tectonics of the Jabal Akhdar and Saih Hatat Domes, Oman Mountains*. Geological Society, London, Memoirs, **54**, 61–66, https://doi.org/10.1144/M54.4

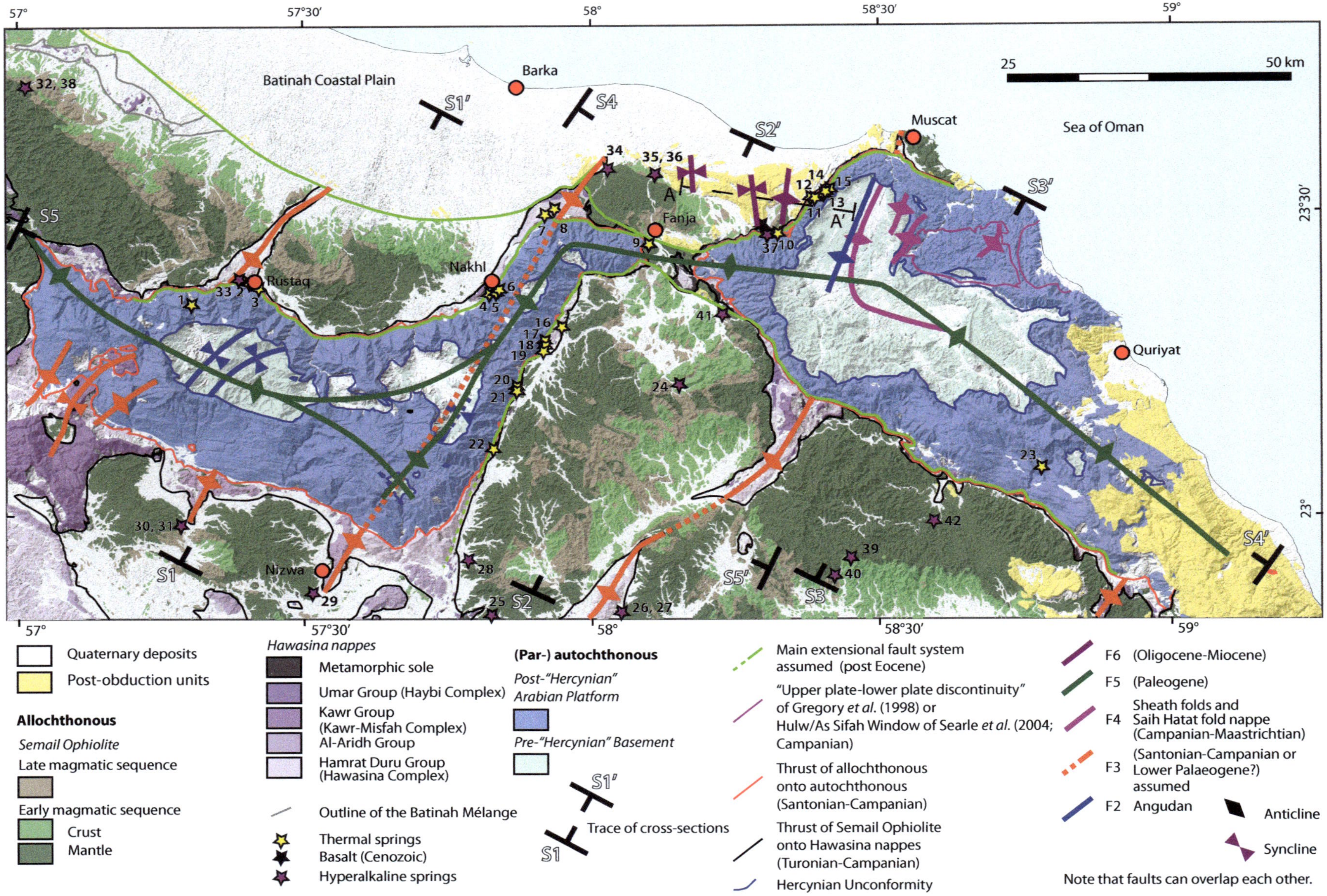

Fig. 4.1. Large-scale structures and tectonic map of the JAD and SHD. Shaded relief: (1) based on data from ASTER GDEM (product of METI and NASA) and (2) compiled with SimpleDEMViewer for Mac (version 4.4.9); modified after Béchennec *et al.* (1993). See section on 'Allochtonous thrusting on autochthonous units' in Scharf *et al.* (2021*b*, Chapter 3, this Memoir) for details, discussion and nomenclature of the Hawasina thrust sheets. Note that the northeastern margin of the SHD consists of high-pressure rocks of the Arabian margin that have been displaced by some tens of kilometres; they are therefore parautochthounous. Thermal springs (1–23) are listed in table A.1 (Scharf *et al.* 2021*a*, Chapter 7, this Memoir). Hyperalkaline springs (24–42) after Paukert *et al.* (2012) and Chavagnac *et al.* (2013*a*, *b*) as listed in table A.2. Mafic intrusion *c.* 20 km east of Fanja is after Al-Harthy *et al.* (1991). F1 folds (Cadomian) are not shown because of the scale; see figure 3.6 (Scharf *et al.* 2021*b*, Chapter 3, this Memoir) for their schematic depictions. A–A' depicts the F6 folds (Fig. 4.3). Note that the Angudan and Hercynian Unconformity are overlapping in the JAD, while they only partly overlap in the SHD.

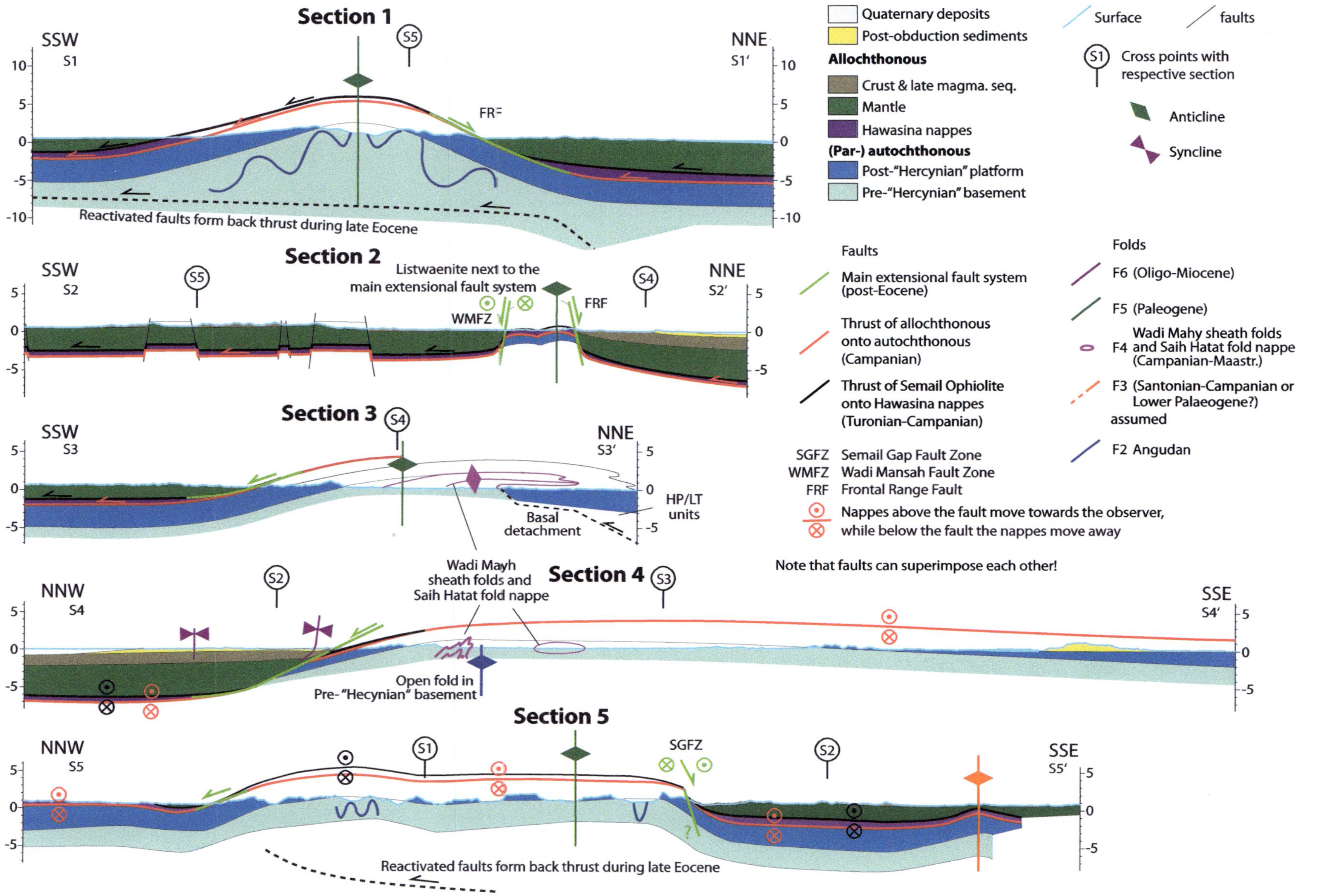

Fig. 4.2. Generalized large-scale orogen-normal and orogen-parallel cross-sections, showing the upper crustal structure of the JAD and SHD at the scale of Figure 4.1. All sections are constructed based on the tectonic map (Fig. 4.1) and own data. The thickness of nappes are from the 1:100 000 scale geological maps. Only faults with a regionally significant displacement could be included in the sections. Note that the northeastern margin of the SHD consists of high-pressure rocks of the Arabian margin that have been displaced by some tens of kilometres; they are therefore not autochthonous. Section 3 depicts the SHD from the NNE to the SSW and is simplified from Searle *et al.* (2004) as well as Warren and Miller (2007). The structures in the eastern SHD are intensely deformed and cannot be depicted in great detail in this section. Searle (2007) provided a schematic section through the same area (fig. 3.2). F1 folds (Cadomian) are not shown because of the scale; see figure 3.6 for schematic depictions. Figure 4.3 depicts more details of the F6 folds in the hanging-wall of the SHD and is parallel to section 4. HP/LT – high-pressure/low-temperature.

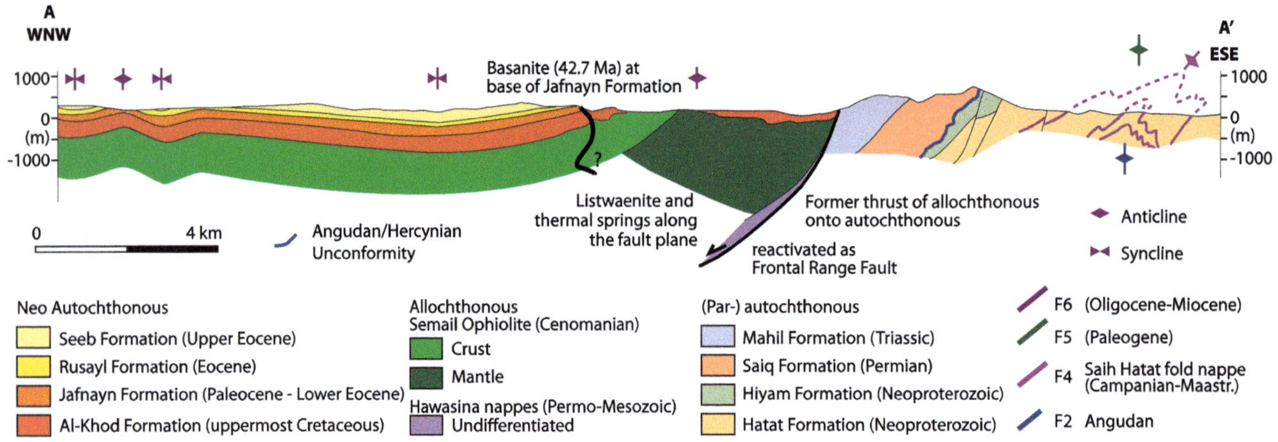

Fig. 4.3. Cross-section across the Frontal Range Fault of the SHD. Along this section the open and gentle F6 folds are shown in the hanging-wall (Figs 4.1 & 4.2). Modified after Villey *et al.* (1986*a*). See Figure 4.1 for trace of section A–A'.

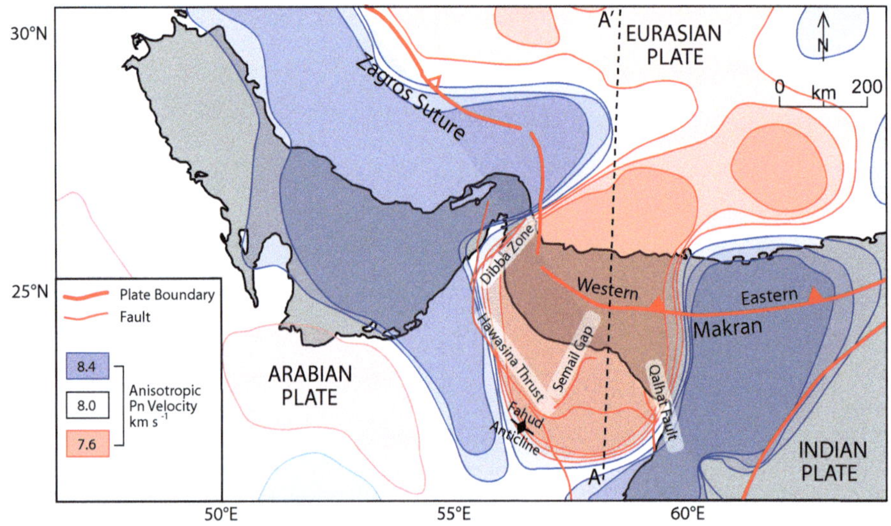

Fig. 4.4. Pn-tomography velocity model of Al-Lazki *et al.* (2014) superimposed on the Arabia–Eurasia collision zone (modified after Hansman *et al.* 2017). Cross-section for the trace (A–A') is depicted in figure 5.11.

the Northern Oman Mountains and the eastern border with the Ja'alan and Qalhat faults near Sur (Al-Lazki *et al.* 2014, fig. 10; Figs 1.1 & 4.4). The mentioned faults and the Dibba Zone are important structural elements which mark the south-easternern and northwesternern end of the Semail Ophiolite, respectively (compare with Moraetis *et al.* 2018; Carminati *et al.* 2020). The southern boundary is probably covered by thick Upper Cenozoic sedimentary rocks (Al-Lazki *et al.* 2014). The border of this anomaly is abrupt to the west, south and east (Fig. 4.4). Al-Lazki *et al.* (2014) interpreted this anomaly as a hot mantle with possible partial melts underlying this region.

Author contributions **AS:** Conceptualization (lead), Writing original draft (lead), Visualization (lead), Supervision (lead), Investigation (lead); **FM:** Conceptualization (equal), Writing – review and editing (equal), Visualization (equal), Investigation (equal); **MAW:** Writing – review and editing (equal); **GF:** Writing – review and editing (equal); **DM:** Writing – review and editing (equal); **BP:** Writing – review and editing (equal); **WB:** Writing – review and editing (equal); **IC:** Writing – review and editing (equal).

Funding This research received no specific grant from any funding agency in the public, commercial, or not-for-profit sectors.

Data availability Data sharing is not applicable to this article as no datasets were generated or analysed during the current study.

References

Agard, P. and Vitale-Brovarone, A. 2013. Thermal regime of continental subduction: the record from exhumed HP-LT terranes (New Caledonia, Oman, Corsica). *Tectonophysics*, **601**, 206–215, https://doi.org/10.1016/j.tecto.2013.05.011

Agard, P., Searle, M.P., Alsop, G. and Dubacq, B. 2010. Crustal stacking and expulsion tectonics during continental subduction: P-T deformation constraints from Oman. *Tectonics*, **29**, 1–19, https://doi.org/10.1029/2010TC002669

Al-Harthy, M.S., Coleman, R.G., Hughes Clarke, M.W. and Hanna, S.S. 1991. Tertiary basaltic intrusions in the Central Oman Mountains. *In*: Peters, T., Nicolas, A. and Coleman, R.J. (eds) *Ophiolite Genesis and Evolution of the Oceanic Lithosphere*. Proceedings of the Ophiolite Conference, Muscat, Oman, Kluwer Academic Publishers, Dordrecht/Boston/London, **1990**, 475–682.

Al-Lazki, A.I., Seber, D., Sandvol, E. and Barazangi, M . 2002. A crustal transect across the Oman Mountains on the eastern margin of Arabia. *GeoArabia*, **7**, 47–78.

Al-Lazki, A.I., Al-Damegh, K.S., El-Hadidy, S.Y., Ghods, A. and Tatar, M. 2014. Pn-velocity structure beneath Arabia-Eurasia Zagros collision and Makran subduction zones. *Geological*

Society, London, Special Publications, **392**, 45–60, https://doi.org/10.1144/SP392.3
Béchennec, F., Le Métour, J., Rabu, D., Bourdillon-de-Grissac, C., de Wever, P., Beurrier, M. and Villey, M. 1990. The Hawasina Nappes: stratigraphy, paleogeography and structural evolution of a fragment of the South-Tethyan passive continental margin. *Geological Society, London, Special Publications*, **49**, 213–223, https://doi.org/10.1144/GSL.SP.1992.049.01.14
Béchennec, F., Roger, J., Le Métour, J. and Wyns, R . 1992. *Geological map of Seeb, sheet NF 40-03, scale 1:250,000, with Explanatory Notes*. Directorate General of Minerals, Oman Ministry of Petroleum and Minerals.
Béchennec, F., Le Métour, J., Platel, J.P. and Roger, J . 1993. *Geological map of the Sultanate of Oman, scale 1:1,000,000, with Explanatory Notes*. Directorate General of Minerals, Oman Ministry of Petroleum and Minerals.
Beurrier, M., Béchennec, F., Rabu, D. and Hutin, G . 1986. *Geological map of Rustaq, sheet NF 40-03D, scale 1:100,000, with Explanatory notes*. Directorate General of Minerals, Oman Ministry of Petroleum and Minerals.
Carminati, E., Aldega, L., Smeraglia, L., Scharf, A., Mattern, A., Albert, R. and Gerdes, A. 2020. Tectonic evolution of the Northern Oman Mountains, part of the Straits of Hormuz Syntaxis: new structural and paleothermal analyses and U–Pb dating of synkinematic calcite. *Tectonics*, **39**(4), https://doi.org/10.1029/2019TC005936
Cawood, P.A., Green, F.K. and Calon, T.J. 1990. Origin of culmination within the southeastern Oman Mountains at Jebel Majhool and Ibra Dome. *Geological Society, London, Special Publications*, **49**, 397–418, https://doi.org/10.1144/GSL.SP.1992.049.01.27
Chavagnac, V., Ceuleneer, G., Monnin, C., Lansac, B., Hoareau, G. and Boulart, C. 2013*a*. Mineralogical assemblages forming at hyperalkaline warm springs hosted on ultramafic rocks: a case study of Oman and Ligurian ophiolites. *Geochemistry, Geophysics, Geosystems*, **14**, 2474–2495, https://doi.org/10.1002/ggge.20146
Chavagnac, V., Monnin, C., Ceuleneer, G., Boulart, C. and Hoareau, G. 2013*b*. Characterization of hyperalkaline fluids produced by low-temperature serpentinization of mantle peridotites in the Oman and Ligurian ophiolites. *Geochemistry, Geophysics, Geosystems*, **14**, 2496–2522, https://doi.org/10.1002/ggge.20147
Coffield, D.Q. 1990. Structures associated with nappe emplacement and culmination collapse in the Central Oman Mountains. *Geological Society, London, Special Publications*, **49**, 447–458, https://doi.org/10.1144/GSL.SP.1992.049.01.28
Cowan, R.J., Searle, M.P. and Waters, D.J. 2014. Structure of the metamorphic sole to the Oman Ophiolite, Sumeini Window and Wadi Tayyin: implications for ophiolite obduction processes. *Geological Society, London, Special Publications*, **392**, 155–175, https://doi.org/10.1144/SP392.8
de Gramont, X., Le Métour, J. and Villey, M. 1986. *Geological Map of Samad, Sheet NF 40-07C, Scale 1:100,000, with Explanatory Notes*. Directorate General of Minerals, Oman Ministry of Petroleum and Minerals.
Glennie, K.W., Boeuf, M.G.A., Hughes Clarke, M.W., Moody-Stuart, M., Pilaar, W. and Reinhardt, B.M. 1974. Geology of the Oman Mountains *Koninklijk Nederlands Geologisch en Mijnbouwkundig Genootschap Transactions*, **31**, 423.
Gray, D.R., Gregory, R.T., Armstrong, R.A., Richards, I.J. and Miller, J.McL. 2005*a*. Age and stratigraphic relationship of structurally deepest level rocks, Oman Mountains: U/Pb SHRIMP evidence for Late Carboniferous Neotethys rifting. *The Journal of Geology*, **113**, 611–626, https://doi.org/10.1086/449325
Gray, D.R., Miller, J.McL. and Gregory, R.T. 2005*b*. Strain state and kinematic evolution of a fold-nappe beneath the Samail Ophiolite, Oman. *Journal of Structural Geology*, **27**, 1986–2007, https://doi.org/10.1016/j.jsg.2005.06.011
Gregory, R.T., Gray, D.R. and Miller, J.M. 1998. Tectonics of the Arabian margin associated with the emplacement of the Oman margin along the Ibra transect: new evidence from northeast Saih Hatat. *Tectonics*, **17**, 657–670, https://doi.org/10.1029/98TC02206
Hanna, S.S. 1990. The Alpine deformation of the central Oman Mountains. *Geological Society, London, Special Publications*, **49**, 341–359, https://doi.org/10.1144/GSL.SP.1992.049.01.21
Hansman, R.J., Ring, U., Thomson, S.N., den Brock, B. and Stübner, K. 2017. Late Eocene uplift of the Al Hajar Mountains, Oman, supported by stratigraphic and low-temperature thermochronology. *Tectonics*, **36**, 3081–3109, https://doi.org/10.1002/2017TC004672
Jolivet, L., Goffé, B., Bousquet, R., Oberhänsli, R. and Michard, A. 1998. Detachments in high-pressure mountain belts, Tethyan examples. *Earth and Planetary Science Letters*, **160**, 31–47, https://doi.org/10.1016/S0012-821X(98)00079-X
Le Métour, J., Villey, M. and de Gramont, X. 1986*a*. *Geological map of Masqat, sheet NF 40-4A, scale 1:100,000, Explanatory notes*. Directorate General of Minerals, Oman Ministry of Petroleum and Minerals.
Le Métour, J., Villey, M. and de Gramont, X. 1986*b*. *Geological map of Quryat, sheet NF 40-4D, scale 1:100,000, Explanatory notes*. Directorate General of Minerals, Oman Ministry of Petroleum and Minerals.
Le Métour, J.F., Rabu, D., Tegyey, M., Béchennec, F., Beurrier, M. and Villey, M. 1990. Subduction and obduction: two stages in the Eo-Alpine tectonometamorphic evolution of the Oman Mountains. *Geological Society, London, Special Publications*, **49**, 285–305, https://doi.org/10.1144/GSL.SP.1992.049.01.20
Le Métour, J.F., Béchennec, F., Roger, J. and Wyns, R. 1992. *Geological map of Muscat, sheet NF40-04, scale 1:250,000, with Explanatory Notes*. Directorate General of Minerals, Oman Ministry of Petroleum and Minerals.
Manghnani, M.H. and Coleman, R.G. 1981. Gravity profiles across the Samail Ophiolite, Oman. *Journal of Geophysical Research*, **86**, 2509–2525, https://doi.org/10.1029/JB086iB04p02509
Michard, A., Bouchez, J.L. and Ouazzani-Touhami, M. 1984. Obduction-related planar and linear fabrics in Oman. *Journal of Structural Geology*, **6**, 39–49, https://doi.org/10.1016/0191-8141(84)90082-8
Miller, J.M., Gray, D.R. and Gregory, R.T. 2002. Geometry and significance of internal windows and regional isoclinal folds in northeast Saih Hatat, Sultanate of Oman. *Journal of Structural Geology*, **24**, 359–386, https://doi.org/10.1016/S0191-8141(01)00061-X
Moraetis, D., Mattern, F., Scharf, A., Frijia, G., Kusky, T.M., Yuan, Y. and Hussain, I.L. 2018. Neogene to Quaternary uplift history along the passive margin of the northeastern Arabian Peninsula eastern Hajar Mountains, Oman. *Quaternary Research*, **90**, 418–434, https://doi.org/10.1017/qua.2018.51
Mount, V.S., Crawford, R.I.S. and Bergman, S.C. 1998. Regional structural style of the Central and Southern Oman Mountains: Jabal Akhdar, Saih Hatat and the Northern Ghaba Basin. *GeoArabia*, **3**, 475–490.
Paukert, A.N., Matter, J.M., Kelemen, P., Shock, E.L. and Havig, J.R. 2012. Reaction path modeling of enhanced in situ CO_2 mineralization for carbon sequestration in the peridotites of the Samail Ophiolite, Sultanate of Oman. *Chemical Geology*, **330–331**, 86–100, https://doi.org/10.1016/j.chemgeo.2012.08.013
Rabu, D., Béchennec, F., Beurrier, M. and Hutin, G. 1986. *Geological map of Nakhl, sheet NF40-3E, scale: 1:100,000, with Explanatory Notes*. Directorate General of Minerals, Oman Ministry of Petroleum and Minerals.
Ravaut, P., Bayer, R., Hassani, R., Rousset, D. and Yahya'ey, A.A. 1997. Structure and evolution of the northern Oman margin: gravity and seismic constraints over the Zagros-Makran-Oman collision zone. *Tectonophysics*, **279**, 253–280, https://doi.org/10.1016/S0040-1951(97)00125-X
Robertson, A.H.F. 1987. The transition from passive margin to Upper Cretaceous foreland basin related to ophiolite emplacement in the Oman Mountains. *Geological Society of America Bulletin*,

99, 633–653, https://doi.org/10.1130/0016-7606(1987)99<633:TTFAPM>2.0.CO;2

Saddiqi, O., Michard, A., Goffé, B., Poupeau, G. and Oberhänsli, R. 2006. Fission-track thermochronology of the Oman Mountains continental widows, and current problems of tectonic interpretation. *Bulletin de la Société Géologique de France*, **177**, 127–134, https://doi.org/10.2113/gssgfbull.177.3.127

Scharf, A., Mattern, F., Moraetis, D., Callegari, I. and Weidle, C. 2019. Postobductional kinematic evolution and geomorphology of a major regional structure – the Semail Gap Fault Zone (Oman Mountains). *Tectonics*, **38**, 2756–2778, https://doi.org/10.1029/2019TC005588

Scharf, A., Mattern, F., Al-Wardi, M., Frijia, G., Moraetis, D., Pracejus, B., Bauer, W. and Callegari, I. 2021*a*. Appendices to: The Geology and Tectonics of the Jabal Akhdar and Saih Hatat Domes, Oman Mountains. *Geological Society, London, Memoirs*, **54**, 113–116, https://doi.org/10.1144/M54.7

Scharf, A., Mattern, F., Al-Wardi, M., Frijia, G., Moraetis, D., Pracejus, B., Bauer, W. and Callegari, I. 2021*b*. Thrusts, extensional faults and fold patterns of the major units. *Geological Society, London, Memoirs*, **54**, 45–56, https://doi.org/10.1144/M54.3

Searle, M.P. 2007. Structural geometry, style and timing of deformation in the Hawasina Window, Al Jabal al Akhdar and Saih Hatat culminations, Oman Mountains. *GeoArabia*, **12**, 99–130.

Searle, M.P. and Cox, J.S. 1999. Tectonic setting, origin and obduction of the Oman ophiolite. *Geological Society of America Bulletin*, **111**, 104–122, https://doi.org/10.1130/0016-7606(1999)111<0104:TSOAOO>2.3.CO;2

Searle, M.P., Warren, C.J., Waters, D.J. and Parrish, R.R. 2004. Structural evolution, metamorphism and restoration of the Arabian continental margin, Saih Hatat region, Oman Mountains. *Journal of Structural Geology*, **26**, 451–473, https://doi.org/10.1016/j.jsg.2003.08.005

Villey, M., de Gramont, X. and Le Métour, J. 1986*a*. *Geological map of Seeb, sheet NF 40-3C, scale: 1:100,000, with Explanatory notes*. Directorate General of Minerals, Oman Ministry of Petroleum and Minerals.

Villey, M., Le Métour, J. and de Gramont, X. 1986*b*. *Geological map of Fanjah, sheet NF 40-3F, scale: 1:100,000, with Explanatory notes*. Directorate General of Minerals, Oman Ministry of Petroleum and Minerals.

Warren, C.J. and Miller, J.M. 2007. Structural and stratigraphic controls on the origin and tectonic history of a subducted continental margin, Oman. *Journal of Structural Geology*, **29**, 541–558, https://doi.org/10.1016/j.jsg.2006.10.006

Weidle, C., Wiesenberg, L., El-Sharkawy, A., Meier, T., Krüger, F., Agard, P. and Scharf, A. 2020. A 3-D crustal model of the eastern Arabian plate margin below the Oman Ophiolite. EGU 2020–17800, **22**, Vienna.

Chapter 5

Tectonic evolution of the Oman Mountains

Andreas Scharf[1]*, Frank Mattern[1], Mohammed Al-Wardi[1], Gianluca Frijia[2], Daniel Moraetis[3], Bernhard Pracejus[1], Wilfried Bauer[4] and Ivan Callegari[4]

[1]Department of Earth Sciences, College of Science, Sultan Qaboos University, PO Box 36, PC 123, Al-Khod, Muscat, Sultanate of Oman

[2]Department of Physics and Earth Sciences, University of Ferrara, Via Saragat 1, 44122, Ferrara, Italy

[3]Department of Applied Physics and Astronomy, College of Sciences, University of Sharjah, PO Box 27272, Sharjah, United Arab Emirates

[4]Department of Applied Geosciences, German University of Technology GUtech, PO Box 1816, PC 130, Halban, Sultanate of Oman

*Correspondence: scharfa@squ.edu.om

Abstract: The tectonic evolution of the Oman Mountains as of the Neoproterozoic begins with a major extensional event, the Neoproterozoic Abu Mahara rifting. It was followed by the compressional Nabitah event, still during the Neoproterozoic, in Oman but possibly not in the study area. During the earliest Cambrian, the Jabal Akhdar area was affected by the Cadomian Orogeny, marked by NE–SW shortening. It is unclear, whether the Saih Hatat area was exposed to the Cadomian deformation, too. Still during the lower Cambrian, the Angudan Orogeny followed, characterized by NW–SE shortening. An episode of rifting affected the Saih Hatat area during the mid-Ordovician. During the mid-Carboniferous, both dome areas were deformed by tilting and large-scale open folding in the course of the 'Hercynian' event. As a consequence, a major unconformity formed. As another Late Paleozoic event, the Permian break-up of Pangaea and subsequent formation of the Hawasina ocean basin, are recorded in the Southeastern Oman Mountains. As a result, a passive margin formed which existed until the mid-Cretaceous, characterized by deposition of mostly shelfal carbonates. This interval of general tectonic quiescence was interrupted during the early Jurassic by uplift and tilting of the Arabian Platform. The platform collapsed during the late Cretaceous, related to the arrival of the obducted allochthonous nappes including the Semail Ophiolite, transforming the passive margin to an active margin.

The Semail Ophiolite formed most likely above a subduction zone within the Neo-Tethys Ocean during the Cenomanian while parts of the Arabian Plate were subducted to the NE. Formation of oceanic lithosphere and SW-thrusting was broadly coeval, resulting in ophiolite obduction onto the Hawasina Basin. The Semail Ophiolite and the Hawasina rocks combined were thrust further onto the Arabian Plate. Their load created a foreland basin and forebulge within the Arabian Platform. Once the continental lithosphere of the Arabian Platform was forced into the subduction zone, a tear between the dense oceanic lithosphere and the buoyant continental lithosphere developed. This led to rapid uplift and exhumation of subducted continental lithosphere of the Saih Hatat area, while obduction was still going on, causing in multiple and intense folding/thrusting within the eastern Saih Hatat Dome. Exhumation of the Saih Hatat Dome was massive. The emplacement of the ophiolite was completed during the Campanian/Maastrichtian. For completeness, we also present alternative models for the developmental history of the Semail Ophiolite.

Immediately after emplacement, the Arabian lithosphere underwent intense top-to-the-NE extensional shearing. Most of the Saih Hatat Dome was exhumed during the latest Cretaceous to Early Eocene, associated with major extensional shearing at its flanks. Further convergence during the late Eocene to Miocene resulted in exhumation of the Jabal Akhdar Dome and some gentle exhumation of the Saih Hatat Dome, shaping the present-day Southeastern Oman Mountains. In the coastal area, east and SE of the Saih Hatat Dome, some late Cretaceous to present-day uplift is evident by, e.g., uplifted marine terraces. The entire Oman Mountains are uplifting today, which is evident by the massive wadi incision into various rock units, including wadi deposits which may form overhangs.

Pre-'Hercynian' deformation

Reconstructions of Gondwana at 540 Ma show that Oman was part of the Arabian–Nubian Shield (e.g. de Wit *et al.* 1988; Allen 2007; Stern and Johnson 2010). The history of Oman before that time is still poorly understood as a result of a lack of palaeomagnetic data (Allen 2007). The Ediacaran to present palaeolatitude positions of Oman have been visualized (Hughes Clarke 2006; Droste 2014; see fig. A.2 in Scharf *et al.* 2021*a*, Chapter 7, this Memoir). During the Late Neoproterozoic, the palaeolatitude of the Muscat region was 13° S (Kilner *et al.* 2005; fig. A.2). Kilner *et al.* (2005) suggested that Oman merged with Gondwana after *c.* 600 Ma, based on the palaeopoles of Oman and Gondwana. According to Allen (2007), Oman was situated between the Arabian–Nubian Shield and an oceanic plate that was subducting, forming a magmatic arc. For a detailed summary of the regional tectonics, basin development and glacial episodes of the greater Oman area during the Neoproterozoic and Cambrian see Allen (2007), Stern and Johnson (2010) and Droste (2014).

Units below the Angudan Unconformity

The sedimentary rocks of the Huqf Supergroup (units 1–7) were deposited above granitic basement of Tonian–Cryogenian age (e.g. Allen 2007), which is not exposed in the study area. Formation of the basement was most likely related to subduction and accretion of terranes and/or island arcs along the eastern margin of the Arabian Platform until *c.* 825 Ma (Allen 2007; Whitehouse *et al.* 2016). In the Jabal Akhdar Dome (JAD), the Ghubrah, Saqlah and Fiq members of the Mistal Formation (Abu Mahara Group) were deposited in N- to NE-striking rift basins (Loosveld *et al.* 1996; fig. A.2). The sediment fill derived from two opposite basin margins (Leather *et al.* 2002, fig. 4; Allen *et al.* 2004). Volcaniclastic sediments, pillow lavas within the Saqlah Member of the Mistal Formation, are related to intracontinental extension and rifting (Le Guerroué *et al.* 2005; Allen *et al.* 2011) at >713 to 635 Ma (Bowring *et al.* 2007). For this timespan, only sparse data, outcrops, wells and poor seismic information are available (Droste, H. 2018, pers. comm.). Extension took place at *c.* 713 Ma (tuff from the Ghubrah

From: Scharf, A., Mattern, F., Al-Wardi, M., Frijia, G., Moraetis, D., Pracejus, B., Bauer, W. and Cellegari, I. 2021. *The Geology and Tectonics of the Jabal Akhdar and Saih Hatat Domes, Oman Mountains*. Geological Society, London, Memoirs, **54**, 67–103, https://doi.org/10.1144/M54.5

Member; Bowring *et al.* 2007) and at *c.* 635 Ma (filling of basin). It is therefore likely that several tectonic events ('Abu Mahara rifting') took place during the respective time interval of *c.* 80 Ma (Droste, H. 2018, pers. comm.).

The rift basin was initially filled with clastic sediments (Fiq Member), and the graben and the surrounding basement were subsequently covered by thin carbonates of the Hadash (Amq) Member of the Mistal Formation (Nafun Group I; Leather 2001; Allen *et al.* 2004; Allen and Leather 2006; Allen 2007). The occurrence of at least two diamictites within the Neoproterozoic Mistal Formation and the fact that Oman was located during that time near the equator encouraged researchers (e.g. Leather *et al.* 2002; Kellerhals and Matter 2003; Kilner *et al.* 2005; Le Guerroué *et al.* 2005) to confirm the Snowball Earth hypothesis of Kirschvink (1992) and Hoffman *et al.* (1998) for Oman.

Tectonics, including an unconformity at the top of the Abu Mahara Group in the Huqf area, are associated with widespread folding, fault reactivation and uplift at 630 to 600 Ma (Droste 2014). This event follows the 'Abu Mahara rifting' and may be related to late phases of the Nabitah Orogeny. The entire Nabitah Orogeny lasted from 690 to 590 Ma or 680 to 600 Ma, according to Genna *et al.* (2002) and Droste (2014), respectively. During this interval, the Oman basement terrane complex collided with the Arabian Shield (Droste 2014). Oman was located in a peripheral zone east of the Nabitah suture zone, which is defined in western/central Arabia (Quick 1991). Thus, only Southern Oman was marginally affected by the Nabitah event, which could explain why it has not been recognized in the JAD or SHD (Droste, H. 2018, pers. comm.). Oman was probably a region of slightly stretched continental lithosphere (cf. Genna *et al.* 2002) during deposition of the Nafun I Group.

The Nafun Group (fig. A.2) is divided into two couplets based on their post-rift lithology (Allen 2007), the Nafun I Group (Amq Member of the Mistal Formation and Hajir Formation) and the Nafun II Group (Mu'aydin and Kharus formations). Condon *et al.* (2005) and Fike *et al.* (2006), both cited in Allen (2007), proposed a stratigraphic gap between both Nafun groups. According to Allen (2007) and based on Al-Husseini (1988), Edgell (1991) and Pandit *et al.* (2001), the Nafun Group can be widely correlated throughout the Arabian/Persian Gulf and Rajasthan of India (Cozzi *et al.* 2012).

The Huqf Supergroup of Oman has a remarkably similar lithostratigraphy and chemostratigraphic signature as the Marwar Supergroup in Rajasthan (Cozzi *et al.* 2012). The latter therefore indicated that Oman was part of India–Pakistan terrain prior to its accretion to the Arabian Shield sometime after 650 Ma. Provenance studies (heavy mineral and zircon geochronology) of sandstones from the Huqf and Haima supergroups in East Oman, combined with sedimentological information, were correlated with coeval formations from western Arabia, India, Pakistan and Iran by Gomez-Perez *et al.* (2019). Based on the provenance studies, Gomez-Perez *et al.* (2019) concluded that East Oman was attached to India and eastern Afghanistan rather than to West Oman and Arabia, supporting the interpretation of Cozzi *et al.* (2012). Eventually, East Oman collided with Arabia during the final Gondwana amalgamation during the Early Cambrian (Angudan Orogeny; Gomez-Perez *et al.* 2019).

The sedimentary rocks of the Nafun Group recorded that Oman underwent an episode of regional subsidence and extension after the Nabitah Orogeny (Droste 2014). Possible reasons for this subsidence could be post-rift thermal relaxation (Loosveld *et al.* 1996; Allen and Leather 2006; Le Guerroué *et al.* 2006; Droste 2014) and/or post-collisional NW-trending crustal extension and collapse of the Arabian–Nubian shield (Al-Husseini 1988, 2000; Blasband *et al.* 2000) and/or westward subduction of an oceanic slab below Arabia along the northeastern margin of Gondwana (Grotzinger *et al.* 2002; Droste 2014). Eastern Oman was located on a passive margin during the time of deposition of the Nafun Group II. The siliciclastic sediments of the Mu'aydin Formation formed in a shallow- to deep-marine environment (Droste 2014).

The analysis of syn-depositional faults between the Hajir and Mu'aydin formations in the JAD identified an extensional fault of similar strike to that of the Semail Gap that may point to extensional Ediacaran block movements along the Proto-Semail Gap (Mattern and Scharf 2019). Vertical block movements of this kind may explain why the Ediacaran Hajir and Mu'aydin formations are present in the JAD but absent from the Saih Hatat Dome (SHD; Mattern and Scharf 2019).

The transition from the uppermost Nafun Group to the Ara Group at the Ediacaran–Cambrian boundary is characterized by a significant basin deepening (Allen 2007), resulting in deposition of pelagic resedimented carbonates, cherts, fine-grained (distal) turbidites and volcaniclastics of the Fara Formation (Béchennec *et al.* 1992). Oblique-dextral ocean closure along the eastern margin of Oman led to initially transtensional deformation (lateral escape tectonics) and reactivation of older structures (Droste 2014; fig. A.2). This event is thought to be the reason for basin deepening (Koopman *et al.* 2007; Droste 2014) and the formation of NE-striking intracontinental salt basins (Ara salt; Hughes Clarke 1988; Mattes and Conway Morris 1990; Al-Husseini 2000). Alternatively, the salt basins may have formed in a retro-arc basin setting (Grotzinger *et al.* 2002; Bowring *et al.* 2007). Droste (2014) correlated this tectonic event with the early stages of the Malagasy Orogeny (East African Orogeny). The sediments of the Ara Group, including carbonates, salt and black shale, display major facies changes, which are characteristics of basins with elevated shoulders and deep troughs (Amthor *et al.* 2003; Schröder *et al.* 2003; Allen 2007). Salt of the Ara Group has been traced from the Arabian/Persian Gulf and the Zagros Mountains to Punjab (northern India; Kent 1979; Gorin *et al.* 1982; Zharkov 1984; Edgell 1991; Bahroudi and Koyi 2003; Allen 2007).

Structural analysis within the Sahtan Bowl of the JAD reveals a folding and thrusting interval that affects the Neoproterozoic to earliest Cambrian Fara Formation (Cadomian Orogeny; Callegari *et al.* 2020; D1 in fig. A.1). These F1 folds are described in the section on 'F1 folds' in Scharf *et al.* 2021*b*, Chapter 3, this Memoir). The Cadomian structures formed during NE–SW-directed compression and are caused by NE–SW-ward convergence of a microplate with Arabia or with a subduction zone. This subduction zone would have consumed oceanic lithosphere of the Proto-Tethys (Callegari *et al.* 2020). Northern Oman was possibly part of a Cadomian fold-and-thrust belt at some time after the deposition of the Fara Formation (*c.* 542 Ma) and before the Angudan Orogeny (525 ± 5 Ma; fig. 3.6).

During the Early Cambrian, a NW–SE-directed (present cardinal directions) compressional event affected northern Oman, associated with the final collision of East and West Gondwana (*c.* 550–510 Ma; Angudan Unconformity; D2 in fig. A.2; Loosveld *et al.* 1996; Al-Husseini 2000; Immerz *et al.* 2000; Koopman *et al.* 2007; Jacobs *et al.* 2008, fig. 1a; Forbes *et al.* 2010; Al-Kindy and Richard 2014; Droste 2014, fig. 12a). The Angudan Unconformity is named after the well Angudan-1 in southern Oman (Droste, H. 2018, pers. comm.). The Angudan Unconformity can be correlated with a major tectonic belt framing the western margin of the South Oman Salt Basin ('Western Deformation Front' of Bowring *et al.* 2007). The Angudan Unconformity has been widely observed as an onlap surface at the base of the Haima Supergroup (Droste, H. 2018, pers. comm.). Al-Husseini (2014) correlated the Angudan Unconformity with the Sub-Siq Unconformity, the 'lower Cambrian peneplain' or 'Afro-Arabian Peneplain' that has been recognized throughout the Middle East and North Africa (e.g. Stern *et al.* 2006; Miller *et al.* 2008, respectively). The age for the Angudan Unconformity is 525 ± 5 Ma according to Al-Husseini (2014). The Angudan Unconformity resulted from transpressional stress during later stages of the Malagasy Orogeny (East African Orogeny) that lasted from 550 to 510 Ma (Immerz *et al.* 2000; Koopman *et al.* 2007; Droste 2014, fig. 6a).

Units above the Angudan Unconformity

In the JAD, no rocks of the Haima Supergroup (above the Angudan Unconformity) occur, but in the SHD the Paleozoic is represented by the *c.* 3.4-km-thick Amdeh Formation, consisting of mostly shallow-marine, coastal deposits (Lovelock *et al.* 1981; figs A.1 & A.2). The Amdeh Formation occurs in the projected extension of the Ghaba Salt Basin (Mount *et al.* 1998; Rickards *et al.* 2010; Heward and Penney 2014) of upper Precambrian–lower Cambrian salt of the Ara Group (Mattes and Conway Morris 1990). However, no evaporites or residues lying below the Amdeh Formation have been mapped in the SHD (Le Métour *et al.* 1986; Villey *et al.* 1986). There may have been salt that was removed before Amdeh Formation deposition, and the Amdeh Formation may have been deposited on the flank of the Ghaba Salt Basin (Droste, H. 2018, pers. comm.).

According to Oterdoom *et al.* (1999), the Huqf area recorded a Middle Ordovician rift pulse or shoulder uplift, and they correlated the rift region with the coeval Tabas Graben of Iran. Such a NNW- to NNE-trending rift zone straddling the Huqf area and the Tabas Graben implies that deposition of the Amdeh Formation was part of this tectonic scenario. Rifting represents a break-up period with local sinistral transtension of eastern Gondwana (Oterdoom *et al.* 1999, fig. 19) and could be one possible mode of subsidence to accumulate the thick Amdeh Formation. Another possible mode could be the withdrawal of the underlying salt of the Ghaba Salt Basin (Heward and Penney 2014). However, no proof of any evaporites within the SHD has been documented (Heward and Penney 2014). In addition, the deep South Oman and Ghaba Salt basins mark a zone of terranes that were accreted during the Neoproterozoic Pan-African Orogeny, consisting of an accretionary wedge, as well as of ophiolitic, magmatic arc and microcontinental terranes (Stuart-Smith *et al.* 2007). According to Stuart-Smith *et al.* (2007), the composition and structural grain, including the terrane boundaries, had a repeated and important influence on the Late Precambrian and Phanerozoic sedimentation and basin evolution (Stuart-Smith *et al.* 2007). We are unaware whether (or what quantity of) dense ophiolite units occur in this suture zone that could cause gravity-driven subsidence.

A possible mid-Paleozoic uplift of Oman was caused by the breakaway of the Cimmeria Superterrane (Sanadaj-Sirjan, NW Iran, Central Iran) from Gondwana during the Mid Silurian–Mid Devonian (Stampfli and Borel 2002; Ruban *et al.* 2007; Torsvik and Cocks 2017). This could explain why so few Middle Paleozoic rocks occur in the study area.

'Hercynian' deformation

Units 1–8 are separated by the Hercynian Unconformity from the younger sequences (fig. 2.2; D3 in fig. A.2). Alternative terms for the Hercynian Unconformity could be 'pre-Late Carboniferous unconformity' or 'mid-Carboniferous unconformity/tectonic event' as applied by Al-Husseini (2004) for the more complete Paleozoic stratigraphy of Saudi Arabian. Nevertheless, we keep the expression 'Hercynian Unconformity'. Three different explanations for this unconformity have been suggested.

According to Beurrier *et al.* (1986), the unconformity correlates with the 'Hercynian Orogeny', based on K/Ar crystallization ages (327 ± 16 Ma, Glennie *et al.* 1974; 329 ± 11 and 321 ± 10 Ma; Beurrier *et al.* 1986) from the Amdeh and Mu'aydin formations, respectively, and NE–SW-oriented folds within the pre-Permian rocks of the JAD (Beurrier *et al.* 1986). Unfortunately, Glennie *et al.* (1974) did not provide the dated mineral (chlorite, white mica or biotite?) and the host rock (Amdeh quartzites or porphyrites?) from the SHD. The dated mineral from the Mu'aydin Formation is chlorite (Beurrier *et al.* 1986). Oman was remotely positioned from the Gondwana–Laurasia collision zone (Konert *et al.* 2001; Guiraud *et al.* 2005). According to Konert *et al.* (2001), the Arabian Plate was affected by the 'Hercynian Orogeny' from the Late Devonian to the Early Carboniferous. This NW–SE-directed compressional event resulted in uplift of several kilometres of sedimentary rocks (Konert *et al.* 2001), which could explain why no pre-Permian Paleozoic rocks are present in the JAD.

On the other hand, Ruban *et al.* (2007) pointed out that any sign of 'Hercynian' deformation is not present in the Arabian Peninsula, since this orogeny is the result of the collision between terranes of the Armorica Group and Laurasia thousands of kilometres away. A correlation between the 'Hercynian Orogeny' and deformation in the Arabian Plate could therefore be misleading (Ruban *et al.* 2007). Furthermore, the major stress direction during both the Angudan and 'Hercynian' orogenies was oriented approximately SE–NW and thus sub-parallel (present cardinal directions; e.g. Stampfli and Borel 2002; Jacobs *et al.* 2008). The regional-scale folds in the western JAD with a NE–SW-orientation (fig. A.1) could therefore have been formed during the Angudan Orogeny and not during the 'Hercynian' Orogeny, as previously thought (e.g. Beurrier *et al.* 1986). A similar tectonic setting could be attributed to the NNE-oriented folds in the Hatat Formation of the SHD (figs 4.1, 4.2 & 4.3).

The 'Hercynian' event within the Arabian Plate may have been a thermal event without orogenic deformation (Abbo *et al.* 2018). Only large-scale Paleozoic arches and basins, as reported by Faqira *et al.* (2009) and Stewart (2016), as well as widespread unconformities, probably resulted from a mantle disturbance (Abbo *et al.* 2018). This view entails that the pre-Permian NE-oriented folds of Neoproterozoic rocks within the JAD and SHD are the result of the East African/Angudan Orogeny (NW–SE-directed compression). A post-East African Orogeny break-off of the Cimmeria Superterrane from Gondwana resulted in deformation during the mid-Carboniferous (Ruban *et al.* 2007). This deformation is caused by compressional subduction–initiation at the margin of the Cimmeria Superterrane (Ruban *et al.* 2007). This event could therefore be responsible for a pre-Permian updoming, causing an unconformity in the Arabian Plate and within the JAD and SHD (Ruban *et al.* 2007), but this cannot explain the mineral ages.

A further possibility for the occurrence of the Hercynian Unconformity may have been tilting and block faulting caused by rift shoulder uplift from the opening of the Neo-Tethys during the Late Paleozoic (Rabu *et al.* 1990; Pillevuit 1993; Al-Belushi *et al.* 1996; Chauvet *et al.* 2009). A combination of block faulting related to Neo-Tethys Ocean opening with one of the two previous explanations is possible.

In any case, it needs to be mentioned that the idea of absent 'Hercynian' folding is not fully accepted, but that it is largely accepted that the unconformity and arch formation on the

Arabian Peninsula is attributed to 'Hercynian' deformation (e.g. Faqira *et al.* 2009; Stewart 2016; Abbo *et al.* 2018). A thermal 'Hercynian' event may explain the abovementioned mineral ages.

Post-'Hercynian' deformation

Pangea rifting and formation of the Neo-Tethys Ocean and Hawasina Basin

Pulsed Neo-Tethys rifting affected the Arabian Plate during the Permian and formed the North Oman passive Neo-Tethys continental margin (Blendinger *et al.* 1990; Robertson and Searle 1990; D4 in fig. A.2). The Neo-Tethys Ocean separated Gondwana from the Cimmerian continental blocks, including the Alborz, Lut and Afghan terranes (e.g. Stampfli and Borel 2002; Torsvik and Cocks 2017). According to different authors, the onset of Neo-Tethys rifting between Arabia and the Cimmerian continental blocks dates as either Late Carboniferous (Gray *et al.* 2005*a*), Early Permian (e.g. Béchennec *et al.* 1990; Blendinger *et al.* 1990; Stampfli *et al.* 1991; Pillevuit *et al.* 1997; Angiolini *et al.* 2003*a*, *b*) or Mid Permian (Murghabian in Béchennec *et al.* 1988; Chauvet *et al.* 2009; fig. A.3).

The timing of this rifting is a complex issue for several reasons. The first reason is the occurrence of several Late Carboniferous–Permian stratigraphic unconformities within southeastern Oman (Blendinger *et al.* 1990; Al-Belushi *et al.* 1996; Angiolini *et al.* 2003*a*, *b*; Chauvet *et al.* 2009). Late Paleozoic doming of the present Oman Mountains is indicated by the northward thinning of the continental clastic beds of the Late Carboniferous and Permian Haushi Group in the subsurface (e.g. Blendinger *et al.* 1990). Much of the Southeastern and Central Oman Mountains were uplifted and eroded during the Early Permian, and the same area had turned into a shallow-marine carbonate shelf during the Mid Permian (Blendinger *et al.* 1990, figs 12 & 13). However, the Late Carboniferous U–Pb crystallization ages of igneous zircons (298 ± 3 Ma) in the eastern SHD could indicate a pre-Permian onset of rifting (Gray *et al.* 2005*a*). These zircons are from felsic schists, infolded with mafic schist from the eastern SHD which underwent Late Cretaceous high-pressure/low-temperature metamorphism. These geological findings and the Mid Permian transgression of the basal Saiq Formation onto Paleozoic rocks are compatible with the palaeogeographic reconstruction of Stampfli and Borel (2002), according to which today's northeastern margin of Arabia was affected by Early Permian (Sakmarian) rifting that led to the opening of the Neo-Tethys Ocean during the Mid Permian (Wordian).

Second, Oman was situated near a triple junction at the southern margin of the Neo-Tethys Ocean and an Early Permian NE-striking rift between the Arabian and Indian plates (e.g. Stampfli *et al.* 1991; Chauvet *et al.* 2009). The break-up of Gondwana resulted in the formation of two basins in Oman. These are the older Batain Basin between the Arabian and Indian plates, and the younger Hawasina Basin between Gondwana and the Cimmerian continental blocks (Immenhauser *et al.* 2000). The sedimentary and volcanic rocks of the Batain Group are of Early Permian to latest Cretaceous age (Immenhauser *et al.* 1998, 2000; see 'Summary of the post-'Hercynian' tectonic evolution'), while the rocks of the Hawasina Basin date as Mid Permian to Upper Cretaceous (e.g, Béchennec *et al.* 1992; see section on 'Hawasina nappes' in Scharf *et al.* (2021*c*, Chapter 2, this Memoir)).

Third, Mid–Late Permian radiolarian-, ammonoid- and conodont-bearing pelagic sedimentary rocks of the Hawasina nappes rest on mafic rocks of the Hamrat Duru Basin and the Baid Formation (e.g. Blendinger *et al.* 1990; Béchennec *et al.* 1992; Kozur *et al.* 2001; Chauvet *et al.* 2009). This indicates that Murghabian thermal subsidence followed pre-Murghabian opening of the Neo-Tethys Ocean (e.g. Angiolini *et al.* 2003*a*, *b*; fig. A.3).

Fourth and finally, Chauvet *et al.* (2009) proposed a Murghabian (Mid Permian) main stretching episode of the Arabian margin, associated with volcanism in the eastern SHD. This interpretation is based on the fact that the post-Murhgabian sedimentary rocks show little facies and thickness variations (fig. A.3). Chauvet *et al.* (2009) therefore suggested that the Arabian margin underwent a short rifting episode of probable ≤ 5 Ma during the Murghabian. The following drift stage started opening the Neo-Tethys Ocean. Chauvet *et al.* (2009) pointed out that volcanism in the SHD is contemporaneous with widespread voluminous magma eruptions on the Arabian and Indian margins (Panjal Traps event).

Extension between Arabia and India led to seafloor spreading of the Proto-Indian Ocean (or SoMoMa) only from the Late Jurassic–Early Cretaceous (Immenhauser *et al.* 2000; Angiolini *et al.* 2003*a*; see 'Summary of the post-' Hercynian' tectonic evolution'). In the SHD and JAD (Saiq Plateau) areas, Early Permian igneous rocks occur (Blendinger *et al.* 1990) within the marine Permian Saiq Formation (Le Métour *et al.* 1986; Béchennec *et al.* 1992). The geochemistry of these volcanic rocks, which include dykes and sills, is mostly alkaline and related to continental rifting (Le Métour *et al.* 1986; Béchennec *et al.* 1992). These volcanic rocks are abundant in the SHD but scarce in the JAD (Le Métour *et al.* 1986; Béchennec *et al.* 1992). This distribution can be explained by a greater distance of the Jabal Akhdar area to the central rift axis than the Saih Hatat area (see Stampfli and Borel 2002, fig. 6). The greater thicknesses of the Saiq and Mahil formations in the Saih Hatat area in comparison to those in the Jabal Akhdar area may also be explained by the different distances to the central rift axis. Being located relatively close to the rift axis, the Saih Hatat area was heated more than the Jabal Akhdar area and therefore had a much more pronounced history of cooling and related thermal subsidence that subsequently resulted in more accommodation space. In addition, horsts and grabens have been identified within the shallow-marine Saiq Formation (e.g. Le Métour *et al.* 1986; Béchennec *et al.* 1992), indicating submarine rifting (Le Métour *et al.* 1986). Generally, conventional rifting of continental lithosphere below a shallow-marine sea-level is not typical because of the vertical extent of the rift dome that raises major parts of the rift area above sea-level (review in Mattern *et al.* 1998); however, in this case the great distance to the central rift axis may provide an explanation as, over a distance, the thermal rift dome diminishes.

The oldest rocks in the Hawasina nappes belong to the lower part of the Al-Jil Formation (unit 18) of the Hamrat Duru Group and are of Wordian age (Murghabian in Béchennec *et al.* 1992; fig. A.3). These rocks are mainly basaltic pillows with dykes and sills of massive dolerite and porphyritic andesite (Béchennec *et al.* 1992). The geochemistry of the volcanic rocks ranges from predominantly alkaline to tholeiitic (Béchennec 1987; Béchennec *et al.* 1992). The oldest deep-water sediments also occur within the Al-Jil Formation. These are radiolarian limestones of Permian–Mid Triassic age (Béchennec *et al.* 1992). Upper Permian radiolarian chert has been found at the base of the Hamrat Duru Group (de Wever *et al.* 1988; Cooper 1990). The type of crust that these cherts accumulated is under debate. The base for the deep-sea sediments is either stretched continental crust (e.g. Béchennec *et al.* 1990; Maury *et al.* 2003), or mid-ocean ridge basalt and therefore a fully-developed oceanic crust (Glennie *et al.* 1974; Pillevuit *et al.* 1997).

The Baid Formation (unit 26, Kawr Group) includes shallow-marine fusulinid-bearing and condensed cephalopod-bearing carbonates (Béchennec *et al.* 1992) that were deposited on a topographic high (fig. 2.6). Béchennec *et al.* (1990) interpreted this topographic high as a horst block of the extended Arabian passive margin (see also Bernoulli *et al.* 1990). The topographic high existed during the Wuchiapingian (Late Permian) as indicated by the shallow-marine limestone (Pillevuit *et al.* 1997). During the Jurassic, deep-water sediments of the Nadan Formation accumulated (Pillevuit *et al.* 1997), indicating thermal subsidence (Searle 2007) of a drowned Triassic platform.

The oldest sediments of the distal Umar Basin (Umar Group of the Hawasina nappes or Haybi Complex thrust sheet) belong to the Triassic Sinni Formation and are red shales, containing scattered blocks of white reefal limestone (Béchennec *et al.* 1992). The shales rest with stratigraphic contact on pillow basalt. These geological findings led Béchennec *et al.* (1990) to propose two stages of rifting and/or extension. A first rifting event opened the Hamrat Duru Sub-basin of the Hawasina Basin during the Permian and a second event affected the Hawasina Basin during the Mid–Late Triassic (Béchennec *et al.* 1990). This Mesozoic rifting event is accompanied by significant magmatic activity, the break-up of the Baid Platform and the formation of the Al-Aridh Trough, Kawr Horst and the Umar Sub-basin (e.g. Béchennec *et al.* 1990; Pillevuit *et al.* 1997). The occurrence of alkaline, transitional and some tholeiitic basalts in the Umar Basin implies that rifting locally reached the stage of ocean-floor formation, which was later aborted (Béchennec *et al.* 1990). Both Hawasina sub-basins were separated by a narrow chain of seamounts (Hanna 1995) or horsts (Béchennec *et al.* 1990), capped with reef and forereef carbonates, known as the 'Oman Exotics' (Searle and Graham 1982). The cause of this Mid–Late Triassic rifting episode is not well understood. Stampfli *et al.* (1991) did not support a Triassic rifting stage in the eastern Oman region. They suggested that Triassic rifting may have occurred in North Oman, assuming the Musandam Peninsula was a transfer zone between the drifting Iranian and Arabian plates.

In summary, the above aspects support the idea of either Late Carboniferous or Early or Mid Permian rifting, followed by the opening of the Neo-Tethys during the Mid Permian. The Arabian passive continental margin and the adjacent oceanic lithosphere of the Neo-Tethys Ocean had formed by then, the Hawasina Basin being part of it (e.g. Béchennec *et al.* 1990, 1991; Bernoulli *et al.* 1990; Stampfli and Borel 2002; Blechschmidt *et al.* 2004; fig. 2.6). As shown by radiolarites, parts of the Hawasina Basin must have already been a deep basin during the Late Permian.

Arabian Platform

Following Neo-Tethys rifting, the Arabian passive continental margin was characterized by a period of tectonic quiescence and accumulation of a thick succession of shallow-marine shelf limestones as can be seen in the JAD and SHD. However, the Mesozoic stratigraphic record is incomplete as there are some Lower Jurassic deposits missing (Béchennec *et al.* 1992; Schlaich *et al.* 2021). Note that Schlaich *et al.* (2021) provided different ages for the Lower Mafraq Formation in the JAD (based on biostratigraphic data from Bendias and Aigner 2015) from those published by Béchennec *et al.* (1992) and Forbes *et al.* (2010). According to Schlaich *et al.* (2021), the Lower Mafraq Formation has an Hettangian–Late Plinsbachian age and not a Pliensbachian–Toarcian age as assigned for the subsurface (Forbes *et al.* 2010; fig. A.2).

The Lower Jurassic stratigraphic gap coincides with a global low sea-level (Haq *et al.* 1987). Furthermore, the hiatus covers some 13 Ma (accepting the ages of Forbes *et al.* 2010) with extensive soil formation and is thought to be related to renewed uplift, most likely of rift shoulders (Droste, H. 2018, pers. comm.). This uplift resulted in NW-ward tilting and deep erosion at the end of the Triassic (Forbes *et al.* 2010). This unconformity progressively truncates Permo-Triassic carbonates, underlying Paleozoic clastics, Pre-Cambrian sedimentary rocks and the granitic basement towards the southeastern margin of Oman (Droste 2014). It is a major tectonic unconformity.

Sedimentation resumed during the Pliensbachian (Béchennec *et al.* 1992) or Hettangian (Schlaich *et al.* 2021). Terrigenous material from an emergent basement was deposited in the Mafraq Formation during the Pliensbachian (or Hettangian) and, within a short time, this succession became marine. The latter seems to correspond to the Pliensbachian transgression (Haq *et al.* 1987).

Marine conditions continued until the Callovian with limestone deposition of the Dhruma Formation (Forbes *et al.* 2010). Shallow-marine deposition was followed by the Middle Callovian–Kimmeridgian hiatus. Within this hiatus, the Oxfordian and Kimmeridgian sequences were probably never deposited in this area because of a lack of accommodation and/or subaerial exposure (Rousseau *et al.* 2005). Moreover, an unconformity related to tilting and erosion had formed between the Upper Jurassic and relatively deep-marine Upper Tithonian–Berriasian sedimentary rocks (Kahmah 1 & 2 Formation; Béchennec *et al.* 1992; Rousseau *et al.* 2005). Permian clasts of the Akhdar Group within the upper Guwaiza Formation of the Hamrat Duru Group are evidence for the erosion of the shelf edge of the Arabian Platform (Rousseau *et al.* 2006). This would suggest incision and/or erosion of the maximum thickness of removed strata following the uplift. Some deep canyons into the northeastern edge of the Arabian Platform may have formed (Rousseau *et al.* 2005), which were associated with the development of high-latitude continental ice sheets and sea-level fall by 40–80 m (Dromart *et al.* 2003).

Rousseau *et al.* (2005) discussed the different causes of the uplift of the margin of the Arabian Platform, including the JAD block. There are some possible causes. Rousseau *et al.* (2005) suggested that an isostatic re-equilibrium of the former thickened shelf edge resulted in the long-lasting uplift of the Arabian Platform in northern Oman. Alternatively, compressive buckling of the lithosphere could explain the uplift of the Arabian margin (Rousseau *et al.* 2005). Rousseau *et al.* (2005) are not in favour of an association between uplift and extension of the Arabian continental margin (Béchennec *et al.* 1992). This extension would correlate with the formation of new oceanic lithosphere of the SoMoMa Ocean which is now exposed on Masirah Island (Béchennec *et al.* 1992). The JAD as a tilted block is 70 km long, which is larger than conventional tilted blocks of 20 km (Rousseau *et al.* 2005). The same authors therefore proposed that a fault cutting deep into the upper mantle must exist. However, no such fault below the JAD has been substantiated yet. Nevertheless, we assume that a major fault at the present Semail Gap (eastern margin of the JAD) existed during the Late Jurassic. This pre-existing fault could have been responsible for the uplift of the eastern JAD by vertical motion.

The mechanical reasons for the downwarping of the platform during the Late Jurassic (Rabu *et al.* 1986; Pratt and Smewing 1990; D5 in fig. A.2) are still under debate. Marine conditions and limestone deposition were re-established during the latest Jurassic (Béchennec *et al.* 1992; fig. A.2). The resulting Tithonian transgression of the Rayda Formation

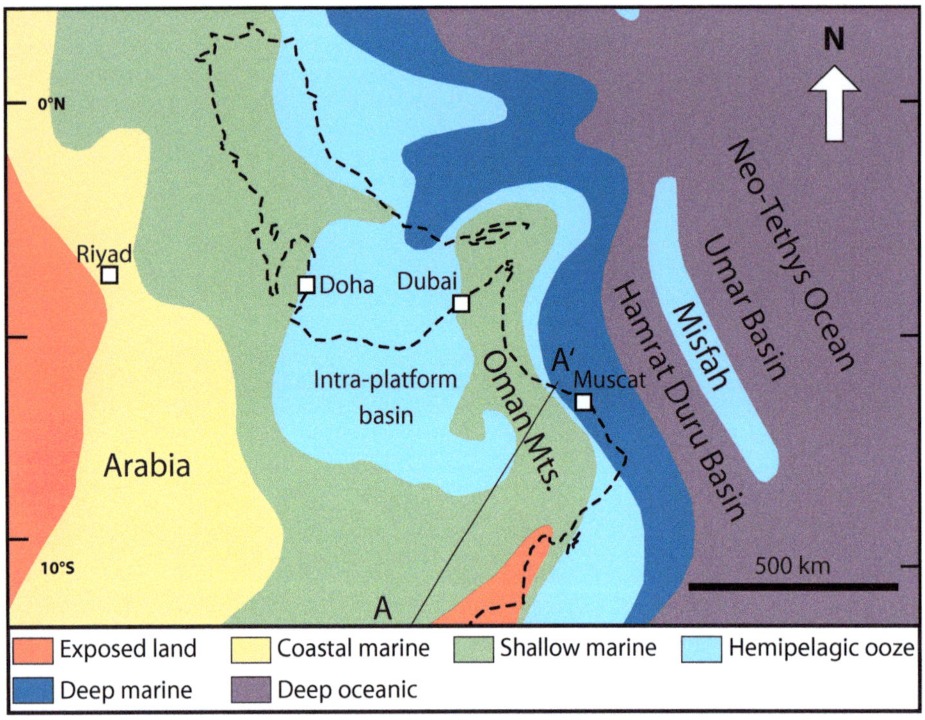

Fig. 5.1. Palaeogeographic map of northeastern Arabia during the Late Albian–Early Turonian (deposition of the Natih Formation; modified after Wohlwend *et al.* 2016). Note the trace of the cross-section A-A' (Fig. 5.2).

onto the Arabian Platform is related to a major sea-level rise (Rousseau *et al.* 2005).

During the entire Early Cretaceous, shallow-marine limestones accumulated (Kahmah Group). Following a sea-level lowstand during the Early Albian, renewed transgression during the Mid–Late Albian resulted in deposition of fine-grained, mixed terrigenous-carbonate sediments (mostly marl) of the Nahr Umr Formation (Béchennec *et al.* 1992).

Droste (2010) analysed the Cretaceous epeiric platform in Oman using seismic data. Parts of the Arabian Platform were an epeiric carbonate platform during the Cretaceous (Fig. 5.1). The following text is a summary of Droste's (2010) findings, including other key references. Epeiric platforms formed in continental shallow waters and displayed a vast lateral extent of hundreds to thousands of kilometres (Tucker and Wright 1990). In Oman, the Shu'aiba Member of the Kahmah 3 Formation and the Natih Formation are considered to have formed within a large carbonate platform system during the Cretaceous (e.g. Hughes Clarke 1988; Pratt and Smewing 1990; Droste and van Steenwinkel 2004; Grélaud *et al.* 2010). The Cretaceous platform sediments prograded *c.* 250 km above the tilted and drowned Middle Jurassic carbonate platform. Progradation ceased during the Hauterivian–Barremian, while an aggradational stage started as a result of a probably rising relative sea-level. This rise resulted in the development of an epeiric Arabian shelf for *c.* 40 Ma until the Turonian. The entire package of carbonates reached up to 700 m in thickness with internal cycles of deepening and shallowing-upward sequences (Fig. 5.2). These cycles exhibit a thickness from several tens of metres to 150 m, and are the response of the overall increase in accommodation space with sea-level fluctuation (van Buchem *et al.* 2002*a*, *b*, Droste and van Steenwinkel 2004; Fig. 5.2). The maximum water depth of deposition of sedimentary rocks during such cycles was 80 m.

Based on seismic data, Droste (2010) could confirm a sequence stratigraphic architecture of the epeiric carbonate platform of earlier studies such as Hughes Clarke (1988), Pratt and Smewing (1990) and Droste and van Steenwinkel (2004). Furthermore, Droste (2010) revealed a pronounced

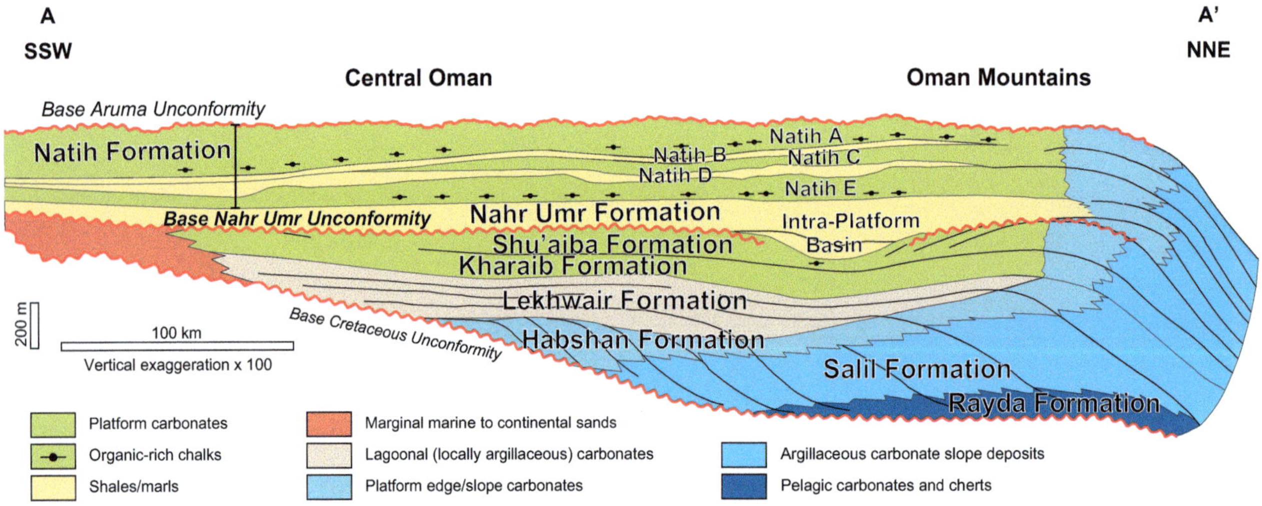

Fig. 5.2. Cross-section of the epeiric Cretaceous platform. See Figure 5.1 for the trace. Modified from Wohlwend *et al.* (2016), after Droste and van Steenwinkel (2004) and Homewood *et al.* (2008).

and complex lateral heterogeneity within the platform. During a transgressive stage, differential vertical carbonate growth was followed by lateral growth during a progradation stage. Intra-platform or intra-shelf basins with restricted bottom-water circulation formed during transgression (Fig. 5.1). These basins were eventually filled with organic-rich sediments. Regression led to exposure and bedrock incisions (Droste 2010; Grélaud *et al.* 2010).

Oblique collision of the Arabian and Indian plates during the Santonian–Campanian (*c.* 85–75 Ma), possibly until the Early Paleogene, created a stress field with NW–SE-directed compression, affecting large parts of northern Oman (Filbrandt *et al.* 2006; Al-Kindy and Richard 2014). In the course of this event, two sets of faults formed south of the Oman Mountains. One set strikes at *c.* 110–120° (WNW–ESE, with dominant right-stepping en echelon faults) and a second set strikes at *c.* 140–150° (NW–SE, with dominant left-stepping en echelon faults; Al-Kindy and Richard 2014). These two fault sets are conjugate and form a fish-net pattern of narrow grabens with small vertical displacements (Al-Kindy and Richard 2014). This fish-net pattern is interpreted to have formed during a thin-skinned transtensional tectonic regime. The resulting overall maximum horizontal stress was oriented 125° during the Late Cretaceous (Al-Kindy and Richard 2014). This event may have caused uplift of parts of the Arabian Platform (e.g. Lekhwair High; Filbrandt *et al.* 2006; Al-Kindy and Richard 2014).

Sets of *c.* 100° and *c.* 130° oriented strike-slip faults within the Natih Formation and the whole underlain sequence throughout Oman and Arabia, seen in both the surface and subsurface, with a stratigraphic offset of up to several hundreds of metres on the southern flank of the JAD (Holland *et al.* 2009*b*) are probably a result of the same NW–SE-directed compressional stress field (Filbrandt *et al.* 2006; Al-Kindy and Richard 2014). Loosveld *et al.* (1996) explained the formation of Late Cretaceous NW-striking transtensional faults with downbending of the Arabian Platform towards the Aruma Foreland Basin, as a result of SW-directed thrusting of the Semail Ophiolite.

Intraoceanic thrusting, formation and obduction of the Semail Ophiolite

Obduction is a process that transports dense oceanic lithosphere on top of light continental lithosphere (e.g. Coleman 1971; Moores and Vine 1971). Such a process occurred over a distance of thousands of kilometres at the northeastern Arabian Plate from Turkey to Oman during the Late Cretaceous (e.g. Ricou 1971). The Semail Ophiolite and the obduction processes have been intensively studied. However, many aspects of the obduction are still under debate and are a subject of ongoing research. The latter provides insights into surprisingly new geodynamic conditions during intraoceanic thrusting and obduction (see below). Intraoceanic thrusting and obduction of the Semail Ophiolite can be divided into four stages (Fig. 5.3) as follows.

Stage I: >104–96 Ma (Fig. 5.3a). The timing and conditions (either spontaneous or induced) of initiation of intraoceanic subduction is under debate. Recent investigations have shown that induced subduction may have started as early as *c.* 104 Ma (Guilmette *et al.* 2018). Other authors constrain the age for the early stage of subduction to *c.* 96.2–94.8 Ma (e.g. Rioux *et al.* 2016). Furthermore, the initial orientation of the subduction zone was probably striking NNE–SSW and not ESE–WNW as previously thought (van Hinsbergen *et al.* 2019).

Stage II: *c.* 96–94 Ma (Fig. 5.3b). Self-sustained subduction is characterized by the formation of new oceanic lithosphere

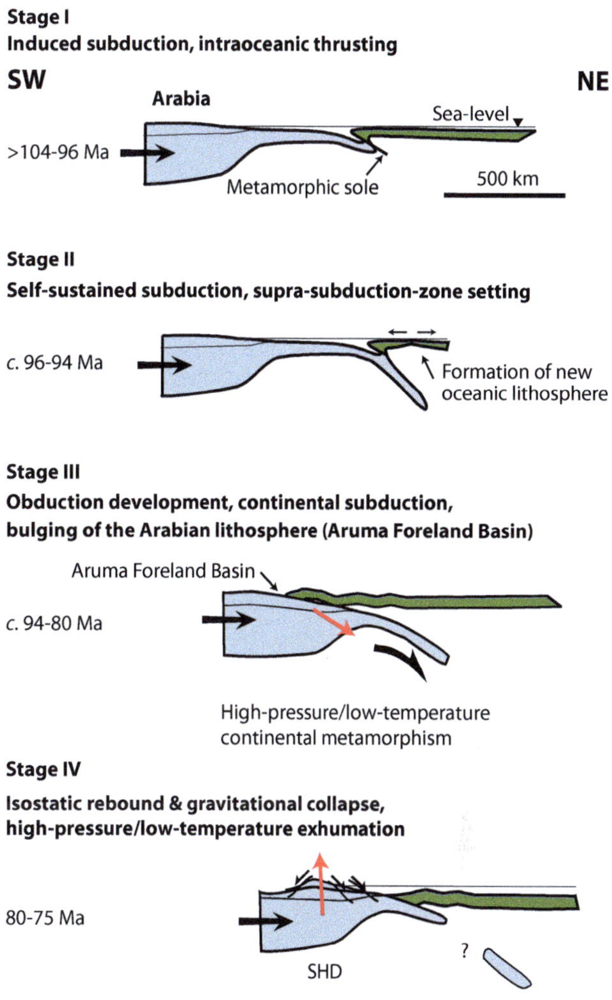

Fig. 5.3. The four stages of intraoceanic thrusting and obduction process during the Cretaceous. Modified from Agard *et al.* (2007, 2010) and Tavani *et al.* (2021).

and a metamorphic sole in a supra-subduction zone setting. This stage also involves intraoceanic thrusting, which eventually led to obduction.

Stage III: *c.* 94–80 Ma (Fig. 5.3c). Eventually, obduction manifested itself with further thrusting of the ophiolite onto Hawasina rocks and later together onto the Arabian Platform. Probably thinned Arabian continental lithosphere underlies the Hawasina Basin. Internal deformation of the Hawasina rocks was thin-skinned. Furthermore, the continental margin was dragged into a subduction zone which resulted in high-pressure/low-temperature metamorphism of the Arabian continental rocks. Recent studies had shown that the ophiolite was rotating in a clockwise fashion during progressing obduction (Morris *et al.* 2016; van Hinsbergen *et al.* 2019). Whether one or two subduction zones existed, and the direction of the subduction (see below), is under debate. The load of the allochthonous units triggered formation of a foredeep and creation of a forebulge exhuming parts of the Arabian lithosphere. As a result, the Aruma Foreland Basin formed between the forebulge and the advancing allochthonous units.

Stage IV: 80–75 Ma (Fig. 5.3d). The last step is the termination of obduction which may be characterized by the rapid exhumation of the continental lithosphere from the subduction zone. A model which provided the reason for the rapid exhumation of the continental lithosphere is the slab break-off (Searle *et al.* 2004), and subsequent isostatic rebound and gravitational collapse (e.g. Mattern *et al.* 2018*b*; Mattern and Scharf 2019). The obduction and exhumation-related structures and formations were sealed by the post-obductional formations of the Aruma Group during the Maastrichtian.

Stage I: subduction initiation. Initiation of subduction, which eventually led to formation of the Semail Ophiolite is a topic under debate (e.g. Stern 2004; Duretz *et al.* 2016; Guilmette *et al.* 2018; Stern and Gerya 2018; Tavani *et al.* 2021). Two concepts have been advocated: 'spontaneous' and 'induced' subduction initiation (e.g. Guilmette *et al.* 2018). In the case of the spontaneous initiation, the subducted material (lower plate) must be contemporaneously replaced by material during extension of the upper plate (Stern and Bloomer 1992; Stern *et al.* 2012). On the other hand, if subduction initiation was induced, subduction started first, followed by extension in the upper plate (Hall *et al.* 2003: Gurnis *et al.* 2004; van Hinsbergen *et al.* 2015). This model requires a time lag of a few million years between subduction and upper-plate extension (Guilmette *et al.* 2018).

Stern (2004) and Whattam and Stern (2011) suggested that the Semail Ophiolite may have formed due to spontaneous subduction initiation. The initiation likely started at a transform fault or fracture zone which separated rocks of major age difference and density contrast. Numerical models provide a timespan of 1.5 Ma for the development of such a subduction zone and formation of a spreading axis (Gerya 2011, fig. 2.2).

Induced subduction requires converging forces between plates (Stern 2004). Eventually, induced subduction becomes self-sustained once a critical amount of oceaninc lithosphere has been consumed (Stern 2004). The transition between induced and self-sustained subduction is reached once the force of the dense, sinking subduction lithosphere exceeds the forces of viscous resistance of the asthenosphere and to lithospheric bending (see Tavani *et al.* 2021). During induced subduction the downgoing slab will be pushed, while it will be pulled during self-sustained subduction. This results in a major reorganization of the forces in the downgoing plate (first compressive, later extensional; see introduction of Tavani *et al.* 2021). Eventually, extensional stresses will also affect distal parts of the downgoing plate, that is, the Arabian margin and platform (Tavani *et al.* 2021).

Duretz *et al.* (2016) modelled obduction dynamics by using two-dimensional thermo-mechanical numerical models. The boundary conditions for their work were adjusted for the obduction process of the Semail Ophiolite, and subduction initiation was assumed to be induced. Their results were tested and verified against existing petrological and structural data, and show that subduction initiation may have started along a proximal thermal anomaly during compression. The thermal anomaly was caused by oceanic magmatism and within the Neo-Tethys Ocean. In this case, subduction would lead to obduction and ophiolite emplacement. Important outcomes of the modelling work of Duretz *et al.* (2016) are that obduction initiation occurred <200 km from the ocean–continent transition and that the amount of convergence amounted to *c.* 400 km.

Based on Lu–Hf geochronology on garnet from the metamorphic sole, Guilmette *et al.* (2018) concluded that subduction initiation was induced and related to far-field forces. Lu–Hf garnet ages reveal ages of *c.* 104 Ma (Guilmette *et al.* 2018). These ages indicate initial burial of the downgoing plate, pre-dating the creation of the Semail Ophiolite (*c.* 96 Ma) by a minimum of 8 Ma (Guilmette *et al.* 2018). The metamorphic sole did not receive any significant thermal or dynamic perturbation between garnet growth at *c.* 104 Ma and zircon crystallization at *c.* 96 Ma (Guilmette *et al.* 2018).

Tavani *et al.* (2021) studied extensional faults in the southern limb of the JAD and in the Northern Oman Mountains. These authors assigned an Albian–earliest Cenomanian age for these faults, which correlate in time with the transition from initiation to self-sustained subduction. This extensional faulting was therefore a product of the reorganization of forces at the transition from induced to self-sustained subduction that also affected the Arabian Platform (Tavani *et al.* 2021).

Subduction zone ridge orientation during the onset of intraoceanic thrusting was NNE-striking, and the slab was WNW-dipping (van Hinsbergen *et al.* 2019). Subduction initiation probably occurred along a NNE-striking fracture zone margin. As a consequence, subduction was induced by (ES)E–(WN)W contraction, and subduction did not initiate near the ridge of the Neo-Tethys Ocean (van Hinsbergen *et al.* 2019). During ongoing subduction after *c.* 95 Ma, the entire setting rotated clockwise to its present-day orientation (van Hinsbergen *et al.* 2019; Fig. 5.4).

Stage II: self-sustained subduction, supra-subduction zone setting, and formation of oceanic lithosphere and metamorphic sole. At *c.* 96 Ma, extension in the hanging wall resulted in the formation of oceanic lithosphere along a spreading centre that correlates with the self-sustaining subduction stage (Guilmette *et al.* 2018). During 96.2–94.5 Ma the metamorphic sole cooled from >850°C to *c.* 700°C (Guilmette *et al.* 2018). Further cooling to 650–600°C has been demonstrated by U–Pb titanite cooling ages (Guilmette *et al.* 2018). Cooling

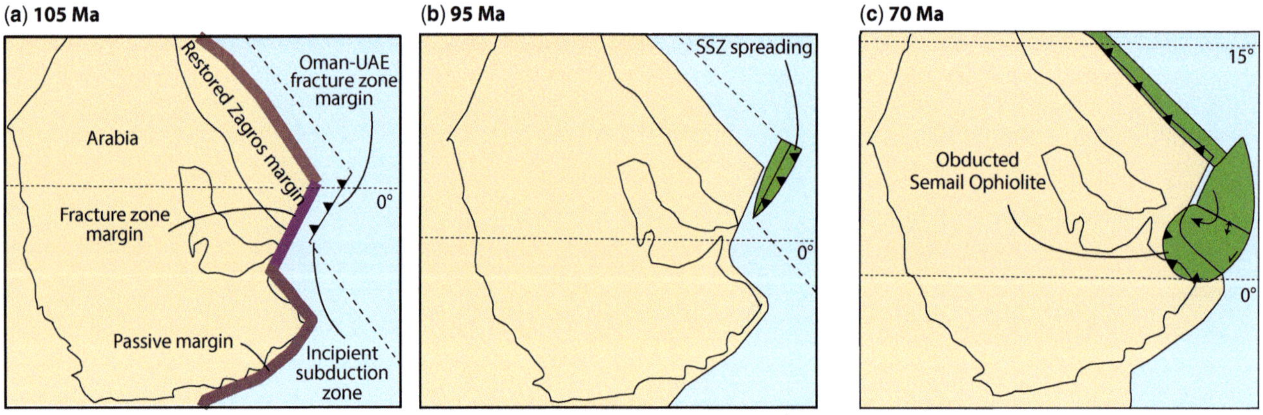

Fig. 5.4. Tectonic model of the subduction initiation along Oman, modified from van Hinsbergen *et al.* (2019). (**a**) Subduction initiation parallel and close to the Oman-UAE fracture zone margin at *c.* 105 Ma. (**b**) Forearc extension at *c.* 96–95 Ma. (**c**) Rotation of the Semail Ophiolite, followed by obduction at *c.* 70 Ma.

continued to <550°C between *c.* 95.5 and *c.* 92 Ma (Hacker *et al.* 1996).

The crustal part of the Semail Ophiolite consists of an early and late magmatic sequence (table 2.2; see section on 'Semail Ophiolite' in Chapter 2). The early magmatic sequence has a U–Pb zircon age in the north and SE of the Oman Mountains of 96.12–95.50 Ma and 96.46–95.50 Ma, respectively (Rioux *et al.* 2013, 2016; table 2.2), suggesting a uniform crustal formation of the Semail Ophiolite along its entire length (Rioux *et al.* 2016) at a spreading centre probably in a marginal basin above an infant subduction zone (e.g. MacLeod *et al.* 2013). According to MacLeod *et al.* (2013), the entire Semail Ophiolite formed above a subduction zone because the protoliths of the ophiolite formed under the presence of a high amount of fluid. This amount of fluid is higher than within conventional open-ocean mid-ocean spreading centres (MacLeod *et al.* 2013).

Rocks from the late magmatic sequence have a U–Pb zircon age of 95.74–95.26 Ma in the Northern (Goodenough *et al.* 2010) and 95.3–94.8 Ma in the Southeastern Oman Mountains (Hacker *et al.* 1996; Warren *et al.* 2005; table 2.2). Rioux *et al.* (2013, 2016) found that late gabbroic, tonalitic and trondhjemitic magmas have a U–Pb zircon age of <95.45 Ma. These rocks cut the Semail Ophiolite 0.25–0.75 Ma after the formation of the crust (Rioux *et al.* 2013). The rocks of the late magmatic sequence formed above a subducting slab (e.g. Goodenough *et al.* 2014).

The U–Pb zircon age of crustal formation (96.12–95.50 Ma) is broadly coeval with the U–Pb zircon age of the metamorphic sole (96.169 ± 0.022 to 96.146 ± 0.035 Ma in Sumeini, Northern Oman Mountains; 94.815 ± 0.030 Ma in Wadi Tayin, Southeastern Oman Mountains, Rioux *et al.* 2016; table 2.2). Some metamorphism along the northern metamorphic sole may have started prior to crustal formation or is contemporaneous with it (Rioux *et al.* 2016). There are three models for intraoceanic thrusting of the future Semail Ophiolite. The first model proposes spreading at a NW-striking, conventional mid-ocean ridge (e.g. Boudier and Coleman 1981; Boudier *et al.* 1985, 1988; Gregory *et al.* 1998) and SW thrusting of the ocean segment that lay on the distal side of the ridge with respect to Arabia (Boudier *et al.* 1985; Hacker *et al.* 1996; Fig. 5.5a).

The second model implies spreading above an intraoceanic NE-directed subduction zone where the spreading centre is located in the proximal hanging wall of the subduction zone (e.g. Searle and Malpas 1980, 1982; Pearce *et al.* 1981; Alabaster *et al.* 1982; Lippard *et al.* 1986; Hacker and Gnos 1997; Searle and Cox 1999, 2002; Warren *et al.* 2005; MacLeod *et al.* 2013; Rioux *et al.* 2013, 2016; Cowan *et al.* 2014; Goodenough *et al.* 2014; Guilmette *et al.* 2018; Fig. 5.5b). Van Hinsbergen *et al.* (2019) assume a different orientation for the subduction zone, one which is WNW-dipping (Fig. 5.4b).

Recently, a third model for thrusting of the ophiolite has been proposed by Nicolas and Boudier (2017) as well as Boudier and Nicolas (2018). Their model suggests that thrusting ensued in a context of collision of an active ridge at the ridge–trench–transform triple junction with a continent-oriented subduction, analogous to the Late Cenozoic evolution of the Gulf of California (Fig. 5.5c).

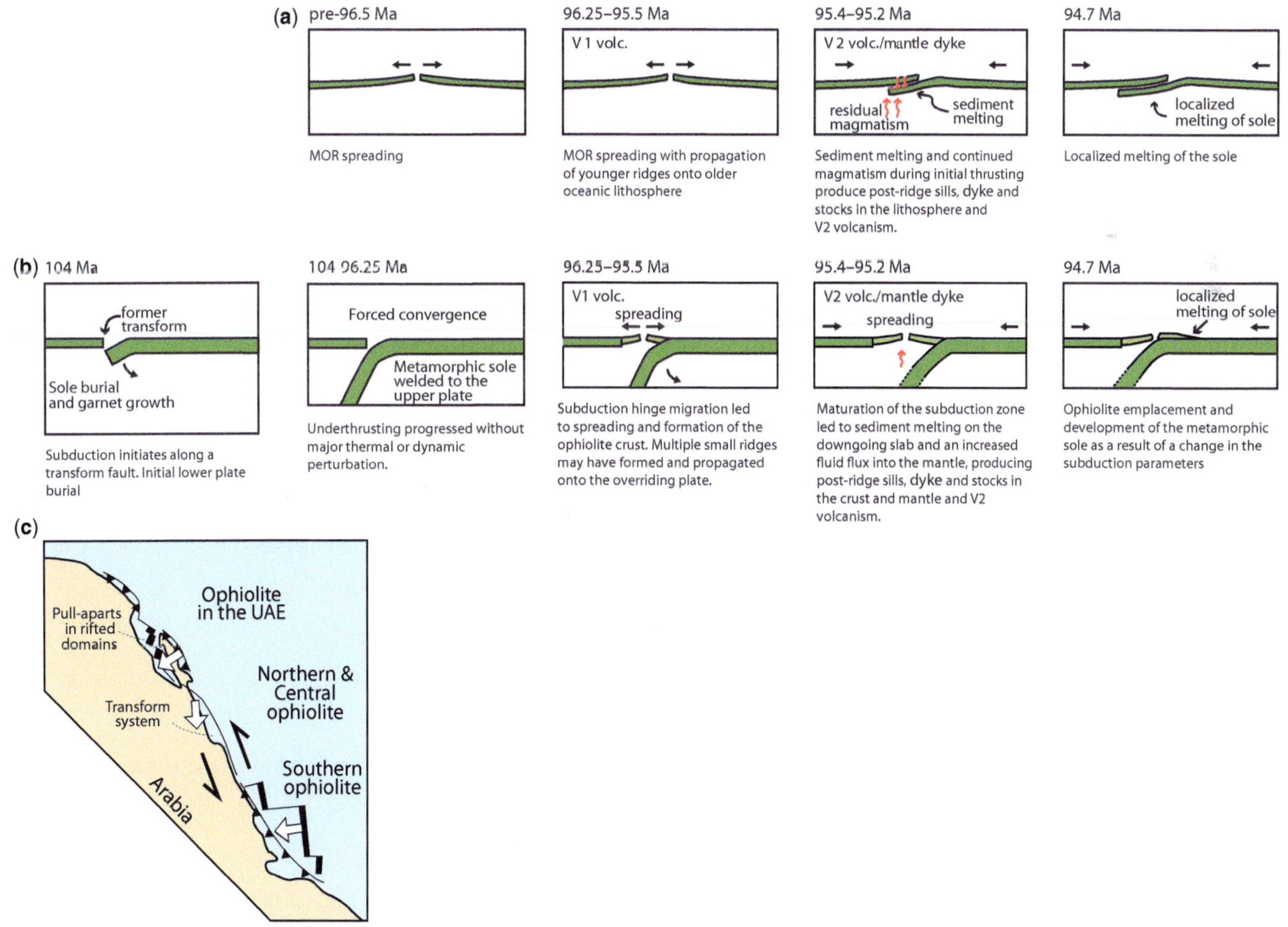

Fig. 5.5. Tectonic model for the Semail Ophiolite formation. (**a**) Formation at a mid-oceanic ridge and later thrusting along the ridge axis (modified from Rioux *et al.* 2013). (**b**) Formation in a supra-subduction zone setting (modified from Rioux *et al.* 2013 and Guilmette *et al.* 2018). (**c**) Ridge–trench–transform model (modified after Boudier and Nicolas 2018).

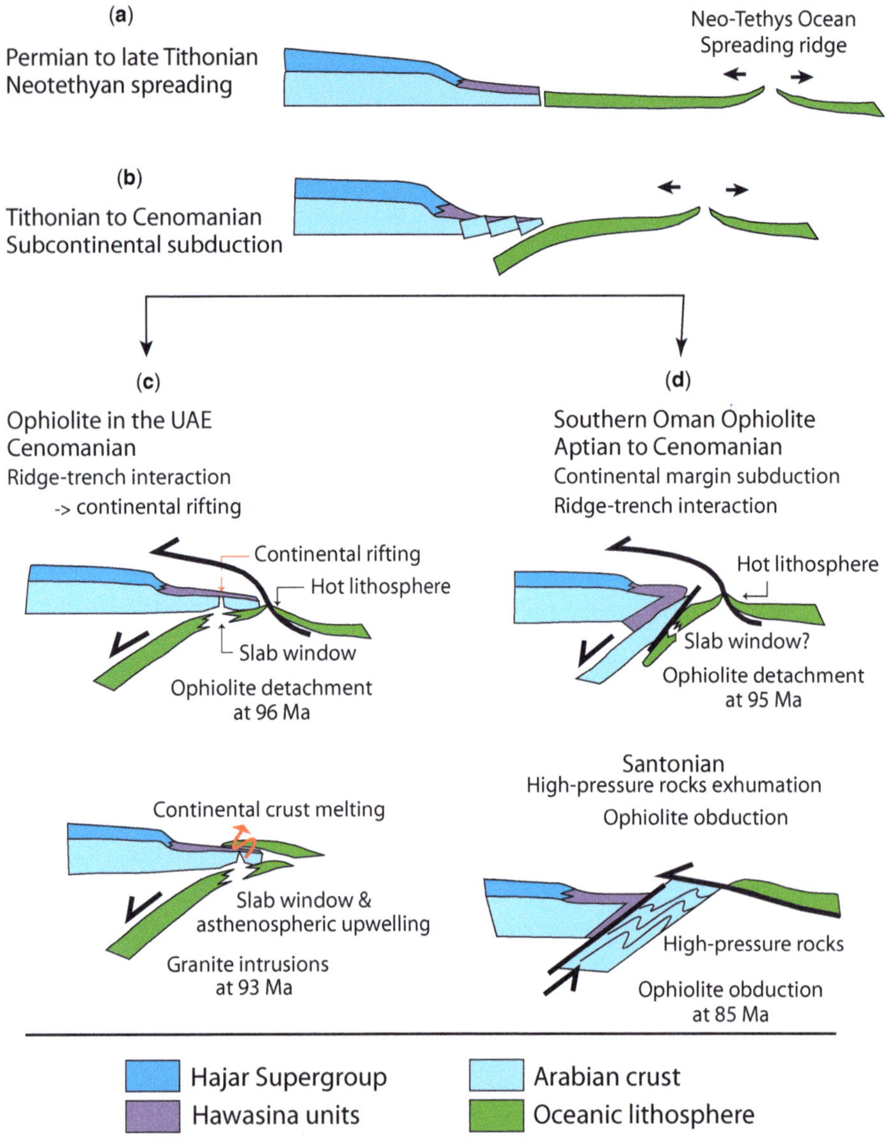

Fig. 5.6. Schematic evolution of the Semail–Arabia convergent margin after Boudier and Nicolas (2018). (**a**) Neo-Tethys oceanic lithosphere formed at impingement upon the Arabian passive margin. (**b**) Initiation of subcontinental subduction along the contact of the thinned basement of the Hamrat Duru and the oceanic lithosphere. (**c**, **d**) Evolution of obduction at the northern and southern sections of the Semail Ophiolite.

The model of Nicolas and Boudier (2017) and Boudier and Nicolas (2018) suggests an Arabian-Platform-directed subduction zone during the Tithonian–Berriasian (Fig. 5.6). The following section is a summary of their model, including other key references. The Early Cretaceous event is reflected in the 110 ± 9 Ma ages of the high-pressure rocks in the SHD (Gray *et al.* 2004*b*). The spreading ridge of the Neo-Tethys Ocean impinged on Arabia during the *c.* 20 Ma duration of subduction (Fig. 5.6). The absence of a volcanic arc could be explained by the limited subduction time. Ophiolite detachment with along-strike variations occurred during the Cenomanian.

Formation of the Semail Ophiolite within an incipient supra-subduction zone setting is favoured by most of the authors today (Fig. 5.5b).

First, only in a supra-subduction zone setting can the *c.* 104 Ma garnet ages of the metamorphic sole be explained (Guilmette *et al.* 2018). It is unlikely for garnets to form in a conventional ocean-spreading centre prior to onset of spreading.

Second, the geochemical and/or isotopic fingerprint of the minerals and rocks from the Semail Ophiolite justify a supra-subduction zone setting. MacLeod *et al.* (2013) compared modern mid-ocean ridge basalts from the East Pacific Rise, Indian Ocean ridges and the Mid-Atlantic Ridge with the V1 (Geotimes) of Oman (fig. 2.13). A clear difference has been observed between the mid-ocean ridge basalts and the V1 lava and dyke composition (MacLeod *et al.* 2013). The geochemistry of the Oman V1 rocks resembles the trend of modern and ancient supra-subduction zone volcanics and not those of mid-ocean rocks. The V1 (Geotimes) lavas and dykes of the Semail Ophiolite have on average a lower concentration of TiO_2 at a given MgO, as well as higher SiO_2 content and lower Cr content, than typical mid-ocean ridge basalts (MacLeod *et al.* 2013; Rioux *et al.* 2016). The reason for the Oman V1 supra-subduction zone geochemistry may be explained by consistently relatively high water concentrations in the magma, more than expected for conventional mid-ocean ridge basalts (MacLeod *et al.* 2013).

Further evidence for the presence of fluids in large amounts during subduction infancy has been discussed by Prigent *et al.* (2018*a*, *b*, *c*). These authors could correlate between fluid migration from the subducting plate into the basal section of the upper plate and the strain localization within a shear zone at the plate interface. The following section is a summary of some of the findings of Prigent *et al.* (2018*b*; see also Prigent *et al.* 2018*a*, *c* for further petrological and microstructural information).

The basal unit of the mantle section of the Semail Ophiolite was located at the contact between the subducting and obducting plates during subduction infancy. These rocks received a significant infiltration of fluids from the lower plate during the early stage of subduction and at temperature conditions of *c.* 900–800°C. As one consequence among others, the grain size of the peridotite was reduced during fluid interaction; shearing resulted in a change of rheology of the basal unit from the upper plate, leading to a decrease in shear strength. The fluids from the subducting plate interacted with the mantle rocks of the upper plate (basal unit) and created a localized high-strain shear zone with a width of <1 km prior to serpentinization (Prigent *et al.* 2018*c*). In turn, this changed the metamorphic sole–Semail Ophiolite mantle interface from a mechanically decoupled to a coupled plate boundary.

Some of the late magmatic rocks (V2) show that during fractionation a high water content must have been present (Alabaster *et al.* 1982), more than is available in a normal mid-ocean spreading centre (MacLeod *et al.* 2013). Moreover, the V2 magmatic rocks have a major- and trace-element geochemistry similar to lavas from an island arc and dissimilar to those of a conventional mid-ocean ridge (fig. 2.13; Pearce *et al.* 1981; Alabaster *et al.* 1982; Ishikawa *et al.* 2002; Rioux *et al.* 2016). In addition, Goodenough *et al.* (2014) showed that the oceanic crust probably formed in a marginal basin above an infant subduction zone (their phase 1; early magmatic sequence; V1), while slightly younger magmas preserved a geochemical signature typical for a supra-subduction zone setting (their phase 2; late magmatic sequence; V2; Fig. 5.7).

The ε_{Nd}(96 Ma) = 7.6 to 8.8 values for the early magmatic sequence volcanics of the ophiolite are comparable with those of modern mid-ocean ridge basalts. The similar but slightly lower ε_{Nd}(96 Ma) = 6.9 to 7.9 values for late magmatic rocks indicate that these rocks derived from a mantle source, similar to the rocks of the early magmatic sequence (Rioux *et al.* 2013). The slightly lower $\varepsilon_{Nd}(t)$ values, however, suggest the addition of a low $\varepsilon_{Nd}(t)$ component by either fluids and/or melts, derived from a subducting slab or underthrust plate (Rioux *et al.* 2013).

Some felsic intrusions cut the mantle section of the Semail Ophiolite in the Northern Oman Mountains. These magmas have a low ε_{Nd}(96 Ma) and ε_{Hf}(96 Ma) signature and formed immediately after ridge magmatism (Rioux *et al.* 2013; Haase *et al.* 2015). The felsic magmas in the Semail Ophiolite probably derived from pelagic sediments and a mixture of sediments and mafic rocks from the downgoing subducting plate (Rioux *et al.* 2013; Haase *et al.* 2015, 2016). Recently, Spencer *et al.* (2017) investigated rare S-type granite dykes from the northernmost Oman Mountains in the UAE intruding the harzburgitic portion of the Semail Ophiolite. These peraluminous rocks have a U–Pb zircon age of 99.8 ± 3.3 Ma and predate the age of the ophiolite formation (Spencer *et al.* 2017). The same rocks have significantly high $\delta^{18}O$ zircon values, which must have derived from melts of pelitic and/or siliceous mud (Spencer *et al.* 2017). According to the same authors, these sediments were subducted, became molten and subsequently intruded the mantle wedge above. All these aspects indicate that felsic material must have been transported (subducted) into the mantle, which supports the model of a supra-subduction zone setting and excludes a conventional mid-ocean ridge setting.

Third, the almost identical age of the ophiolite's crust and the metamorphic sole (e.g. Rioux *et al.* 2016) favours a supra-subduction zone setting. In a conventional mid-ocean ridge model, the formation of the ophiolite crust would pre-date thrusting.

Fourth, if intraoceanic thrusting began at a mid-ocean spreading centre, the hot basal mantle from one side of the ridge would have been thrust onto the basalt of the other side of the ridge to form a metamorphic sole (Searle and Cox 2002). In this case, the proto-basalts of the metamorphic sole would have a geochemistry similar to that of the basalts from the Semail Ophiolite. However, since the geochemistry of the metamorphic sole is close to that of the Hawasina sediments and volcanics, and is dissimilar to that of the ophiolite suite, the model of thrusting along the spreading centre must be ruled out (see Rioux *et al.* 2016). Indeed, the model of NE-ward subduction of the Hawasina nappes would best explain the observation (Searle and Cox 2002).

Fifth, the pressure conditions of the metamorphic sole of 1.1–1.3 GPa (e.g. Cowan *et al.* 2014; Soret *et al.* 2017) imply a depth of >35 km. This depth exceeds the thickness of the entire ophiolite (maximum 20 km). The metamorphic sole therefore probably formed within a subduction zone (Searle and Cox 2002; Searle *et al.* 2015; Rioux *et al.* 2016). Nicolas (1989) and Hacker and Gnos (1997) suggested that the ophiolite was initially 30–40 km thick and that the metamorphic sole was exhumed to its final position along normal faults. Most of the recent literature advocates metamorphic sole formation within a subduction zone.

The vergence of subduction is debated. A SW-dipping subduction zone towards the Arabian continent was proposed by Gregory *et al.* (1998), Gray *et al.* (2000), Gray and Gregory (2003, 2004), Searle (2007) and Boudier and Nicolas (2018), but is difficult to substantiate with field evidence (e.g. lack of subduction-related igneous rocks in the overriding Arabian Plate). The alternative idea of a NE-directed subduction has been suggested and/or supported by Searle and Malpas (1980), Lippard *et al.* (1986), Goffé *et al.* (1988), Searle *et al.* (1994, 2003), Searle and Cox (1999, 2002), El-Shazly *et al.* (2001), Breton *et al.* (2004), Agard *et al.* (2010), Rioux *et al.* (2012, 2013, 2016), MacLeod *et al.* (2013), Cowan *et al.* (2014), Goodenough *et al.* (2014) and van Hinsbergen *et al.* (2019). The NE-directed subduction interpretation fits with the absence of a magmatic arc within the Arabian Plate. It also takes into account the geochemistry of

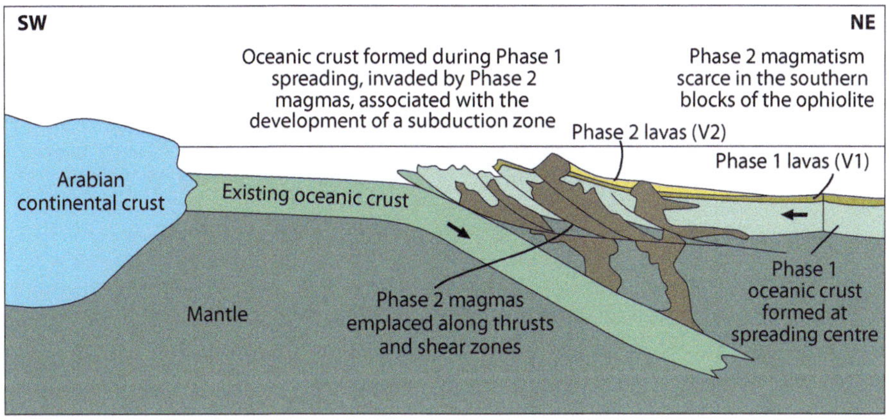

Fig. 5.7. Schematic model depicting formation of the V2 magmas above a subduction zone during the Cenomanian. Redrawn after Goodenough *et al.* (2014).

the ophiolite lavas (e.g. Pearce *et al.* 1981), and the similar ages of ophiolite formation and the metamorphic sole, which are best reconciled with ophiolite formation above a NE-dipping subduction zone (Searle and Malpas 1980; Lippard *et al.* 1986).

On the basis of $^{40}Ar/^{39}Ar$ hornblende cooling ages from the ophiolite's metamorphic sole, Hacker *et al.* (1996) calculated the timespan of 1–2 Ma for the 200-km-wide intraoceanic thrust range, which corresponds to a spreading rate of 100–200 mm a^{-1}. Rioux *et al.* (2012) quantified a half rate of spreading as 50–100 mm a^{-1}, confirming the results of Hacker *et al.* (1996). These calculations are compatible with the proposed spreading rate of >65–100 mm a^{-1}, deduced on the basis of variations in U–Pb zircon ages of plagiogranites (belonging to the ophiolite) within the outcrop zone of the Semail Ophiolite (Tilton *et al.* 1981; Hacker *et al.* 1996).

Stage III: obduction development, subduction of the Arabian margin and bulging of the Arabian lithosphere. Based on palaeomagnetic data from the volcanic and lower crustal rocks of the ophiolite, Perrin *et al.* (2000) and Morris *et al.* (2016, fig. 2) suggested clockwise rotation by as much as *c.* 150° along a vertical axis of the entire Semail Ophiolite. Morris *et al.* (2016) suggested that, during the beginning of ophiolite formation, the subduction zone hinge migration resulted in supra-subduction zone spreading and early rotation of the newly formed crust at 95–90 Ma. Ongoing N-ward movement of Arabia and hinge migration led to further clockwise rotation and probably thrusting of the rotated ophiolite onto Arabia. At *c.* 75 Ma, the emplacement in the Saih Hatat area is associated with remagnetization and anticlockwise back rotation resulting from exhumation of the SHD (Morris *et al.* 2016). Clockwise rotation of the Semail Ophiolite above a supra-subduction zone setting during obduction has been confirmed by van Hinsbergen *et al.* (2019; Fig. 5.4) on the basis of palaeomagnetic data of sheeted dykes along the entire Oman Mountains.

The downgoing oceanic slab eventually pulled the Arabian passive continental margin down to depths of high-pressure/low-temperature metamorphism ($D6_c$ in fig. A.2). Corresponding blueshist- and eclogite-facies rocks are exposed at the northeastern flank of the SHD (As-Sifah eclogites; fig. A.1).

High-pressure/low-temperature metamorphism may have occurred during the late stage of the ophiolite obduction. The interpretations of the tectonic setting during subduction and exhumation of the Arabian margin differ. Three models have been offered as follows.

First, some authors suggest that one NE-dipping subduction zone existed during the Late Cretaceous (e.g. Goffé *et al.* 1988; Searle *et al.* 2004; Agard *et al.* 2010; Rioux *et al.* 2016). The U–Pb zircon crystallization and $^{40}Ar/^{39}Ar$ phengite cooling ages for the peak high-pressure/low-temperature metamorphism of the eclogites has been determined as 78.95 ± 0.13 Ma (U–Pb zircon; Warren *et al.* 2003, 2005) and 80–70 Ma ($^{40}Ar/^{39}Ar$ phengites; Miller *et al.* 1999), respectively. Garber *et al.* (2016) dated garnets using U–Pb zircon thermal ionization mass spectrometry (TIMS) from the high-pressure units of the SHD. Their data suggest that garnet was growing between *c.* 85 and 74 Ma. Garber *et al.* (2016) concluded a subduction rate of 7 ± 2 mm a^{-1} and an exhumation of 3 ± 2 mm a^{-1} for the Arabian continental margin. The differences between subduction and exhumation rates are explained with subduction of small continental slivers followed by exhumation of a larger continental body, which is in line with subduction of extended continental crust followed by thickening (Garber *et al.* 2016).

Second, other researchers proposed a further subduction zone. This second NE-dipping intracontinental subduction zone would be part of a 'North Muscat microplate' (El-Shazly *et al.* 2001; Breton *et al.* 2004).

Third, it has been suggested that the high-pressure/low-temperature event was part of a intracontinental SW-dipping subduction at *c.* 130–90 Ma (e.g. Gregory *et al.* 1998; Gray *et al.* 2000; Gray and Gregory 2003). Miller *et al.* (1999) and Gray *et al.* (2004*a*, *b*) arrived at ages of *c.* 110 Ma for the peak pressure during subduction, using the $^{40}Ar/^{39}Ar$ method on mica and the Sm–Nd technique on garnet. Metamorphism started some 15–25 Ma before the onset of obduction.

Most researchers today advocate a single NE-dipping subduction zone during the Late Cretaceous because of the majority of Late Cretaceous ages within the subduction and/or exhumation zone, deformation pattern and pressure–temperature data (see Agard *et al.* 2010). Furthermore, Garber *et al.* (2016) concluded that the *c.* 110 Ma ages were likely inaccurate, and therefore considered these data as outdated.

During the Late Cretaceous, the Oman sector of the Arabian Platform experienced major deformation by the obduction of the Semail Ophiolite (see the section on 'Intraoceanic thrusting, formation and obduction of the Semail Ophiolite'; $D6_b$ in fig. A.2). The passive margin changed into a convergent margin, and progressive loading by the Semail Ophiolite and associated Hawasina thrust units triggered the formation of a foreland basin (Robertson 1987*a*, *b*) known as the 'Aruma Basin' or 'Aruma Foreland Basin' (e.g. Glennie *et al.* 1974; Ravaut *et al.* 1997). The thickness of the Santonian–Campanian (Searle 2007) Aruma Basin fill (Fiqa Formation), which is largely clastic (Béchennec *et al.* 1992), has been estimated to reach up to 3.5 km, with the greatest thickness towards the NE of the Oman Mountains (Searle 2007). Warburton *et al.* (1990) reported a Late Coniacian age for the oldest sediments in the foredeep based on planktonic microfaunas. The Coniacian–Santonian Muti Formation represents the lower fill of the Aruma Foreland Basin. It interfingers with the Fiqa Formation (Glennie *et al.* 1974; section on 'Aruma Group' in Chapter 2). The top of the basin is occupied by the Juweiza and Simsima formations. The foredeep was completely filled with clastic detritus during the Campanian and Early Maastrichtian (Warburton *et al.* 1990; Searle 2007, fig. 4). The first appearance of igneous ophiolite clasts in the sediments on the Arabian Platform occurred during the Mid–Late Campanian (78–71 Ma) in the Juweiza Formation (Glennie *et al.* 1974; Rabu *et al.* 1993; Hacker *et al.* 1996), which temporally coincides with the age of the high-pressure/low-temperature metamorphism.

The onset of deposition in the Aruma Foreland Basin and the collapse of the Arabian margin are marked by uplift and erosion of the forebulge (Searle 2007). Erosion of the collapsing platform succession related to the flexural uplift amounted to 200 m south of the Oman Mountains (Aldega *et al.* 2017). The uplift is a consequence of flexural bending of the Arabian Foreland (Robertson 1987*a*; Warburton *et al.* 1990; Loosveld *et al.* 1996; Filbrandt *et al.* 2006). Uplift of the flexural bend is manifested by exhumation of the northern JAD and SHD flanks (Béchennec *et al.* 1992) where the Natih and Nahr Umr formations are missing. Even greater uplift and exhumation is indicated in the realm of the SHD since the Kahmah Group is also missing there (fig. A.1).

Stage IV: exhumation of high-pressure/low-temperature rocks. Agard *et al.* (2010) introduced a model for the origin, preservation and exhumation of the high-pressure rocks in the SHD within the context of the Semail Ophiolite obduction (Fig. 5.8). This section summarizes a few key aspects of the work by Agard *et al.* (2010). Figure 5.8a depicts the palaeogeography immediately before obduction during the Late Cretaceous. Figure 5.8b presents the obduction of the Semail

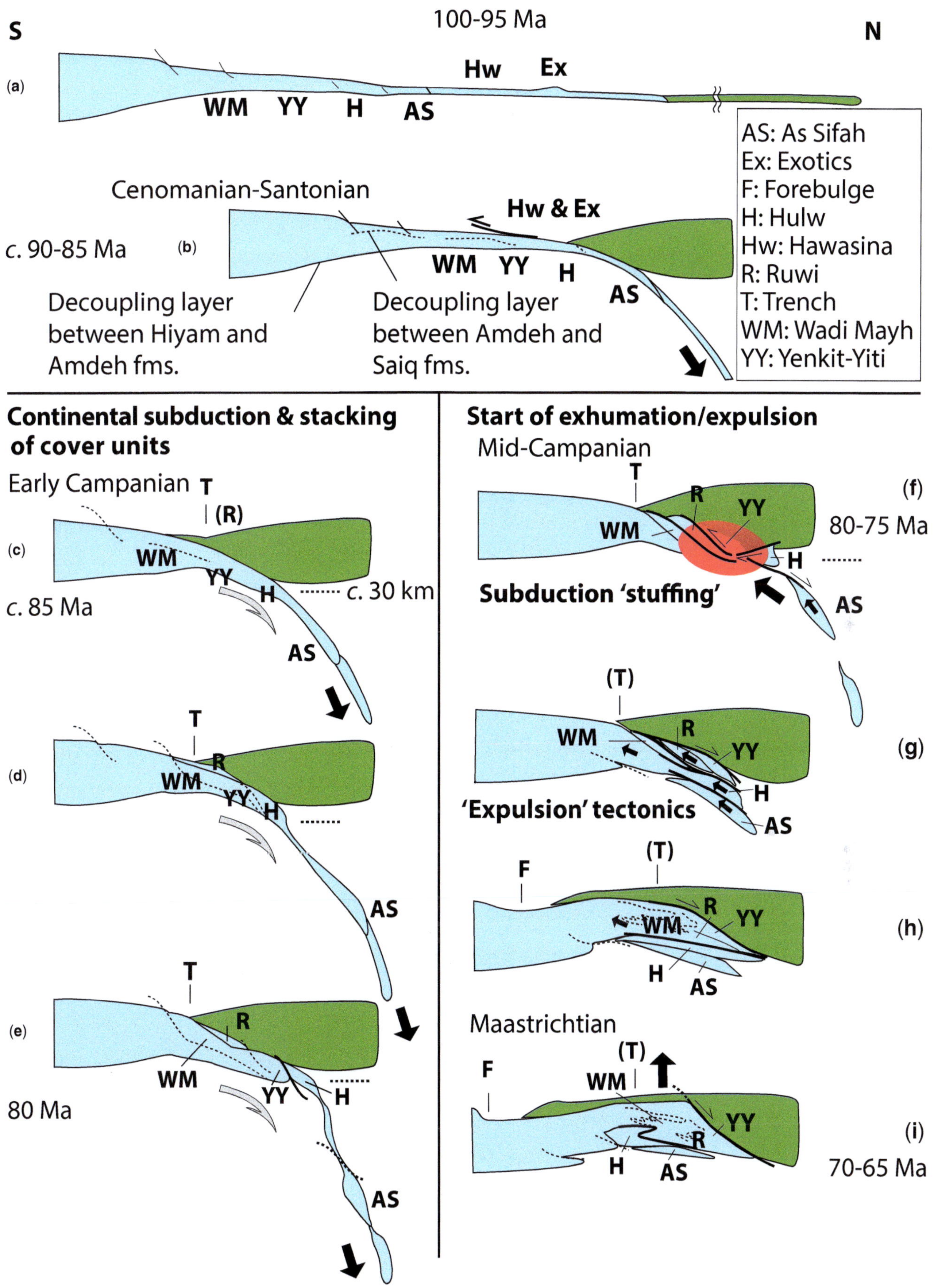

Fig. 5.8. Schematic sketch of the response of the Arabian margin to the obduction of the Semail Ophiolite in the Saih Hatat area. Pre-Amdeh formations are not depicted. Modified after Agard *et al.* (2010). (**a**) Initial palaeogeography; (**b**) arrival of the ophiolite at the Arabian margin; (**c–e**) subduction of the Arabian margin; and (**f–i**) beginning of rapid exhumation or 'expulsion' of HP nappes, crustal stacking and 'choking' or 'stuffing' of the subduction zone.

Ophiolite, Haybi and Hawasina thrust sheets. Thrusting ensued towards the SW along the Hawasina Basin; coeval with the onset of thrusting, the metamorphic sole formed. The nappes below the ophiolite evolved into a thin-skinned thrust belt that was transported onto the Arabian margin and platform. The most distal part of the Arabian margin was situated in a subduction zone at a later phase of the thrusting during the Early Campanian (Fig. 5.8c). Subduction depth of the most distal Arabian margin (As Sifah) reached at least *c.* 80–90 km (Searle *et al.* 1994, 2004) at *c.* 79 Ma (Warren *et al.* 2003, 2005), while more proximal parts did not go down to such great depths. This difference in subduction depth between different parts of the continental margin triggered the formation of major shear zones (e.g. between Ruwi and Yenkit-Yiti units; Searle *et al.* 1994, 2004; Yamato *et al.* 2007; Fig. 5.8d, e). When the Arabian passive continental margin was subducted, its Late Paleozoic and Mesozoic cover was partly decoupled. This decoupling resulted in complex folding and faulting as well as in high-pressure metamorphism in the northeastern SHD (figs 3.2 & 5.8).

Rocks of the passive continental margin were exhumed and intensely folded (Gregory *et al.* 1998; Searle 2007), including the formation of sheath folds (Searle and Alsop 2007; see sections on 'Upper–lower plate discontinuity or Hulw–As Sifah Window' and 'F4 folds' in Chapter 3). The As Sifah unit was rapidly exhumed beneath a passive roof fault immediately after the high-pressure/low-temperature metamorphism (Fig. 5.8f–h). During this SSW-directed footwall exhumation process, the prominent NNE-directed folds developed. The Hulw Unit was exposed to 1–1.2 GPa and 250–300°C and later underwent a re-equilibration stage at 0.9–1 GPa at 450°C (Yamato *et al.* 2007). The other units (Yenkit-Yiti, Ruwi and Wadi Mayh) were exposed to similar depths as the Hulw Unit. Furthermore, the sheath folds formed at depths of 25–30 km (>0.8–1 GPa). The pressure and temperature differences in the different units across shear zones were negligible. The vertical displacement across the shear zones that were between the units is therefore small. However, the shear zones indicate an intense strain. The high-pressure/low-temperature metamorphism rocks within the SHD are interpreted to have progressively 'choked' the subduction zone during the Late Cretaceous (Fig. 5.8). The subducted rocks were exhumed under highly ductile conditions during bottom-to-SSW exhumation beneath passive roof faults (Agard *et al.* 2010; Fig. 5.8). Zircon and apatite fission tracks allow the cooling process of the eclogite-facies rocks from 550°C (at 80 Ma) via *c.* 260°C (at 70 Ma) to *c.* 100°C (at 60–58 Ma; Saddiqi *et al.* 2006) to be dated.

The high-strain deformation in the SHD, including the sheath folds, formed coevally with the Hawasina Foreland Fold-and-Thrust Belt south of the Oman Mountains (Searle and Alsop 2007). The possible slab break-off at the transition between the subducted buoyant continental Arabian lithosphere and the dense Neo-Tethys ocean floor triggered fast exhumation of Arabia during the Late Cretaceous (Searle *et al.* 2004; compare with Frisch *et al.* 2011). The intensity of metamorphism and deformation increases from the JAD to the SHD (Breton *et al.* 2004). No subduction-related metamorphism has been observed in the southern JAD. This observation is consistent with a NE-directed subduction zone of the Arabian margin and platform (Breton *et al.* 2004; Grobe *et al.* 2019). During the Maastrichtian, the pre-Permian rocks of the SHD were exhumed. Clasts of pre-Permian formations within the Maastrichtian post-obductional Al-Khod Formation are evidence of this impressive exhumation (e.g. Nolan *et al.* 1990). This indicates a massive upward movement of the rocks of the SHD with the entire exhumation of the allochthonous nappes, the Muti Formation and the Hajar Supergroup, as well as some pre-Permian formations.

Peak-temperature estimates based on Raman spectroscopy on solid bitumen and organic-rich matrix particles in autochthonous Mesozoic formations in the southern JAD reveal *c.* 225–240°C (Fink *et al.* 2016), reconstructed by one-dimensional numerical basin modelling, which corresponds to 7.6–8.2 km of overburden of allochthonous ophiolitic and (meta)sedimentary rocks (Grobe *et al.* 2016). Recent Raman spectroscopy data on carbonaceous matter show peak temperatures of 305 to *c.* 266°C within the Cretaceous to Jurassic rocks of the Hajar Supergroup of the JAD (Grobe *et al.* 2019). The lower and higher temperatures were reached at the base of the Natih Formation, and within the Sahtan and Kahmah groups, respectively (Grobe *et al.* 2019). Peak temperatures within the Muti Formation and Hawasina nappes from the JAD are in the range of 178–208 ± 30°C and 193–213 ± 30°C, respectively (Grobe *et al.* 2019). These peak-temperature estimates indicate that the JAD was blanketed by 8–10 km of rocks of the Semail Ophiolite and 2 km of rocks of Hawasina nappes, corresponding to a lithostatic pressure of 0.28–0.34 GPa (Grobe *et al.* 2019).

Temperatures at the northern flank of the JAD are slightly higher than at the southern side (Aldega *et al.* 2017; Grobe *et al.* 2019). This thermal overprint was reached after emplacement of the Semail Ophiolite immediately prior to exhumation at *c.* 55 Ma (Grobe *et al.* 2019). Holland *et al.* (2009*b*) obtained minimum temperature conditions in the Natih Formation at the southern flank of the JAD of 134–221°C, obtained from fluid inclusions in quartz. According to Breton *et al.* (2004, fig. 4b), the peak-temperatures in the JAD ranged from <200°C in the SW to >200°C in the NE. Based on illite content in mixed illite–smectite layers, deformed ooids and one-dimensional numerical modeling, Aldega *et al.* (2017) obtained peak temperatures of *c.* 150–200°C in Mesozoic autochthonous rocks from the northern part of JAD and *c.* 120–150°C from the southern JAD area. These temperatures correspond to 4.5 km of overburden of allochthonous rocks in the northern JAD with decreasing overburden towards the south (Aldega *et al.* 2017). The discrepancy between the obtained peak-temperatures of *c.* 100°C between Fink *et al.* (2016) and Grobe *et al.* (2016, 2019) on one side and Aldega *et al.* (2017) on the other side was discussed by (Grobe *et al.* (2019). Either clay minerals did not have enough time to recrystallize because of a possible lack of potassium or the reflectance of vitrinite increased, or clay minerals may have been deformed during top-to-the-NNE shearing following the Semail Ophiolite emplacement (see 'Post-obduction deformation') and after peak-burial conditions (Grobe *et al.* 2019). This might explain why the estimated temperatures of Aldega *et al.* (2017) are lower than those of Fink *et al.* (2016) and Grobe *et al.* (2016, 2019).

Summary of thrusting, obduction and response of the Arabian crust. Stage I: Between 104 Ma and 96.5 Ma, WNW-dipping intraoceanic subduction was either spontaneously initiated (Stern 2004; Whattam and Stern 2011) or induced (Guilmette *et al.* 2018; Tavani *et al.* 2021) along a NNE-striking transform fault or fracture zone (e.g. van Hinsbergen *et al.* 2019). The orientation of the subduction zone may have been parallel to a transform fault or fracture zone. During that time, garnets formed in the subducted lower plate, indicating burial of the lower plate (Guilmette *et al.* 2018). Spencer *et al.* (2017) indicated that subducted sediments (pelitic and/or siliceous mud) were molten and intruded the intact section of mantle lithosphere at 99.8 ± 3.3 Ma to form peraluminous granite dykes in the Northern Oman Mountains.

Stage II: The probably induced subduction zone became self-sustained at *c.* 96 Ma, which resulted in extension in the upper plate (Guilmette *et al.* 2018; Tavani *et al.* 2021). Eventually, a spreading centre formed in the upper plate which produced new oceanic lithosphere (e.g. Rioux *et al.* 2013, 2016). The slab was subducted to >35 km depth (e.g. Cowan *et al.* 2014). Most researchers advocate that the ophiolite crust (early magmatic sequence) formed along a submarine spreading centre above a subduction zone (e.g. MacLeod *et al.* 2013). Metamorphic sole formation commenced immediately before or coeval with formation of the new oceanic lithosphere (Rioux *et al.* 2016). Contemporaneous with metamorphic sole formation, the Semail Ophiolite was obducted above either thinned continental crust or oceanic crust of the Hawasina Basin. Fluids derived from the subducted plate interacted with the basal section of the newly formed Semail Ophiolite under temperature conditions of *c.* 900–800°C (Prigent *et al.* 2018*b*). This interaction created a narrow and <1 km thick ductile shear zone at the plate interface. As a consequence, a transition zone from a mechanically decoupled to a coupled plate interface formed (Prigent *et al.* 2018*b*).

The late magmatic sequence, which has a geochemical supra-subduction zone setting fingerprint, intruded the newly formed oceanic lithosphere during the late Cenomanian (*c.* 96.4–94.7 Ma; U–Pb on zircon; Warren *et al.* 2005; Styles *et al.* 2006; Goodenough *et al.* 2010). Peraluminous granitic dykes of the late magmatic sequence provide possible indirect evidence for melting of sedimentary rocks during subduction (e.g. Haase *et al.* 2015). Molten sediments, together with released fluids, interacted with the overlying mantle to form the late magmatic rocks of the ophiolite (e.g. Goodenough *et al.* 2014).

Stage III: Between *c.* 95 and 70 Ma, the Semail Ophiolite and the metamorphic sole were thrust to the SW over distal Haybi Complex rocks and structurally lower Hawasina units for a minimum distance of 450 km. The entire Semail Ophiolite underwent significant clockwise rotation around a vertical axis from its formation to its emplacement onto Arabia with coeval N-ward motion of the Arabian Plate (e.g. Morris *et al.* 2016; van Hinsbergen *et al.* 2019). The Hawasina nappes were imbricated and partly truncated by later out-of-sequence motion along the thrust together with the ophiolite (e.g. Glennie *et al.* 1974; Searle 2007). The lower Arabian continental margin at the Saih Hatat area was subducted to a minimum depth of 80 km at 78.95 ± 0.13 Ma (Miller *et al.* 1999; Warren *et al.* 2003, 2005).

Stage IV: Buoyant crustal rocks jammed the subduction zone, causing possible slab break-off (e.g. Searle *et al.* 2004; Agard *et al.* 2010). The Arabian Crust within the subducting slab was highly folded multiple times during exhumation and ongoing convergence (e.g. Miller *et al.* 2002; Gray *et al.* 2004*a*; Agard *et al.* 2010). Emplacement of the Semail Ophiolite on top of Arabia was completed during the Maastrichtian (e.g. Glennie *et al.* 1974). Pre-Permian rocks of the SHD were exhumed towards the surface and were eventually deposited in the upper sequence of the Maastrichtian Al-Khod Formation (e.g. Nolan *et al.* 1990).

The Turonian–Maastrichtian Aruma Foreland Basin formed as a result of flexural bending of the Arabian Platform in the course of the SW-ward moving allochthonous units (e.g. Glennie *et al.* 1974; Robertson 1987*a*, *b*; Warburton *et al.* 1990). The onset of deposition in the basin coincided with the uplift and erosion of a forebulge. The forebulge was either stationary with basin narrowing during thrusting (Warburton *et al.* 1990), or migrating towards the continent (Robertson 1987*a*).

Post-obduction development

Uplift and cooling history

Zircon fission-track cooling ages from the SHD reveal that the exhumed high-pressure/low-temperature rocks cooled from 550°C at 80 Ma to 260°C at 70 Ma at a rate of *c.* 30°C Ma^{-1} (Poupeau *et al.* 1998; Saddiqi *et al.* 2006). These rocks cooled even more down to *c.* 100°C at 60–58 Ma at a cooling rate of *c.* 16–13°C Ma^{-1} (apatite fission track of Poupeau *et al.* 1998 and Saddiqi *et al.* 2006). The same authors explained the different cooling rates with different exhumation mechanisms during the Late Cretaceous to Cenozoic. Fast cooling was caused by extensional tectonics during the Late Cretaceous, and slow cooling was the result of the isostatic rebound and erosion of the Arabian crust during the Cenozoic (Poupeau *et al.* 1998; Saddiqi *et al.* 2006). Weak reheating of apatite fission tracks to 70°C during the Late Miocene–Early Pliocene is explained by hydrothermal events related to the collision between Arabia and Eurasia (Poupeau *et al.* 1998; Saddiqi *et al.* 2006). However, based on their new apatite helium ages Hansman *et al.* (2017) exclude such reheating in the SHD. The final Pliocene–Quaternary cooling occured at rates of nearly 10°C Ma^{-1}, which is explained with isostatic rebound and erosion (Poupeau *et al.* 1998).

In our western study area, rapid exhumation of the JAD started at *c.* 40–30 Ma (Hansman *et al.* 2017, 2018; fig. A.2), previously thought to have occurred at 30–25 Ma (Mount *et al.* 1998), resulting from roughly SSW-directed shortening ($D6_d$ in fig. A.2). Hansman *et al.* (2017, 2018) studied the Late Eocene uplift of the JAD and SHD, based on low-temperature thermochronology, using the apatite (U–Th)/He (AHe), apatite fission track (AFT) and zircon (U–Th)/He (ZHe) techniques as well as U–Pb dating of calcite veins. They concluded that uplift of the JAD could be the result of SSW-oriented horizontal shortening, which was controlled by a blind NNE-dipping basal thrust (Hansman *et al.* 2017). This Late Eocene uplift amounted to 4–6 km at the JAD and 1 km at the SHD (Hansman *et al.* 2017). In agreement with the Eocene cooling data of the JAD from Hansman *et al.* (2017, 2018), Grobe *et al.* (2019) obtained 48.7 ± 1.8 to 39.8 ± 3.0 Ma cooling ages for the centre of the JAD based on zircon (U–Th)/He data. These ages indicate an Eocene cooling of the JAD below *c.* 170 to 130°C (Grobe *et al.* 2019). Furthermore, Corradetti *et al.* (2019) recently provided Late Eocene–Early Miocene AHe cooling ages from the JAD that, in terms of age, overlap with the results of the former studies.

Doming can be studied at a spectacular outcrop at the westernmost end of the JAD in the approximately NE-striking Wadi Shakhar (fig. 2.1). At this outcrop, the fold axis of the JAD plunges moderately–shallowly towards the NW below the allochthonous rocks (Fig. 5.9; Muri Anticline).

Hansman *et al.* (2017) modelled one cooling event for the SHD since the Cretaceous with cooling from 500°C to 60 ± 20°C during 80–50 Ma. Since 50 Ma, less than 60°C of cooling has occurred (Hansman *et al.* 2017; Fig. 5.10).

The cooling history in the JAD has also been modelled by Hansman *et al.* (2017) using low-temperature thermochronology (Fig. 5.10). They modelled two versions of rapid cooling. Version 1 includes two distinct cooling events since the Late Cretaceous. Cooling began at *c.* 80 Ma with cooling from 250 ± 50°C to near-surface temperatures at 75–65 Ma (Fig. 5.10). This first cooling episode was followed by reheating to 180 ± 20°C at some time during 65–40 Ma (Hansman *et al.* 2017). Version 2 entails a cooling event from 250 ± 50°C to 180 ± 20°C during 80–65 Ma with a constant temperature until *c.* 40 Ma (Fig. 5.10). The second cooling event (from 180 ± 20°C to 45 ± 15°C) occurred during 40–30 Ma. The rocks of the JAD underwent cooling of only 40°C during the past 30 Ma (Hansman *et al.* 2017; Fig. 5.10). The work of Grobe *et al.* (2019) supports the theory that the JAD underwent two steps of rapid cooling. The first cooling event to below *c.* 260°C occurred during 96–70 Ma, followed by

Fig. 5.9. Westernmost outcrop of the JAD near the village of Muri (Muri Anticline; fig. 2.1). The anticline is part of the set of F5 folds of the JAD (fig. 4.1; section 'F5 folds' in Chapter 3). The wadi cuts >500 m into the Natih and Nahr Umr formations. The NW-plunging anticline is framed by the Muti Formation Hawasina nappes and the Semail Ophiolite. The light-coloured mountains in the background are part of the Hawasina Window. Folding occurred during the Paleogene–Neogene. Courtesy of Christopher M. Bailey (see Appendix C for further information). Width of the anticline in Wadi Shakhar is *c.* 2 km.

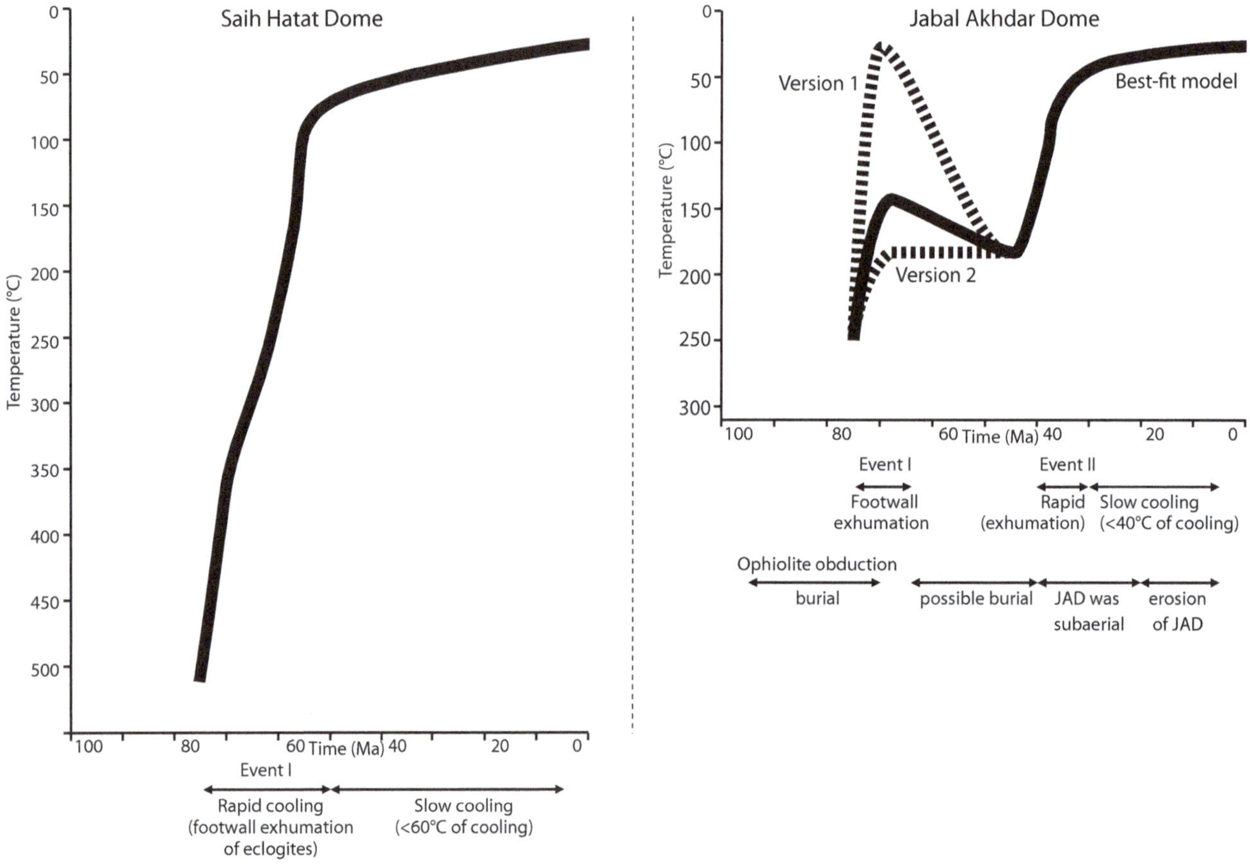

Fig. 5.10. Temperature–time paths for the Saih Hatat and Jabal Akhdar domes, redrawn from Hansman *et al.* (2017). Possible versions for the cooling path in the Jabal Akhdar Dome are possible (see text). Note that the Saih Hatat Dome underwent one unsteady cooling event, and the Jabal Akhdar Dome two separate events of rapid cooling.

slow cooling of 100°C during 70–55 Ma (Saddiqi *et al.* 2006; Grobe *et al.* 2019). A rapid exhumation phase of the JAD began at 55 Ma, resulting in rapid cooling during the Eocene (Grobe *et al.* 2019).

Post-obductional deformation

Following emplacement of the Semail Ophiolite onto Arabia during the Late Cretaceous, northeastern Oman underwent a phase of multiple extensions and a subsequent phase of compression, associated with the Arabia–Eurasia convergence (e.g. Fournier *et al.* 2006; Grobe *et al.* 2018, 2019; Mattern and Scharf 2018). The tectonic phases are discussed in the following subsections. Hansman *et al.* (2017) provided a hypothetical tectonic model for the Oman Mountains from the Late Cretaceous until recent (Fig. 5.11), from which large parts of our interpretation are based on.

While obduction of the Semail Ophiolite was completed during the Late Cretaceous, some million years later a second ophiolite obduction onto Arabia took place to the SE of Sur. WNW-directed obduction of the Masirah Ophiolite onto the Arabian margin occurred at the Mesozoic–Cenozoic transition (e.g. Smewing *et al.* 1991; Peters *et al.* 1995; Gnos *et al.* 1997; Peters and Mercolli 1997; Marquer *et al.* 1998; Schreurs and Immenhauser 1999; fig. A.2; 'Tectonic framework' in Scharf *et al.* 2021*d*, Chapter 1, this Memoir).

Late Cretaceous: rapid exhumation and cooling in the SHD. Rapid exhumation and cooling of the SHD during late stages of obduction may have been caused by slab break-off between dense and eclogitized oceanic lithosphere of the Neo-Tethys Ocean and the subducting Arabian continental margin (e.g. Searle *et al.* 2004; Fig. 5.11a, b). No such rapid exhumation and cooling have been reported from the JAD; the JAD may therefore have been distal to the subduction zone (e.g. Breton *et al.* 2004; Grobe *et al.* 2019).

Late Cretaceous to Paleogene–Early Eocene: top-to-the-NNE shearing immediately after Semail Ophiolite emplacement. Immediately after the emplacement of the Semail Ophiolite, the ophiolite underwent a period of tectonic thinning above a NNE-directed major shear zone during the Early Campanian (Grobe *et al.* 2018, 2019; fig. A.2). Shearing ensued under ductile conditions within Mesozoic limestone of the northern JAD as proven by a new microscopic shear-sense indicator (Scharf *et al.* 2019*b*). Top-to-the-NNE shear strain affected the Arabian Platform at the northern and southern realms of the JAD, although deformation is more pronounced in the northern part (Grobe *et al.* 2018). In addition, the same authors concluded that this extensional tectonic event was associated with a relatively slow plate convergence during the Early Campanian and with the formation of a precursor JAD. The respective ductile extensional structures were overprinted by brittle extensional faults during the Campanian–Maastrichtian (Grobe *et al.* 2018).

Additionally, Fournier *et al.* (2006) worked on the post-obduction deformation history of the study area and established three main phases of deformation after the emplacement of the Semail Ophiolite. The first stage ('Tectonic Stage 1') is characterized by ENE-directed extension in the SHD and its vicinity of Maastrichtian–Early Eocene age (fig. A.2). Fournier *et al.* (2006) explained the origin of this extension with crustal thinning after exhumation of high-pressure/low-temperature rocks in the northern SHD. The Frontal Range Fault that bounds the northern margin of the SHD and JAD has two phases of activity (section on 'Main extensional fault systems' in Chapter 3). The older phase lasted from the Late Cretaceous to the Paleocene–Eocene boundary (Mattern and Scharf 2018).

Structural investigations of the contact of listwaenite bodies with the post-obductional latest Cretaceous Al-Khod Formation near Fanja were carried out by Scharf *et al.* (2019*c*, 2021). The contact between the tabular listwaenite bodies and the Al-Khod Formation is an extensional fault contact where carbonate veins, extending from the listwaenite, occur within the first few metres of the Al-Khod Formation. This might indicate that the listwaenite formed after the obduction within an overall extensional tectonic regime (e.g. Wilde *et al.* 2002; Mattern and Scharf 2018; Bailey *et al.* 2019; fig. 4.1; section on 'Main extensional fault systems' in Chapter 3) and not during Late Cretaceous subduction as suggested by Falk and Kelemen (2015). Carbonate veins, forming a complex network, are common in the listwaenite. Scharf *et al.* (2021) assumed that those veins with a thickness of $\leq$ 0.5 cm formed at a late stage of listwaenitization. Two samples of such veins were dated using the U–Pb system and applying the laster ablation – sector field – inductively coupled plasma – mass spectrometry (LA-SF-ICP-MS) technique. The obtained ages of both samples are 55 ± 4 Ma (Scharf *et al.* 2019*c*, 2021); the listwaenite therefore probably formed at the late stage of the first phase of activity of the Frontal Range Fault.

In the vicinity of the SHD, sedimentation of fluvial and/or deltaic sediments in the lower part and shallow-marine limestones in the upper part of the post-obduction succession occurred during the Maastrichtian–Late Eocene (e.g. Nolan *et al.* 1990; Skelton *et al.* 1990; Hersi and Al-Harthi 2010; Fig. 5.11c). During the Late Eocene–Early Miocene, the Arabian Platform was largely emergent (Fournier *et al.* 2006).

Braathen and Osmundsen (2019) reported that metamorphic core complexes formed with extension towards the NE at the JAD and SHD after the Semail Ophiolite emplacement. Their interpretation is based on: (1) the significant post-obductional extension of northeastern Oman; (2) the shape of the SHD and JAD, with steeper limbs towards the NNE (i.e. they might be asymmetric monoclines); (3) the formation of Maastrichtian–Paleogene extensional basins; and (4) the consistent top-to-the-NNE shearing at the base of the allochthonous nappes, interpreted as detachments by Braathen and Osmundsen (2019).

Eocene–Miocene: compression within the Oman Mountains. The origin of the compressional structures within the Oman Mountains during the Arabia–Eurasia convergence is still under debate. Potential causes of the formation of compressional structures within northern Oman are the Makran Subduction Zone and the Zagros Mountains, to the north of Oman. The Makran Subduction Zone in the Sea of Oman was initiated by the Arabia–Eurasia convergence from the Eocene (McQuarrie and van Hinsbergen 2013). The Zagros Mountains formed by continent–continent collision from the Oligocene (e.g. Gavillot *et al.* 2010; McQuarrie and van Hinsbergen 2013).

However, the low-temperature geochronological data from the JAD and SHD of Hansman *et al.* (2017) do not support the idea that the Zagros Mountains and/or Makran Subduction Zone are responsible for compression in the Oman Mountains of the study area; this is because the uplift and rapid cooling in the JAD started *c.* 10 Ma before the formation of the Zagros Mountains (Hansman *et al.* 2017). Furthermore, Miocene cooling along the JAD and SHD was slow while recent seismicity is negligible, especially along the JAD (El-Hussain *et al.* 2012), but as the Zagros–Makran convergence is active, active exhumation and/or seismicity in the JAD would be expected (Hansman *et al.* 2017).

Instead of the Zagros–Makran convergence, Hansman *et al.* (2017) speculated that exhumation of the JAD and SHD is

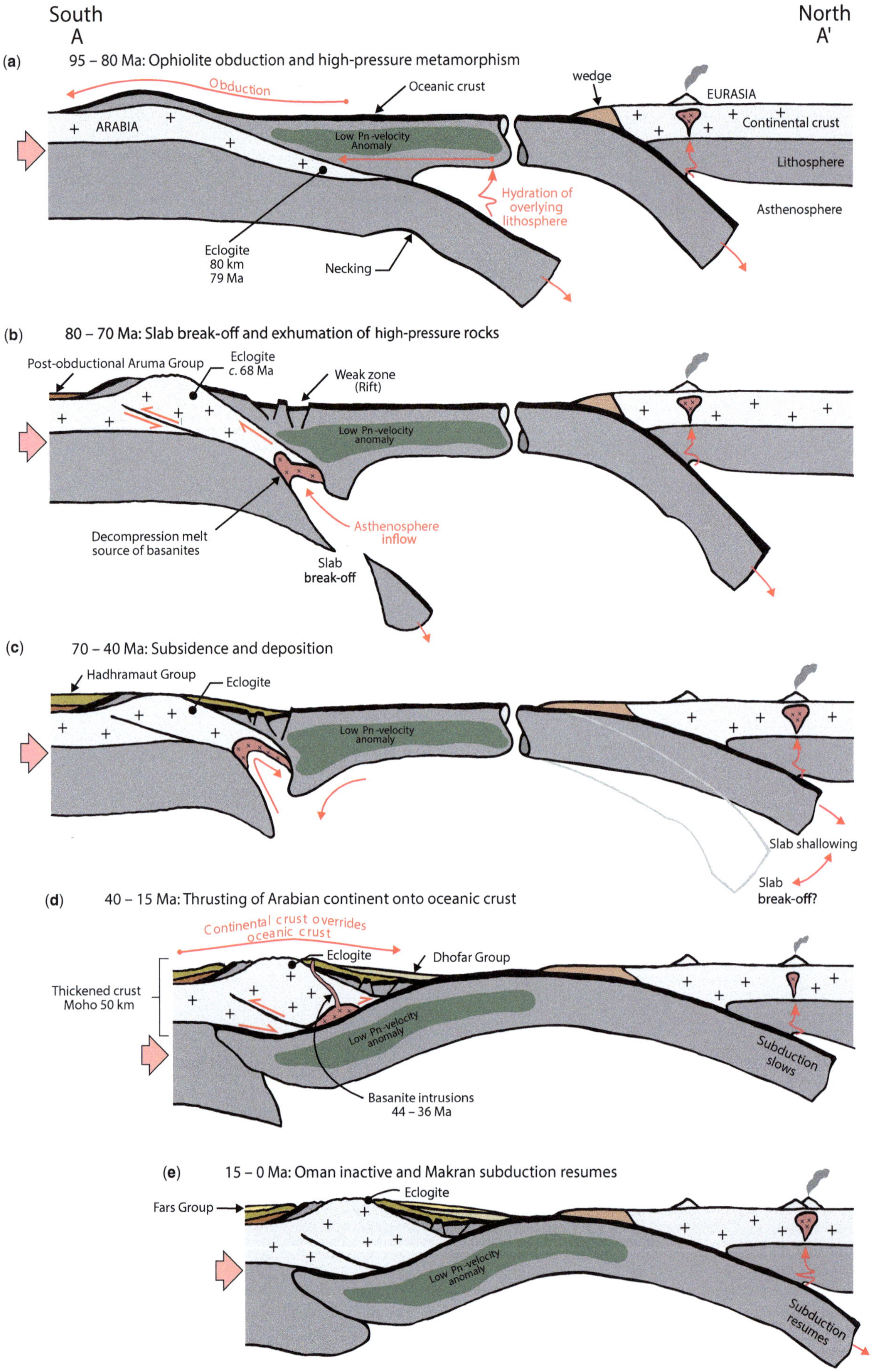

Fig. 5.11. Schematic cross-section depicting the tectonic evolution of the Oman Mountains modified after Hansman *et al.* (2017). Trace of cross-section is depicted in figure 4.4.

related to a slowdown of the downgoing slab in the Makran Subduction Zone. The following paragraph is a summary of their work.

The slowdown may be a result of shallowing of the subduction angle during the Eocene as indicated by craton-ward migration of the volcanic arcs in Iran (Fig. 5.11c). The shallower angle would be expected to have increased friction and therefore reduced subduction velocity. As a consequence of the slowdown in subduction, Arabian continental crust was thrust for 300 km onto oceanic lithosphere with the low Pn-tomography velocity anomaly (figs 4.4 & 5.11d). Thrusting would have initiated along the lithospheric weak area that formed during the Late Cretaceous slab break-off in NE Oman (Fig. 5.11d).

The Arabian continental crust below the Semail Ophiolite was thickened to 50 km (Al-Lazki *et al.* 2002). Thickening may have been due to reactivation of obduction-related structures in the continental crust as back thrusts (Fig. 5.11d). The model of Hansman *et al.* (2017) may explain the occurrence of basanites intrusions during 44–36 Ma along the Batain area and near Muscat (Fig. 5.11d; Worthing and Wilde 2002; Scharf *et al.* 2021*b*). These intrusions may have been the result of deformation of the structural anomaly of the asthenosphere–lithosphere boundary from the slab break-off in the tearing slab. At 15 Ma, the Arabian lithosphere below the Semail Ophiolite was mechanically rigid; convergence decreased at the Arabian margin while it increased in the Makran Subduction Zone (Fig. 5.11e).

E- to NE-directed shortening during the Late Oligocene–Early Miocene (Fournier *et al.* 2006; F6 in fig. 4.1) is marked by large-scale open folds or box folds with kink band sets and local brittle faults within the Paleogene sediments (e.g.

Fig. 5.12. NNE-dipping normal fault near Muscat at Wadi Kabir (fig. 2.1), which separates the Mahil Formation of the SHD to the WSW (footwall) from the hanging wall to the ENE, represented by the Semail Ophiolite, which is overlain by Paleocene–Eocene shallow-marine limestones (Jafnayn and Rusayl formations). The hanging wall forms a rollover anticline. Vertical displacement of the Cenozoic rocks is *c.* 150 m (Mattern and Scharf 2018).

Fournier *et al.* 2006; Searle 2007; Scharf *et al.* 2016; F6 in section on 'F6 folds' in Chapter 3). During this time and/or during the Late Eocene, Paleogene limestones were also uplifted to >2000 m SE of the SHD (e.g. Wyns *et al.* 1992; Searle *et al.* 2004; Moraetis *et al.* 2018).

Rifting in the Gulf of Aden began during the Oligocene–Early Miocene (Roger *et al.* 1989; Watchorn *et al.* 1998; Fournier *et al.* 2006, 2010) with an onset of ocean spreading at *c.* 18 Ma (Leroy *et al.* 2004). The onset of compression in the study area was therefore older than the start of rifting in the Gulf of Aden. To the SE of the SHD, compression caused inversion of the Abat Basin and Qalhat Fault (both near Sur) during the Late Burdigalian (Wyns *et al.* 1992). Molasse-type deposits of the Barzaman Formation started to accumulate during the Mid Miocene as a result of the uplift of the Oman Mountains (Wyns *et al.* 1992). This uplift probably continued until the Pliocene (Wyns *et al.* 1992).

Late Eocene–Oligocene: extension. A second extensional interval occurred, probably during the Late Eocene–Oligocene (Fig. 5.12). Fournier *et al.* (2006) divided this second stage into two substages with extension along N20° E and N150° E, and assigned a probable Oligocene age to them ('Tectonic Stage 2A and 2B'; fig. A.2). Scharf *et al.* (2019*a*) assumed a Late Eocene age for the onset of an extensional stage along the Semail Gap. Further major post-Eocene normal or extensional faults exist in the Batinah Coastal Plain (Mann *et al.* 1990).

The origin of the forces leading to the second extensional interval are interpreted to be the combined effect of the bending of the Neo-Tethys slab beneath Eurasia, which is part of the Arabian Plate, and of uplift of the Oman Mountains (Mann *et al.* 1990; Fournier *et al.* 2006). Hansman *et al.* (2017) pointed out that the onset of uplift and exhumation of the Oman Mountains occurred as early as during the Late Eocene. Gravitational collapse of the JAD along the Semail Gap may therefore have already started during that time or slightly later (Scharf *et al.* 2019*a*). Applying the same reasoning to the Frontal Range Fault, its second stage of extensional activity may also have started during the Late Eocene (Scharf *et al.* 2021*b*). Alternatively, Fournier *et al.* (2006) pointed out that far-field effects of the rifting and/or doming of the Gulf of Aden–Red Sea region could have caused patterns of extensional structures within the Arabian Plate, including our study area.

Oligocene–Miocene: E-to-NE-directed compression. The third phase of Fournier *et al.* (2006) reveals E- to NE-directed compression, lasting from the Oligocene–Miocene boundary into the Miocene ('Tectonic Stage 3'). The direction of compression later rotated towards N–S and NNE–SSW during the Pliocene (Fournier *et al.* 2006; fig. A.2). Scharf *et al.* (2016) confirmed the post-Eocene NNE-directed compressional phase west of Muscat that was expressed by left-lateral strike-slip. According to Fournier *et al.* (2006), the apparent stress direction during the Mio-Pliocene seems to have changed from transpressional in the Oman Mountains to strike-slip south of the Oman Mountains. This could be related to decreasing intensity of compression with increasing distance from the Zagros Mountains (Fournier *et al.* 2006). However, this cannot explain the occurrence of the N-oriented compressional structures (e.g. F6 in Fig. 4.1) parallel to the N-oriented shortening direction between Arabia and Eurasia (Vernant *et al.* 2004; Regard *et al.* 2005).

Roughly E-directed post-Eocene compression formed open folds with approximately N-trending and gently dipping fold axes in Cenozoic shallow-marine sedimentary rocks in the Rusayl Embayment (fig. 2.1). This embayment is the triangular-shaped outcrop zone of the Cenozoic rocks north of the Nakhl Subdome and SHD (e.g. Mann *et al.* 1990; fig. 4.1; section on 'F6 folds' in Chapter 3). The folds occur on the hanging-wall block of an extensional fault that was active during gravitational and/or culmination collapse of the Nakhl Subdome and the SHD (see Coffield 1990; Hanna 1990; Mann *et al.* 1990; Fig. 5.13) and belongs to the Frontal Range Fault System (Mattern and Scharf 2018; 'Main extensional fault systems' in Chapter 3). Extensional tectonic transport ensued to the NW from the SHD and to the NE from the Nakhl Subdome (Mann *et al.* 1990; Fournier *et al.* 2006); transport therefore had a convergent aspect to it on the downthrown block (Fig. 5.13). With progressing transport, crowding and constriction of the Cenozoic rocks increased. The resulting compressional forces within this downsliding block created the mentioned folds (Fig. 5.13). This model explains the deformation at a moderate regional scale north of the Nakhl Subdome and the SHD. The origin of the Oligo-Miocene E- to NE-directed compression for large parts of the Oman Mountains is not well understood. Fournier *et al.* (2006) pointed out that this compression may be a result of interaction between the Arabian and Indian plates.

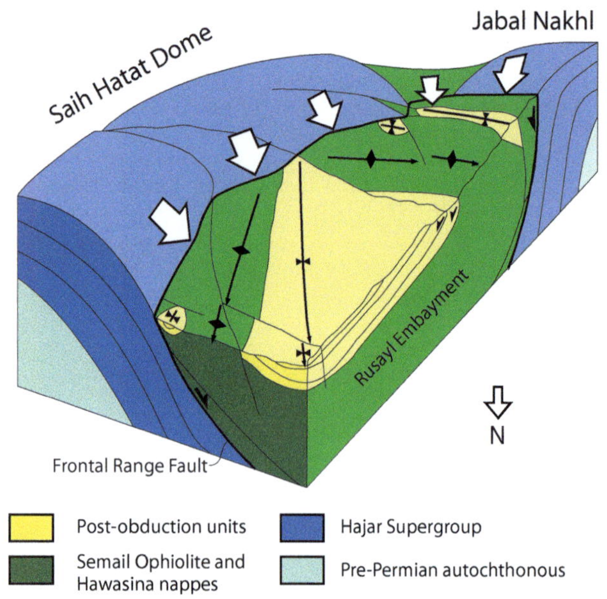

Fig. 5.13. Model of Eocene–Miocene extension at the northwestern margin of the SHD, the Rusayl Embayment. Extension direction is towards the NW from the SHD and to the NE from the Nakhl Subdome, resulting in E–W compressional structures. Redrawn after Mann *et al.* (1990).

Nine generations of calcite veins and stylolites within the Mesozoic limestones reflect the geodynamic evolution of the JAD from the Late Cretaceous to the Pliocene–Holocene (Gomez-Rivas *et al.* 2014; Grobe *et al.* 2018). Hansman *et al.* (2018) dated the different vein generations. The reader is referred to Gomez-Rivas *et al.* (2014, table 3) and Grobe *et al.* (2018, table 1) for details on the veins, their formation and their correlation with tectonic events.

Although Hansman *et al.* (2017) provided a hypothetical tectonic model for the exhumation of the Oman Mountains from the Late Cretaceous to the present, which includes reactivation of back thrusts in the subsurface (Fig. 5.11d), the deep structure of the JAD and its role during Cenozoic exhumation is still under debate. The JAD represents a single fold either associated with (Searle 1985; Bernoulli and Weissert 1987; Cawood *et al.* 1990; Hanna 1990; Hansman *et al.* 2017) or without thrusting (Glennie *et al.* 1973, 1974; Al-Lazki *et al.* 2002) of basement nappes beneath the JAD. A respective part of the thrust is emerging south of the JAD at the Salakh Range, while other parts of the thrust are blind (Cawood *et al.* 1990; Hanna 1990). Mount *et al.* (1998) proposed that steep basement faults beneath the JAD are the reason for its exhumation.

Quaternary deformation

Active convergence between Arabia and Eurasia created an active forebulge in the Oman Mountains (Rodgers and Gunatilaka 2002; fig. 1.2). Their model is supported by: (1) an outstanding spatial coincidence between a NNE-oriented modern topographic cross-section and a theoretical flexural profile across northern Oman; (2) the spatial distribution of Pliocene–Pleistocene sediments; (3) the inferred Late Miocene–Pliocene age of the bajadas south of the Oman Mountains; and (4) the possible recent uplift of the Oman Mountains (Fig. 5.14; Rodgers and Gunatilaka 2002). In this tectonic model, the southern Sea of Oman would correlate with the foredeep, the Oman Mountains with the subaerial forebulge, the vast bajada south of the Oman Mountains with the distal forebulge, and the sabkha sediments south of the bajadas with the backbulge basin (Fig. 5.14).

The largest percentage of the Quaternary deposits is represented by conglomerates that mainly occur in the interior of Oman (south of the Oman Mountains) and in the Batinah Coastal Plain (north of the Oman Mountains). The evolution of the conglomerates is directly related to pluvial periods. Vast bajadas south of the Oman Mountains comprise Miocene–Pliocene sedimentary rocks of the Barzaman Formation which are blanketed by weakly cemented Quaternary gravels (Blechschmidt *et al.* 2009). The enormous amount of sedimentary rocks formed during pluvial periods correlates with monsoonal precipitation (Blechschmidt *et al.* 2009). The same authors attributed the formation of Quaternary coalescing alluvial fans to the interglacial and glacial periods between Marine Isotope Stage (MIS) 11 and 3 (424–30 ka), based on the optically stimulated luminescence (OSL) technique. Rapid speleothem growth from the Hoti cave (southern JAD) occurred at 325–300 ka BP, 200–180 ka BP, 135–120 ka BP, 82–78 ka BP and 10.5–6 ka BP (Burns *et al.* 1998, 2001). These aspects, and the fact that the respective calcite is highly depleted in ^{18}O with respect to modern speleothems, led Burns *et al.* (2001) to date the pluvial periods in northern Oman.

Stepped fluvial and/or wadi terraces and deformed Quaternary alluvium in the JAD are attributed to activities of reactivated faults (Beurrier *et al.* 1986; Rabu *et al.* 1986). All wadis in and around the JAD and SHD exhibit a (sub) recent incision of several metres to tens of metres or more (Fig. 5.15). These significant incisions testify ongoing uplift of the Oman Mountains, probably related to isostatic rebound in response to erosion of thick and dense allochthonous units and/or the flexural bulging (Rodgers and Gunatilaka 2002), providing the Oman Mountains with its final shape ($D6_e$ in fig. A.2).

Additional evidence for recent uplift is manifested in the Southeastern Oman Mountains between Quriyat and Sur (Le Métour *et al.* 1986; figs 1.1 & 2.1). Uplift of the coastal area between Quriyat and Sur started during the Early Miocene after deposition of the Shama Formation and probably continues today (e.g. Moraetis *et al.* 2018). Wyns *et al.* (1992), Kusky *et al.* (2005), Hoffmann *et al.* (2013, 2019), Yuan *et al.* (2016), Mattern *et al.* (2018*a*), Moraetis *et al.* (2018), Pavlopoulos *et al.* (2018) and Falkenroth *et al.* (2019) showed uplifted marine terraces of presumably neotectonic activity in

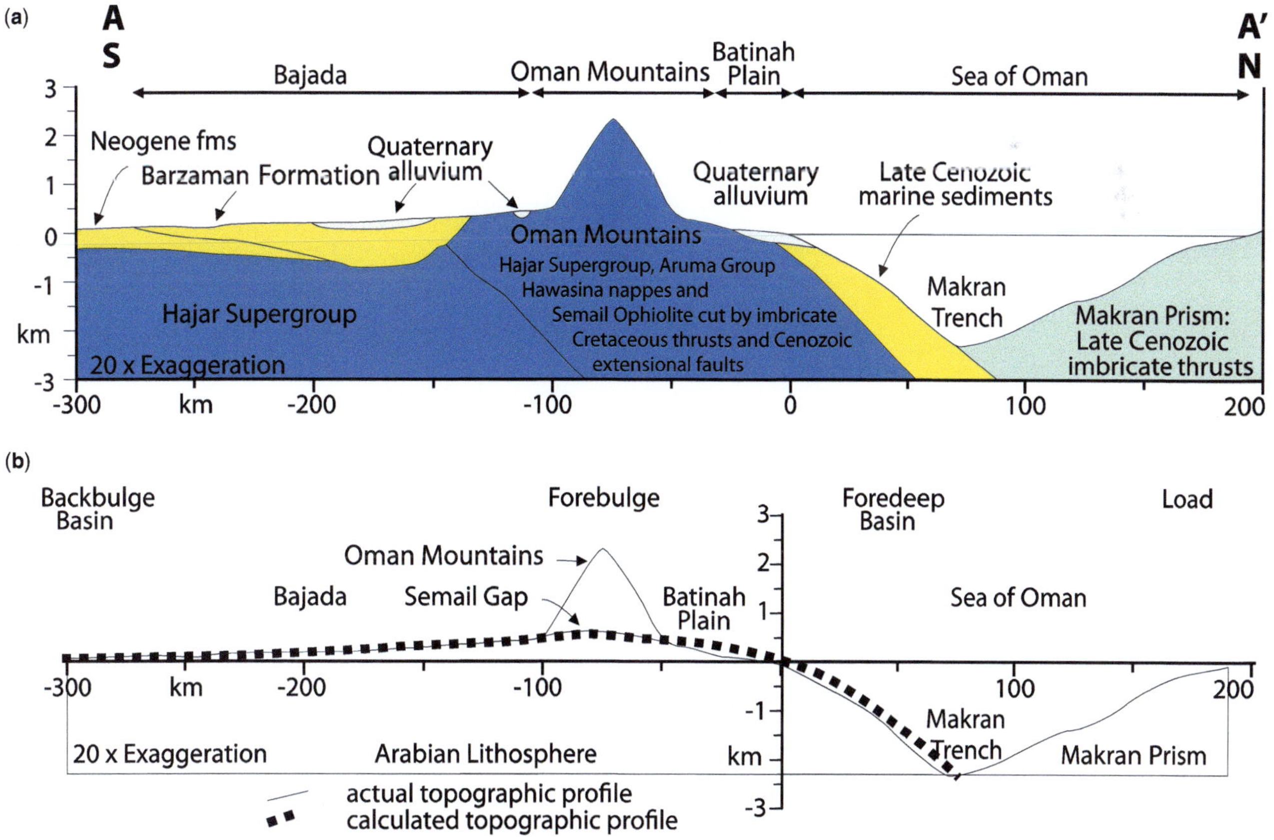

Fig. 5.14. (**a**) Schematic cross-section through northern Oman, the Sea of Oman up to the Makran area. (**b**) Topographic and/or bathymetric cross-section across the same area as for (a). The actual topography and/or bathymetry matches the calculated flexural profile very well. Redrawn from Rodgers and Gunatilaka (2002). Traces of the section are depicted in figure 1.2.

Fig. 5.15. Wadi incision of ≥ 10 m. This type of young incision (note overhangs) as shown from Wadi Al-Abyad (between Rustaq and Nakhl; fig. 2.1) is common within the Oman Mountains

the same area. In addition to a tectonic reason for the uplift, Hoffmann *et al.* (2019) hypothesized that ongoing serpentinization of the Semail Ophiolite below the uplifted Cenozoic carbonates may be a contributor to the uplift. Serpentinization results in volume increase and density decrease and therefore possible uplift in the Southeastern Oman Mountains (Hoffmann *et al.* 2019). Al-Hatrushi (2017) investigated marine terraces *c.* 20 km SE of Quriyat (fig. 2.1; between Daghmar and Dhabab). The presented marine mollusc ^{14}C ages of Al-Hatrushi (2017, table 1; Hannss 1991; Al-Hatrushi 1995) range between 2605 ± 85 and 5960 ± 100 a BP, and the respective terrace has an elevation of <10 m. According to Al-Hatrushi (2017) this terrace formed during the last interglacial sea-level high and therefore non-tectonically.

Fig. 5.16. Four marine terraces (T1 to T4), NW of Sur near Tiwi, bear witness to (sub)recent uplift. Ages of dated terraces T1 and T2 are from Moraetis *et al.* (2018).

However, his interpretation conflicts with the work of Moraetis *et al.* (2018), Pavlopoulos *et al.* (2018) and Hoffmann *et al.* (2019), which conclude ongoing tectonic uplift in this coastal area.

Gardner (1988) dated mollusc shells with the ^{14}C technique at elevations of 12–15 m with an age of 31.11 ± 0.53 to 21.28 ± 0.28 ka BP. The site is located at the coastline 136 km south of Sur. Considering the eustatic sea-level of 80 m below the present sea-level at 31 ka BP (Siddall *et al.* 2003), an uplift rate of *c.* 3–4.5 mm a^{-1} has been calculated (Moraetis *et al.* 2018). A similar magnitude of uplift rates south of Sur have been presented by Moraetis *et al.* (2018; figs 1.1 & 5.16) for the area between Fins and Tiwi (*c.* 40 km NW of Sur). However, slower uplift rates (0.01–0.89 mm a^{-1}) have been calculated by Hoffmann *et al.* (2019). The latter rates are between MIS 5a and MIS 19, with the lowest rates towards Quriyat and the highest rates towards the Qalhat Fault (Hoffmann *et al.* 2019).

The WNW-striking Ja'alan and NNE-striking Qalhat reverse faults are possibly related to recent tectonic activity in the Sur area (fig. 1.1; Moraetis *et al.* 2018). The Qalhat Fault deformed Middle Miocene sediments in its footwall and partly lifted Neoproterozoic basement rocks to the surface (Wyns *et al.* 1992). The accumulated total vertical displacement at the Ja'alan and Qalhat faults amounts to 3.5 and 5 km, respectively (Ali *et al.* 2016). According to Roger *et al.* (1991), the northern Ja'alan Fault cuts through terraces of Quaternary alluvium.

The Ja'alan and Qalhat faults as well as the Coastal Parallel Shear Zone bound a wedge of uplifted area: the Ja'alan–Qalhat Wedge of Moraetis *et al.* (2018; fig. 10). The uplifted marine terraces are within this wedge. Uplift of the Ja'alan–Qalhat Wedge is possibly caused by NE–SW compression along the Zagros collision zone (Moraetis *et al.* 2018) and/or the slowdown of subduction in the Zagros–Makran Subduction Zone (compare to Hansman *et al.* 2017). Rodgers and Gunatilaka (2002), Kusky *et al.* (2005) and Yuan *et al.* (2016) suggested that the Late Cenozoic uplift of the Oman Mountains is related to the position of the area within an active forebulge on the downgoing Arabian Plate beneath Eurasia. This setting would reactivate faults within the Ja'alan–Qalhat Wedge and caused uplift of the wedge (Moraetis *et al.* 2018). Furthermore, the wedge uplift during the Pleistocene–Mid Holocene may be attributed to a slab slowdown in the subduction zone beneath Eurasia (compare with Hansman *et al.* 2017; Moraetis *et al.* 2018). This interpretation is based on the uniform uplift of the eastern Arabian Peninsula, which requires a uniform uplift mechanism. Hoffmann *et al.* (2019) speculated that ongoing serpentinization below the Cenozoic carbonates may have contributed to uplift. This uplift mechanism is unrelated to any tectonics.

The area between Sur and Qalhat is characterized by some mild tectonic activity including several earthquakes with Richter magnitudes of <3 during the years 2000 to 2015 (e.g. El-Hussain *et al.* 2012; Moraetis *et al.* 2018). Nevertheless, a major earthquake occurred in 1483 along the Qalhat Fault, and possibly destroyed the historical town of Qalhat (Aubin 1973; Ambraseys *et al.* 1994; Musson 2009; El-Hussain *et al.* 2012; Rougeulle 2017; Ermertz *et al.* 2019). The remains of this city have been part of the UNESCO World Heritage site since 2018 (https://whc.unesco.org/en/list/1537).

At least 22 thermal springs with water temperatures of 35–66°C occur along the northern and eastern margin of the JAD and on the northwestern flank of the SHD (fig. 4.1; table A.1), indicating a heat souce (see the following paragraph). Another thermal spring is located in the gorge of Wadi Daiqa in the southern SHD. This spring is probably related to deep faults, cutting through the Hajar Supergroup. The location of the northern 22 thermal springs at or near main extensional faults suggests a relationship between extensional faulting and the origin of the thermal springs (section on 'Main extensional fault systems' in Chapter 3).

The thermal source(s) of the springs is/are still unidentified, and could include: Cenozoic mafic intrusions (alkali-olivine) into extended post-obduction sedimentary rocks near the Frontal Range Fault at the northwestern margin of the SHD (Al-Harthy *et al.* 1991; fig. 4.1); ongoing serpentinization, which is an exothermal reaction facilitated by water able to percolate into and circulate within the extensional fault system; and/or the heating of the water by ambient rocks due to a regionally elevated heat flow.

Hyperalkaline warm springs are hosted within ultramafic rocks of the Semail Ophiolite (e.g. Chavagnac *et al.* 2013*a*, *b*) with a pH value of >11 (table A.2). Hyperalkaline water forms by percolation of meteoric water along fractures within the mantle sequence of the Semail Ophiolite by low-temperature serpentinization reactions at depth (Chavagnac *et al.* 2013*a*, *b*; Miller *et al.* 2016). Recent research targeted the geochemistry and microbial habitability of the hyperalkaline spring water (e.g, Paukert *et al.* 2012; Chavagnac *et al.* 2013*a*, *b*; Miller *et al.* 2016).

Hyperalkaline fluids (pH >10) contain a high concentration of Ca^{2+}, OH^{-} and dissolved H_2 and CH_4 (Miller *et al.* 2016). The high content of gas in the aqueous fluids may be the product of ongoing low-temperature serpentinization (Miller *et al.* 2016). Extracted 16 s rRNA data of DNA from microbial organisms prove the existence of aerobic and anaerobic life in hyperalkaline water (Miller *et al.* 2016). The same authors identified abundant microorganisms such as *Meiothermus*, *Thermodesulfovibrionaceae* (sulfate reducers), Clostridia (fermenters) and *Hydrogenophaga* (H_2 oxidizers). The organism *Methanobacterium* (methanogen) and candidate phylum OP1 have also been found in the hyperalkaline water (Miller *et al.* 2016). The coupling between microorganisms, aqueous geochemistry and mineralogy in hyperalkaline water is not well understood (Miller *et al.* 2016). However, novel research based on the International Continental Drilling Program (ICDP) as part of the Oman Drilling Project (Oman DP), with more than a hundred participating scientists led by Peter Kelemen (Columbia University, USA), Jürg M. Matter (University of Southampton, UK) and Damon A.H. Teagle (National Oceanography Centre Southampton, UK), will shed light on this interesting topic.

Summary of the post-'Hercynian' tectonic evolution

We summarize and simplify or generalize the tectonic evolution of the Arabian margin and adjacent areas at the lithospheric scale, starting from continental rifting that postdates the Hercynian Unconformity to the present. The data and sources for construction of the schematic sketches of Figure 5.17 are provided in the previous section. Schematic and general geodynamic reconstructions on a lithospheric scale are provided in 10 steps in Figure 5.17a–j. See the previous section or Breton *et al.* (2004, fig. 8) for alternative interpretations and details (e.g. the origin of the Semail Ophiolite and the metamorphic sole, or different orientations of the Arabian margin during subduction). Note that the future JAD and SHD are shown in unnatural projections in favour of a schematic depiction; note also the different horizontal scale between the figure parts. The legend in Figure 5.17a applies to Figure 5.17a–j. Figure 5.18a–j provides the overall palaeogeography from the Neoproterozoic and for approximately the same timespan covered by Figure 5.17: (a) 270 Ma,

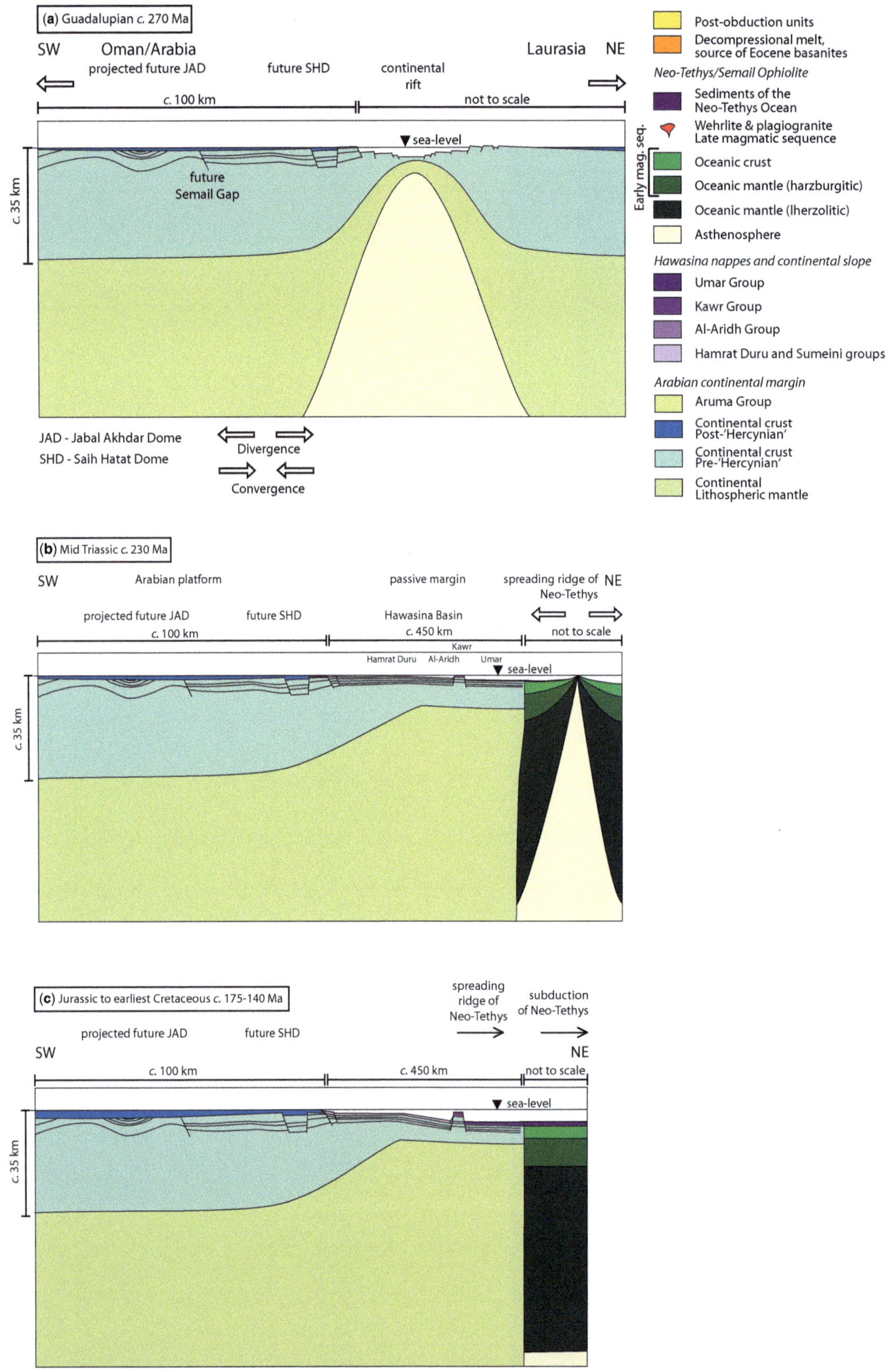

Fig. 5.17. (**a**) Continental rifting NE of the future SHD. Formation of a horst and graben system. The calcareous and partly marine synrift Saiq Formation started receiving siliciclastic input, due to uplift and erosion of the southwestern rift shoulder. Local rift volcanism and intrusions of sills and dykes occurred mainly in the Saiq Formation within the SHD. The pre-Permian units were separated by a fault between the locations of the future JAD and SHD (Semail Gap; section 'Main extensional fault systems' in Chapter 3). This fault may have played a pivotal role in the localization of the Amdeh Formation, which is restricted to the SHD in our study area ('Units above the Angudan Unconformity'). Also shown are the East African Orogeny / Angudan folds of the JAD. The position of the future JAD is projected into the section. (**b**) Deposition of the thick, shallow-marine calcareous Mahil Formation along the Arabian margin, following the transition from the rift to the drift stage. The deep-marine Hawasina Basin had formed (fig. 2.6; 'Pangea rifting and formation of the Neo-Thethys Ocean and Hawasina Basin'). The crust beneath the Hawasina Basin was thinned continental Arabian crust. The position of the future JAD is projected into the section. For legend see (a).

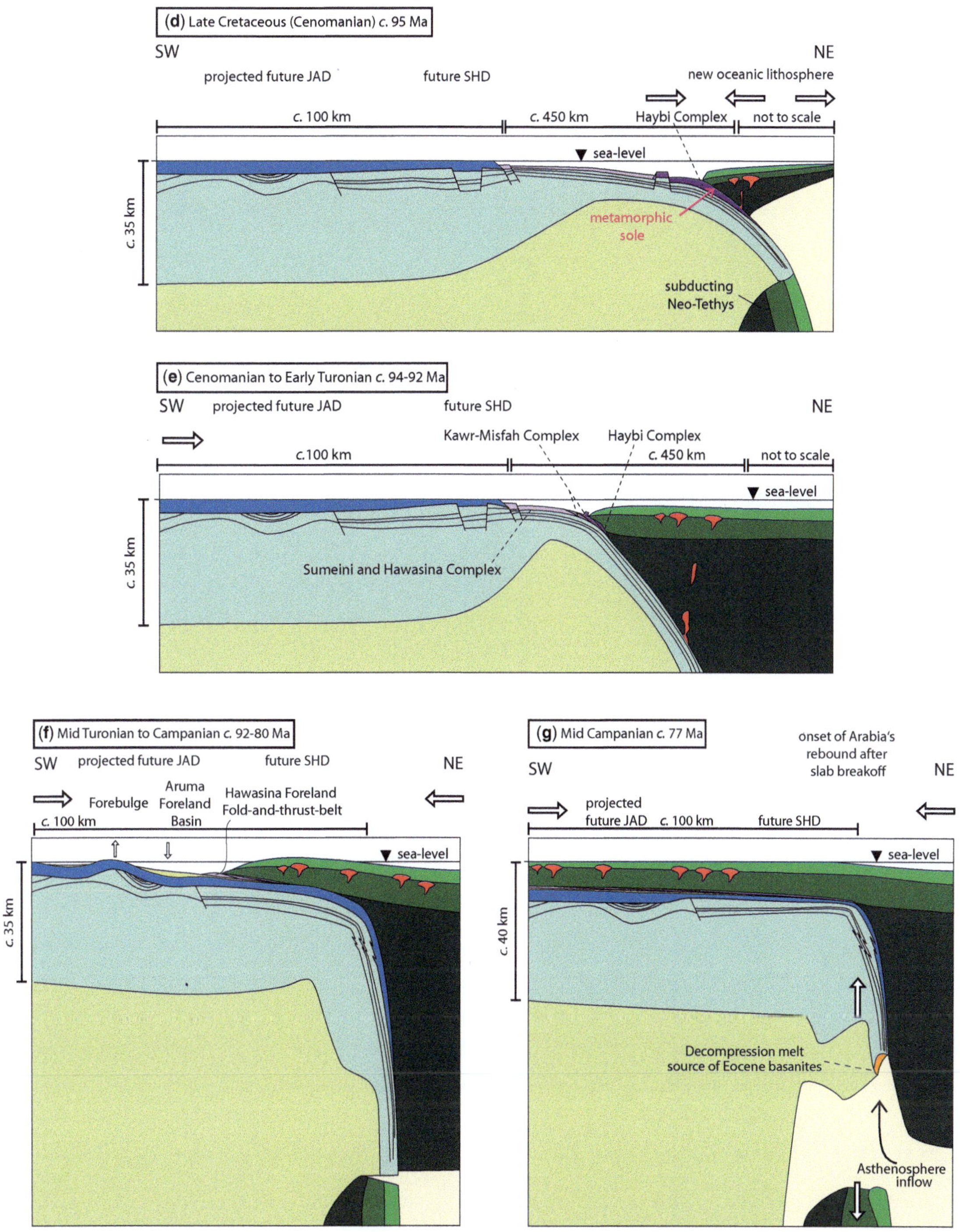

Fig. 5.17. *Continued.* (**c**) Neo-Tethys lithosphere was subducted beneath Laurasia (Fig. 5.18e). Local gentle uplift affected the Arabian margin and the Hawasina Basin during the Late Jurassic–Early Cretaceous because of the opening of the SoMoMa Ocean, creating the protoliths of the Masirah Ophiolite east of the Arabian margin (Béchennec *et al.* 1992; 'Tectonic framework' in Chapter 1). Local tilting and deposition of some terrigenous material at the Arabian Platform occurred during the Mid Jurassic (Mafraq Formation; Béchennec *et al.* 1992). The position of the future JAD is projected into the section. For legend see (a). (**d**) Subduction initiation either spontaneously initiated along a transform fault/fracture zone or was induced at *c.* 104 Ma. The induced subduction zone matured and a self-sustaining subduction zone formed at *c.* 96 Ma. A supra-subduction zone spreading centre formed, producing new oceanic lithosphere. The transform fault/fracture zone may have been striking NNE. The entire subduction zone therefore rotated clockwise after the formation of the new oceanic lithosphere and during obduction onto the Hawasina Basin (van Hinsbergen *et al.* 2019). Above the initial subducting slab, a back-arc spreading centre was established. Along the spreading axis, the early magmatic rock suite of the ophiolite crust formed at 96.25–95.5 Ma ('Intraoceanic thrusting, formation and obduction of the Semail Ophiolite'). Shortly afterwards, the subduction zone matured in which the late magmatic rocks formed during 95.4–95.2 Ma. This plate tectonic configuration led to the spreading above a subduction zone within the Neo-Tethys Ocean. In the course of the opening of the South Atlantic Ocean, Africa and Eurasia converged and Africa and Arabia rotated anticlockwise (Ricou 1994; Breton *et al.* 2004). The new young, hot and buoyant oceanic lithosphere was thrust ($D6_a$ in fig. A.2) onto the older Hawasina Basin (Haybi Complex). The crust of the Hawasina Basin subducted to a depth of *c.* ?35 km during plate convergence (section on 'Metemorphic sole' in Chapter 2). Segments of the subducting crust were exhumed to the base of the Semail Ophiolite to form the metamorphic sole at 96.1–94.8 Ma ('Intraoceanic thrusting, formation and obduction of the Semail Ophiolite'). Generation of the new oceanic lithosphere and thrusting of the ophiolite were coeval. For more details on tectonic modeling and formation of the Semail Ophiolite and the metamorphic sole see sections 'Metamorphic sole' and 'Semail Ophiolite' in Chapter 2 and 'Intraoceanic thrusting, formation and obduction of the Semail Ophiolite' in this chapter. Intraoceanic thrusting occurred at a possible rate of *c.* 150 mm a^{-1} for 1–2 Ma at *c.* 95 Ma (Hacker *et al.* 1996). The position of the future JAD is projected into the section. For legend see (a).

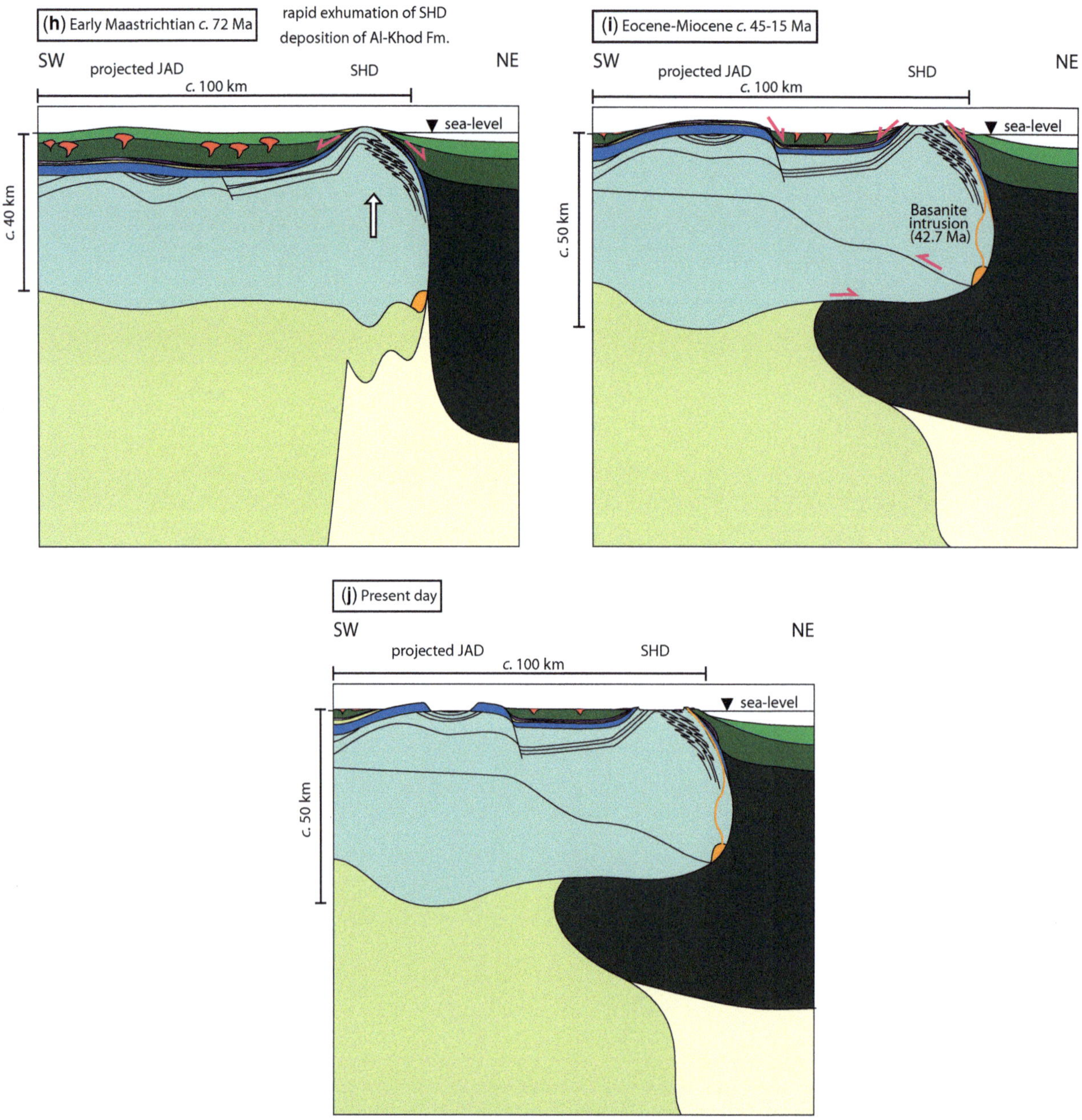

Fig. 5.17. *Continued.* (**e**) Ongoing obduction of the Semail Ophiolite onto the Hawasina Basin. The gentle dip of the slab is an assumption. The subduction angle could have steepened as the ophiolite progressed to the SW. Hawasina and continental slope deposits were deformed into nappes in the course of thrusting of the Semail Ophiolite, starting at *c.* 95 Ma (Sumeini Complex, Hawasina Complex, Kawr-Misfah Complex and Haybi Complex; see Searle 2007; section on 'Hawasina nappes' in Chapter 2 and 'Allochtonous thrusting on autochthonous units' in Chapter 3). The position of the future JAD is projected into the section. For legend see (a). (**f**) The Semail Ophiolite was thrust over the entire Arabian margin and platform from the Mid Turonian to Early Campanian. The load of the allochthonous units buried the underlying Arabian continental crust and caused the formation of the Aruma Foreland Basin and a forebulge (marked by arrows; $D6_b$ in fig. A.2; see 'Arabian Platform'). The Hawasina Foreland Fold-and-Thrust Belt formed as the leading part of the allochthonous units overrode the Aruma Group sediments of the foreland basin. The Arabian crust was too buoyant to be subducted to greater depth. The extended continental crust below the Hawasina Basin subducted down to ≥ 80 km (see section on 'Akhdar Group' in Chapter 2 and 'Intraoceanic thrusting, formation and obduction of the Semail Ophiolite'). A tear started to form in the subducted Arabian Plate at the boundary between the continental and the oceanic lithosphere (incipient slab break-off). Thrusting of the Semail Ophiolite onto the Hawasina Basin lasted for *c.* 17 Ma (from *c.* 95 to 78 Ma) over a distance of ≥ 450 km (width of the Hawasina Basin) at an average thrust velocity of *c.* 25–33 mm a^{-1} ('Arabian Platform'). The steep dip of the subducting Arabian slab is assumed, considering the downpull of the dense eclogitic lower part of the slab. Note that the Aruma Foreland Basin was formed in front of the Hawasina Fold-and-Thrust-Belt. The position of the future JAD is projected into the section. For legend see (a). (**g**) The Semail Ophiolite and Hawasina nappes obducted above the Arabian Platform and the Aruma Foreland Basin. Detailed sketches of synobduction thrusting based on field, seismic and well-log data are shown in Warburton *et al.* (1990, fig. 2.6). Arabia was pressed down by the weight of the allochthonous units. Slab break-off was completed. The loss of the oceanic slab pull triggered rapid isostatic uplift of the continental lithosphere and exhumation as well as deformation. Deformation included formation of crustal-scale sets of sheath folds (F4) and thrusting within and around the Hulw–As Sifah Window in the SHD ($D6_c$ in fig. A.2; sections 'Upper–lower plate discontinuity or Hulw–As Sifah Window' and 'F4 folds' in Chapter 3). Slab break-off facilitated an inflow of asthenospheric material between the sinking slab and the exhuming lithosphere (Hansman *et al.* 2017). This asthenospheric rise may be the source of the Eocene basanite intrusions (Hansman *et al.* 2017). The position of the future JAD is projected into the section. For legend see (a). (**h**) Continental rocks from a depth of ≥ 80 km were rapidly exhumed. The rocks of the Arabian crust were folded and sheared during exhumation. This deformation can be studied along the outcrops of the mega-sheath folds and the upper–lower

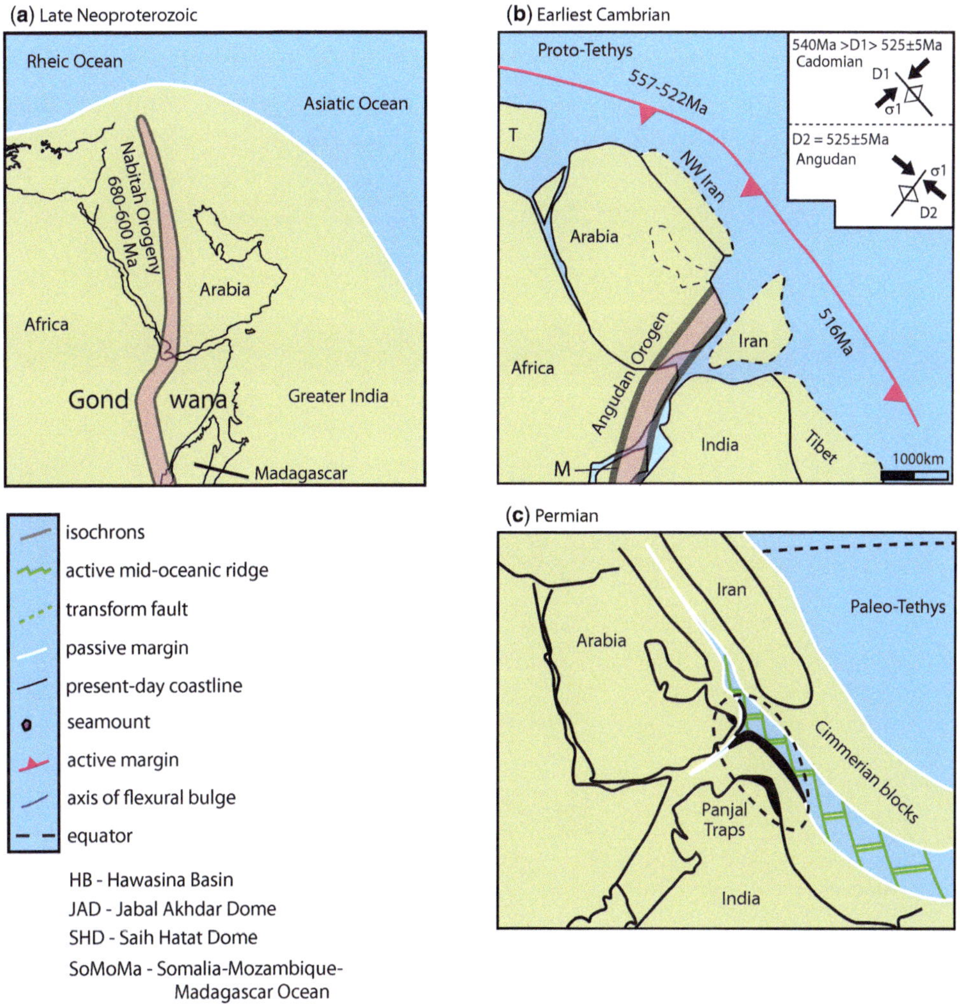

Fig. 5.18. (a–j). Palaeogeography of the continents and oceans of the greater Arabian realm. (**a**) The final amalmagation of Gondwana at the end of the Neoproterozoic, after Droste (2014) and Boger *et al.* (2015, fig. 1). (**b**) Cambrian, after Callegari *et al.* (2020). (**c, d**) Break-up of Pangea and development of the passive margin ((c) after Chauvet *et al.* (2009)).

Fig. 5.17. *Continued.* plate discontinuity or Hulw–As Sifah Window in the northeastern SHD. See figure A.3 for visualization of these high-strain structures. Coevally, the northern part of the SHD was forming. Major normal faults bound the SHD. Pre-Carboniferous autochthonous Arabian rock clasts of the Amdeh Formation in the post-obductional Al-Khod Formation indicate that exhumation at the northern SHD was considerable (several kilometres; sections on 'Main extensional fault systems' in Chapter 3 and 'Post-obduction deformation' in this chapter). No shallow-marine limestones were deposited during that time, indicating that the obducted Semail Ophiolite was above sea-level. Some exhumation of the Oman Mountains could have occurred due to erosion of the Semail Ophiolite and isostatic rebound of the Arabian Plate (section on 'Post-obduction deformation'). The entire ophiolite rotated *c.* 90° clockwise along a vertical axis from its formation to its emplacement (Morris *et al.* 2016). The Arabian margin at the site of the JAD was affected by extensional shearing towards the NNE, resulting in thinning of the Semail Ophiolite and formation of a precursor of the JAD (Grobe *et al.* 2018). The position of the JAD is projected into the section. For legend see (a). (**i**) The interval between the Maastrichtian and the Late Eocene is marked by stable tectonic conditions. Some parts (close to the present coastline) were covered with thick, shallow-marine limestone (≤ 2000 m in thickness). At *c.* 40 Ma, the final uplift of the JAD and SHD started (F5 folds). The exhumation at the southern SHD amounted to ≤ 2000 m, considering the uplift of Middle Eocene shallow-marine limestone. Synchronously with the Late Eocene–Early Miocene uplift of the two domes, gravitational collapse resulted in extension at their northern flanks ($D6_d$ in fig. A.2; F6 folds of fig. 4.3; Frontal Range Fault). Deep thrusting within the Arabian crust was possibly associated with the doming of the JAD and SHD (e. g. Hanna 1990; Hansman *et al.* 2017). Furthermore, some basanite intrusion occured during this episode (Scharf *et al.* 2021*b*). During doming and/or slightly after, gravitational collapse played a major role in the shaping of the Oman Mountains. Extensional fault zones at the margins of the two domes have a throw of a few to several kilometres (e. g. Mattern and Scharf 2018; Scharf *et al.* 2019*a*; sections on 'Main extensional fault systems' in Chapter 3 and 'Post-obduction deformation' in this chapter). Because of the scale of this figure, no details of the shallow crustal deformation could be drawn. The reader is referred to figure 4.3, which depicts shallow-crustal Cenozoic deformation in the hanging wall of the Frontal Range Fault of the Rusayl Embayment. The position of the JAD is projected into the section. For legend see (a). (**j**) Most of the major Neogene faults were inactive. However, some faults were active at the easternmost margin of the Arabian Plate (e. g. Ja'alan and Qalhat faults). Uplift of large parts of the Oman Mountains is evident by the incision of wadis into their own Quaternary deposits (often with overhangs) and into recent to subrecent marine terraces. Thermal springs occur along faults at the northern and eastern margins of the JAD and the northwestern margin of the SHD, some minor faults displace Quaternary deposits (see section on 'Quaternary deformation' in this chapter; $D6_e$ in fig. A.2), and earthquakes in the Sea of Oman and the SHD point to active geodynamics. The depth of the present-day Moho has been estimated to be *c.* 50 km below the JAD and *c.* 42 km below the SHD (section 4.1). Northeastern Oman is underlain by a lithosphere with an Pn-velocity anomaly (Al-Lazki *et al.* 2014; Hansman *et al.* 2017; Fig. 5.11). The thickness of the remaining Arabian crust amounts to *c.* 40 km (Scharf *et al.* 2021*e*, Chapter 4, this Memoir). The position of the JAD is projected into the section. For legend see (a).

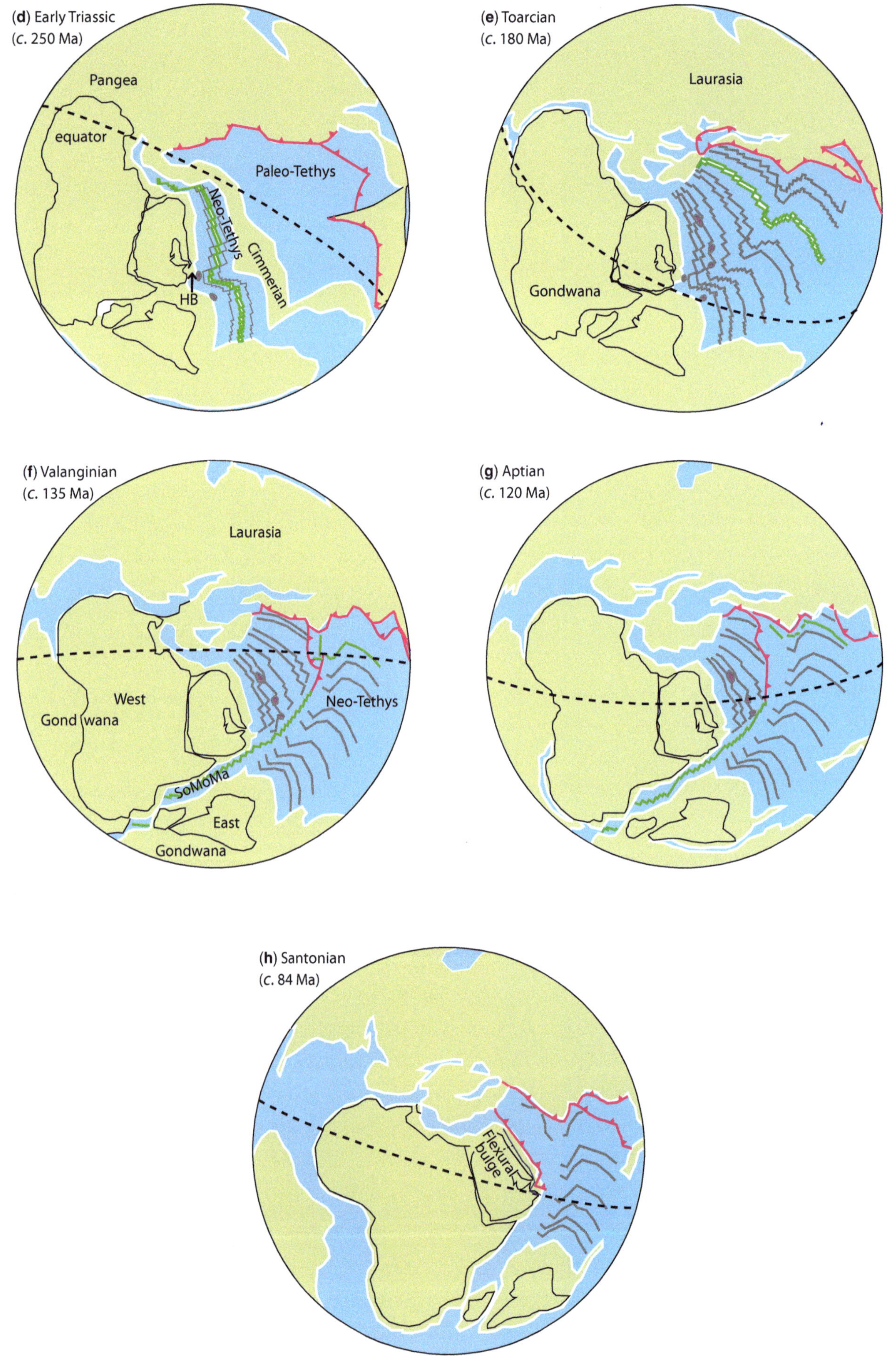

Fig. 5.18. *Continued.* (**e**) Toarcian passive margin. (**f**) Valanginian passive margin. (**g**) Aptian passive margin. (**h**) Santonian convergence between Arabia and Eurasia, obduction of the Semail Ophiolite and formation of Aruama Foreland Basin and forebulge.

(i) Paleocene (*c.* 60 Ma)

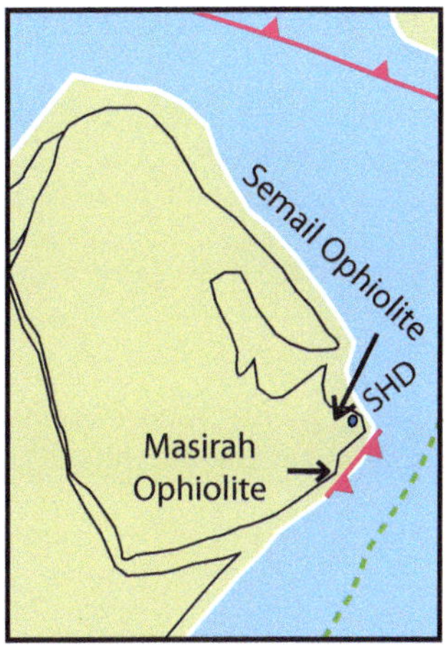

(j) Oligo-Miocene (*c.* 23 Ma)

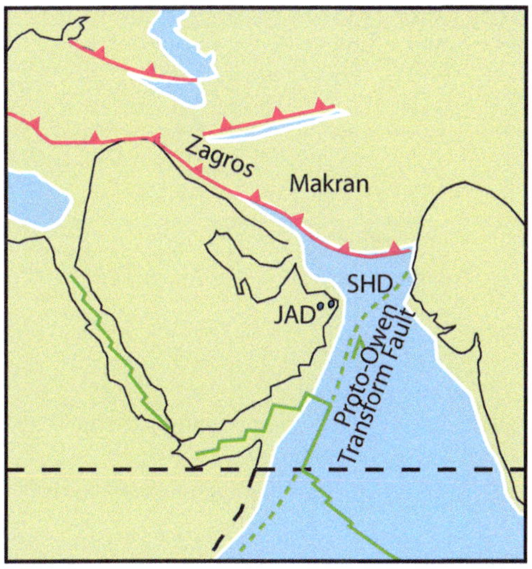

Fig. 5.18. *Continued.* (**i**) Paleocene obduction of the Masirah Ophiolite, modified after Gaina *et al.* (2015, fig. 9i). (d–h) Redrawn from Stampfli and Borel (2002). (**j**) Oligo-Miocene boundary. The JAD and SHD developed to their present-day situation as depicted in figure 1.2. Redrawn after Fournier *et al.* (2006) and Droste (2014), showing the sinistral Proto-Owen Transform Fault. Note that the Toarcian age of *c.* 180 Ma is after the timescale of Gradstein *et al.* (2012), whereas in Stampfli and Borel (2002) the Aalenian is shown for plate tectonic stage at 180 Ma.

Guadalupian; (b) 230 Ma, Mid Triassic; (c) 175 Ma, Early–Mid Jurassic; (d) 95 Ma, Cenomanian; (e) 94–92 Ma, Cenomanian–Early Turonian; (f) 92–85 Ma, Mid Turonian–Early Campanian; (g) 77 Ma, Mid Campanian; (h) 72 Ma, Early Maastrichtian; (i) 40–23 Ma, Late Eocene–Early Miocene; and (j) present day.

Author contributions **AS:** Conceptualization (lead), Writing original draft (lead), Visualization (lead), Supervision (lead), Investigation (lead); **FM:** Conceptualization (equal), Writing – review and editing (equal), Visualization (equal), Investigation (equal); **MAW:** Writing – review and editing (equal); **GF:** Writing – review and editing (equal); **DM:** Writing – review and editing (equal); **BP:** Writing – review and editing (equal); **WB:** Writing – review and editing (equal); **IC:** Writing – review and editing (equal).

Funding This research received no specific grant from any funding agency in the public, commercial, or not-for-profit sectors.

Data availability Data sharing is not applicable to this article as no datasets were generated or analysed during the current study.

References

Abbo, A., Avigad, D. and Gerdes, A. 2018. The lower crust of the Northern broken edge of Gondwana: evidence for sediment subduction and syn-Variscan anorogenic imprint from zircon U-Pb-Hf in granulite xenoliths. *Gondwana Research*, **64**, 84–96, https://doi.org/10.1016/j.gr.2018.08.002

Agard, P., Jolivet, L., Vrielynck, B., Burov, E. and Monié, P. 2007. Plate acceleration: the obduction trigger? *Earth and Planetary Science Letters*, **258**, 428–441, https://doi.org/10.1016/j.epsl.2007.04.002

Agard, P., Searle, M.P., Alsop, G. and Dubacq, B. 2010. Crustal stacking and expulsion tectonics during continental subduction: P-T deformation constraints from Oman. *Tectonics*, **29**, 1–19, https://doi.org/10.1029/2010TC002669

Alabaster, T., Pearce, J.A. and Malpas, J. 1982. The volcanic stratigraphy and petrogenesis of the Oman ophiolite complex. *Contribution to Mineralogy and Petrology*, **81**, 168–183, https://doi.org/10.1007/BF00371294

Al-Belushi, J.D., Glennie, K.W. and Williams, B.J.P. 1996. Permo-Carboniferous Al Khlata Formation, Oman: a new hypothesis of origin of its glaciation. *GeoArabia*, **1**, 389–404.

Aldega, L., Carminati, E., Scharf, A., Mattern, F. and Al-Wardi, M. 2017. Estimating original thickness and extent of the Semail Ophiolite in the eastern Oman Mountains by paleothermal indicators. *Marine and Petroleum Geology*, **84**, 18–33, https://doi.org/10.1016/j.marpet.geo.2017.03.024

Al-Harthy, M.S., Coleman, R.G., Hughes Clarke, M.W. and Hanna, S.S. 1991. Tertiary basaltic intrusions in the Central Oman Mountains. *In*: Peters, T., Nicolas, A. and Coleman, R.J. (eds) *Ophiolite Genesis and Evolution of the Oceanic Lithosphere. Proceedings of the Ophiolite Conference*, Muscat, Oman, Kluwer Academic Publishers, Dordrecht/Boston/London, **1990**, 475–682.

Al-Hatrushi, S.M. 1995. *Morphology and Late Tertiary and Quaternary Evolution of the Oman Coastline Between Muscat and Ras Al Hadd.* Unpublished Ph.D. thesis, University College of Swansea, UK.

Al-Hatrushi, S.M. 2017. Morphology of the raised shore platforms along the coastline between Daghmar and Dhabab, Sultanate of Oman. *Journal of Arts and Social Sciences [JASS],* Sultan Qaboos University, **8**, 13–24, https://doi.org/10.24200/jass.vol8iss2pp13-24

Al-Husseini, M.I. 1988. The Arabian Infracambrian extensional system. *Tectonophysics*, **148**, 93–103, https://doi.org/10.1016/0040-1951(88)90163-1

Al-Husseini, M.I. 2000. Origin of the Arabian Plate Structures: Amar Collision and Najd Rift. *GeoArabia*, **5**, 527–542.

Al-Husseini, M.I. 2004. Pre-Unayzah unconformity, Saudi Arabia. Carboniferous, Permian and Early Triassic Arabian Stratigraphy. *GeoArabia Special Publication*, **3**, 15–59.

Al-Husseini, M.I. 2014. Ediacaran-Cambrian Middle East geologic time scale 2014, proposed correlation of Oman's Abu Mahara Supergroup and Saudi Arabia's Jibalah Group. *GeoArabia*, **19**, 107–132.

Ali, M., Al Bahri, A.O. and Al-Harthy, A. 2016. Gravity experiment in Oman for exposing a basin. *Bahria University Research Journal of Earth Sciences*, **1**, 24–27.

Al-Kindy, M.H. and Richard, P.D. 2014. The main structural styles of the hydrocarbon reservoirs in Oman. *Geological Society, London, Special Publications*, **392**, 409–445, https://doi.org/10.1144/SP392.20

Al-Lazki, A.I., Seber, D., Sandvol, E. and Barazangi, M. 2002. A crustal transect across the Oman Mountains on the eastern margin of Arabia. *GeoArabia*, **7**, 47–78.

Al-Lazki, A.I., Al-Damegh, K.S., El-Hadidy, S.Y., Ghods, A. and Tatar, M. 2014. Pn-velocity structure beneath Arabia-Eurasia Zagros collision and Makran subduction zones. *Geological*

Society, London, Special Publications, **392**, 45–60, https://doi.org/10.1144/SP392.3

Allen, P.A. 2007. The Huqf Supergroup of Oman: basin development and context for Neoproterozoic glaciation. *Earth-Science Reviews*, **84**, 139–185, https://doi.org/10.1016/j.earscirev.2007.06.005

Allen, P.A. and Leather, J. 2006. Post-Marinoan marine siliciclastic sedimentation: the Masirah Bay Formation, Neoproterozoic Huqf Supergroup of Oman. *Precambrian Research*, **144**, 167–198, https://doi.org/10.1016/j.precamres.2005.10.006

Allen, P.A., Leather, J. and Brasier, M.D. 2004. The Neoproterozoic Fiq glaciation and its aftermath, Huqf Supergroup of Oman. *Basin Research*, **16**, 507–534, https://doi.org/10.1111/j.1365-2117.2004.00249.x

Allen, P.A., Leather, J. *et al.* 2011. Chapter 20 The Abu Mahara Group (Ghubrah and Fiq formations), Jabal Akhdar, Oman. *Geological Society of London, Memoirs*, **36**, 251–262, https://doi.org/10.1144/M36.20

Ambraseys, N.N., Melville, C.P. and Adams, R.D. 1994. *The Seismicity of Egypt, Arabia and Red Sea*. Cambridge University Press, Cambridge.

Amthor, J.E., Grotzinger, J.P., Schröder, S., Bowring, S.A., Ramezani, J., Martin, M.W. and Matter, A. 2003. Extinction of Clodina and Namacalathus at the Precambrian-Cambrian boundary in Oman. *Geology*, **31**, 431–434, https://doi.org/10.1130/0091-7613(2003)031<0431:EOCANA>2.0.CO;2

Angiolini, L., Balini, M., Garzanti, E., Nicora, A. and Tintori, A. 2003*a*. Gondwanan deglaciation and opening of Neotethys: the Al Khlata and Saiwan Formation of Interior Oman. *Palaeogeography. Palaeoclimatology, Palaeoecology*, **196**, 99–123, https://doi.org/10.1016/S0031-0182(03)00315-8

Angiolini, L., Balini, M., Garzanti, E., Nicora, A., Tintori, A., Crasquin, S. and Muttoni, G. 2003*b*. Permian climatic and paleogeographic changes in Northern Gondwana: the Khuff Formation of Interior Oman *Palaeogeography, Palaeoclimatology, Palaeoecology*, **196**, 269–300, https://doi.org/10.1016/S0031-0182(02)00668-5

Aubin, J. 1973. Le Royaume d'Ormuz au début du XVIe Sie'cle. *Mare Luso-Indicum*, **2**, 77–179.

Bahroudi, A. and Koyi, H.A. 2003. Effect of spatial distribution of Hormuz salt on deformation style in the Zagros folds and thrust belt: an analogue modeling approach. *Journal of Geological Society, London*, **160**, 719–733, https://doi.org/10.1144/0016-764902-135

Bailey, C.M., Hurtado, C., Scharf, A. and Mattern, F. 2019. Structural controls on listwaenite genesis in the Semail Ophiolite, Northern Oman. EGU 2019–5794, Vienna, Austria.

Béchennec, F. 1987. Géologie des Nappes Hawasina dans les parties orientale et central des Montagnes d'Oman. Thesis, Docteur des-Sciences, Pierre and Marie Curie Université, Paris 6. Bureau de Recherches Géologiques et Minières, Document No. 127, Orléans, France, 427.

Béchennec, F., Le Métour, J., Rabu, D., Villey, M. and Beurrier, M. 1988. The Hawasina Basin: a fragment of a starved passive continental margin, thrust over the Arabian Platform during obduction of the Sumail Nappe. *Tectonophysics*, **151**, 323–343, https://doi.org/10.1016/0040-1951(88)90251-X

Béchennec, F., Le Métour, J., Rabu, D., Bourdillon-de-Grissac, C., de Wever, P., Beurrier, M. and Villey, M. 1990. The Hawasina Nappes; Stratigraphy, paleogeography and structural evolution of a fragment of the South-Tethyan passive continental margin. *Geological Society, London, Special Publications*, **49**, 213–223, https://doi.org/10.1144/GSL.SP.1992.049.01.14

Béchennec, F., Tegyey, M., Le Métour, J., Lemière, B., Lescuyer, J.L., Rabu, D. and Milesi, J.P. 1991. Igneous rocks in the Hawasina Nappes and the Hajar Supergroup, Oman Mountains: their significance in the birth and evolution of the composite extensional Margin of the eastern Tethys. *In*: Peters, T., Nicolas, A. and Coleman, R.J. (eds) *Ophiolite Genesis and Evolution of the Oceanic Lithosphere. Proceedings of the Ophiolite Conference*, Muscat, Oman. Kluwer Academic Publishers, Dordrecht/Boston/London, **1990**, 593–611.

Béchennec, F., Roger, J., Le Métour, J. and Wyns, R. 1992. *Geological map of Seeb, sheet NF 40-03, scale 1:250,000, with Explanatory Notes*. Directorate General of Minerals, Oman Ministry of Petroleum and Minerals.

Bendias, D. and Aigner, T. 2015. Facies, sequence stratigraphy, reservoir and seal potential of the Mafraq Formation, Sultanate of Oman: an integrated outcrop analogue study. *GeoArabia*, **20**, 17–94.

Bernoulli, D. and Weissert, H. 1987. The Upper Hawasina nappes in the central Oman mountains stratigraphy, palinspastics and sequence of nappe emplacement. *Geodinamica Acta*, **1**, 47–58, https://doi.org/10.1080/09853111.1987.11105124

Bernoulli, D., Weissert, H. and Blome, C.D. 1990. Evolution of the Triassic Hawasina Basin, Central Oman Mountains. *Geological Society, London, Special Publications*, **49**, 189–202, https://doi.org/10.1144/GSL.SP.1992.049.01.12

Beurrier, M., Béchennec, F., Rabu, D. and Hutin, G. 1986. *Geological map of Rustaq, sheet NF 40-03D, scale 1:100,000, with Explanatory notes*. Directorate General of Minerals, Oman Ministry of Petroleum and Minerals.

Blasband, B., White, S., Brooijmans, P., De Boorder, H. and Visser, W. 2000. Late Proterozoic extensional collapse in the Arabian-Nubian Shield. *Journal of the Geological Society, London*, **157**, 615–628, https://doi.org/10.1144/jgs.157.3.615

Blechschmidt, I., Dumitrica, P., Matter, A., Krystyn, L. and Peters, T. 2004. Stratigraphic architecture of the northern Oman continental margin – Mesozoic Hamrat Duru Group Hawasina Complex, Oman. *GeoArabia*, **9**, 81–132.

Blechschmidt, I., Matter, A., Preusser, F. and Rieke-Zapp, D. 2009. Monsoon triggered formation of Quaternary alluvial megafans in the interior of Oman. *Geomorphology*, **110**, 128–139, https://doi.org/10.1016/j.geomorph.2009.04.002

Blendinger, W., van Vliet, A.T. and Hughes Clarke, M.W. 1990. Updoming, rifting and continental margin development during the Late Paleozoic in northern Oman. *Geological Society, London, Special Publications*, **49**, 27–37, https://doi.org/10.1144/GSL.SP.1992.049.01.03

Boger, S.D., Hirdes, W., Ferreira, C.A.M., Jenett, T., Dallwig, R. and Fanning, C.M. 2015. The 580-520 Ma Gondwana suture of Madagascar and its continuation into Antarctica and Africa. *Gondwana Research*, **28**, 1048–1060, https://doi.org/10.1016/j.gr.2014.08.017

Boudier, F. and Coleman, R.G. 1981. Cross sections through the peridotite in the Samail Ophiolite, southeastern Oman Mountains. *Journal of Geophysical Research*, **86**, 2573–2592, https://doi.org/10.1029/JB086iB04p02573

Boudier, F. and Nicolas, A. 2018. Synchronous seafloor spreading and subduction at the paleo-convergent margin of Semail and Arabia. *Tectonics*, **37**, 2961–2982, https://doi.org/10.1029/2018TC005099

Boudier, F., Bouchez, J.L., Nicolas, A., Cannat, M., Ceuleneer, G., Miseri, M. and Montigny, R. 1985. Kinematics of oceanic thrusting in the Oman ophiolites: model of plate convergence. *Earth and Planetary Science Letters*, **75**, 215–222, https://doi.org/10.1016/0012-821X(85)90103-7

Boudier, F., Ceuleneer, G. and Nicolas, A. 1988. Shear zones, thrusts and related magmatism in the Oman ophiolite: initiation of thrusting on an oceanic ridge. *Tectonophysics*, **151**, 275–296, https://doi.org/10.1016/0040-1951(88)90249-1

Bowring, S.A., Grotzinger, J.P., Condon, D.J., Ramezani, J., Newall, M. and Allen, P.A. 2007. Geochronologic constraints of the chronostratigraphic framework of the Neoproterozoic Huqf Supergroup, Sultanate of Oman. *American Journal of Science*, **307**, 1097–1145, https://doi.org/10.2475/10.2007.01

Braathen, A. and Osmundsen, P.T. 2019. Extensional tectonics rooted in orogenic collapse: long-lived disintegration of the Semail Ophiolite, Oman. *Geology*, **48**, https://doi.org/10.1130/G47077.1

Breton, J.-P., Béchennec, F., Le Métour, J., Moen-Maurel, L. and Razin, P. 2004. Eoalpine (Cretaceous) evolution of the Oman Tethyan continental margin: insights from a structural field study in Jabal Akhdar (Oman Mountains). *GeoArabia*, **9**, 41–58.

Burns, S.J., Matter, A., Frank, N. and Mangini, A. 1998. Speleothem-based paleoclimate record from northern Oman. *Geology*, **26**, 499–502, https://doi.org/10.1130/0091-7613(1998)026<0499:SBPRFN>2.3.CO;2

Burns, S.J., Fleitmann, D., Matter, A., Neff, U. and Mangini, A. 2001. Speleothem evidence from Oman for continental pluvial events during interglacial periods. *Geology*, **29**, 623–626, https://doi.org/10.1130/0091-7613(2001)029<0623:SEFOFC>2.0.CO;2

Callegari, I., Scharf, A. *et al.* 2020. Gondwana accretion tectonics and implications for the geodynamic evolution of eastern Arabia: first structural evidence of the existence of the Cadomian Orogeny in Oman (Jabal Akhdar Dome, Central Oman Mountains). *Journal of Asian Earth Sciences*, **187**, 104070, https://doi.org/10.1016/j.jseaes.2019.104070

Cawood, P.A., Green, F.K. and Calon, T.J. 1990. Origin of culmination within the southeastern Oman Mountains at Jebel Majhool and Ibra Dome. *Geological Society, London, Special Publications*, **49**, 397–418, https://doi.org/10.1144/GSL.SP.1992.049.01.27

Chauvet, F., Dumont, T. and Basile, C. 2009. Structures and timing of Permian rifting in the central Oman Mountains (Saih Hatat). *Tectonophysics*, **475**, 563–574, https://doi.org/10.1016/j.tecto.2009.07.008

Chavagnac, V., Ceuleneer, G., Monnin, C., Lansac, B., Hoareau, G. and Boulart, C. 2013*a*. Mineralogical assemblages forming at hyperalkaline warm springs hosted on ultramafic rocks: a case study of Oman and Ligurian ophiolites. *Geochemistry, Geophysics, Geosystems*, **14**, 2474–2495, https://doi.org/10.1002/ggge.20146

Chavagnac, V., Monnin, C., Ceuleneer, G., Boulart, C. and Hoareau, G. 2013*b*. Characterization of hyperalkaline fluids produced by low-temperature serpentinization of mantle peridotites in the Oman and Ligurian ophiolites. *Geochemistry, Geophysics, Geosystems*, **14**, 2496–2522, https://doi.org/10.1002/ggge.20147

Coffield, D.Q. 1990. Structures associated with nappe emplacement and culmination collapse in the Central Oman Mountains. *Geological Society, London, Special Publications*, **49**, 447–458, https://doi.org/10.1144/GSL.SP.1992.049.01.28

Coleman, R.G. 1971. Plate tectonic emplacement of upper mantle peridotites along continental edges. *Journal of Geophysical Research*, **76**, 1212–1222, https://doi.org/10.1029/JB076i005p01212

Condon, D., Zhu, M., Bowring, S., Wang, W., Yang, A. and Jin, Y. 2005. U-Pb ages from the Neoproterozoic Doushantuo Formation, China. *Science*, **308**, 95–98, https://doi.org/10.1126/science.1107765

Cooper, D.J.W. 1990. Sedimentary evolution and paleogeographical reconstruction of the Mesozoic continental rise in Oman: evidence from the Hamrat Duru Group. *Geological Society, London, Special Publications*, **49**, 161–187.

Corradetti, A., Spina, V. *et al.* 2019. Late-stage tectonic evolution of the Al-Hajar Mountains, Oman: new constraints from Palaeogene sedimentary units and low-temperature thermochronometry. *Geological Magazine*, 1–14, https://doi.org/10.1017/S0016756819001250

Cowan, R.J., Searle, M.P. and Waters, D.J. 2014. Structure of the metamorphic sole to the Oman Ophiolite, Sumeini Window and Wadi Tayyin: implications for ophiolite obduction processes. *Geological Society, London, Special Publications*, **392**, 155–175, https://doi.org/10.1144/SP392.8

Cozzi, A., Rea, G. and Craig, J. 2012. From global geology to hydrocarbon exploration: Edicaran-Early Cambrian petroleum plays of India, Pakistan and Oman. *Geological Society, London, Special Publications*, **366**, 131–162, https://doi.org/10.1144/SP366.14

de Wever, P., Bourdillon-de-Grissac, C.H. and Béchennec, F. 1988. A Permian age from radiolarites of the Hawasina nappes, Oman Mountains. *Geology*, **16**, 912–914, https://doi.org/10.1130/0091-7613(1988)016<0912:PAFROT>2.3.CO;2

de Wit, M.J., Jeffry, M.B. and Nicolayson, L. 1988. *Geological Map of Sectors of Gondwana, Reconstruction to their Disposition c. 550 Ma*. American Association of Petroleum Geologists and the University of Witwatersrand, Tulsa.

Dromart, G., Garcia, J.-P., Picard, S., Atrops, F., Lécuyer, C. and Sheppard, S.M.F. 2003. Ice age at the Middle-Late Jurassic transition? *Earth and Planetary Science Letters*, **213**, 205–220, https://doi.org/10.1016/S0012-821X(03)00287-5

Droste, H. 2010. High-resolution seismic stratigraphy of the Shu'aiba and Natih formations in the Sultanate of Oman: implications for Cretaceous epeiric carbonate platform systems. *Geological Society, London, Special Publications*, **329**, 145–162, https://doi.org/10.1144/SP329.7

Droste, H. 2014. Petroleum geology of the Sultanate of Oman. *In*: Marlow, L., Kendall, C. and Yose, L. (eds) *Petroleum Systems of the Tethyan Region*. American Association of Petroleum Geologists, Memoirs, **106**, 713–755, https://doi.org/10.1036/12431870M1063039

Droste, H. and van Steenwinkel, M. 2004. Stratal geometries and patterns of platform carbonates: the Cretaceous of Oman. *In*: Eberli, G., Massaferro, J.L. and Sarg, J.F.R. (eds) *Seismic Imaging of Carbonate Reservoirs and Systems*. American Association of Petroleum Geologists, Memoirs, **81**, 185–206.

Duretz, T., Agard, P., Yamato, P., Ducassou, C., Burov, E.B. and Gerya, T.V. 2016. Thermo-mechanical modeling of the obduction process based in the Oman Ophiolite case. *Gondwana Research*, **32**, 1–10, https://doi.org/10.1016/j.gr.2015.02.002

Edgell, H.S. 1991. Proterozoic salt basins of the Persian Gulf area and their role in hydrocarbon generation. *Precambrian Research*, **54**, 1–14, https://doi.org/10.1016/0301-9268(91)90065-I

El-Hussain, I., Deif, A. *et al.* 2012. Probabilistic seismic hazard maps for the Sultanate of Oman. *Natural Hazards*, **64**, 173–210, https://doi.org/10.1007/s11069-012-0232-3

El-Shazly, A.K., Broker, W., Hacker, B. and Calvert, A. 2001. Formation and exhumation of blueshists and eclogites from NE Oman: new perspectives from Rb-Sr and ^{40}Ar/^{39}Ar dating. *Journal of Metamorphic Geology*, **19**, 233–248, https://doi.org/10.1046/j.1525-1314.2001.00309.x

Ermertz, A.M., Kázmér, M., Adolphs, S.K., Falkenroth, M. and Hoffmann, G. 2019. Geoarchaelogical evidence for the decline of the Medieval City of Qalhat, Oman. *Open Quaternary*, **5**, 1–14, https://doi.org/10.5334/oq.56

Falk, E.S. and Kelemen, P.B. 2015. Geochemistry and petrology of listvenite in the Samail ophiolite, Sultanate of Oman: complete carbonation of peridotites during ophiolite emplacement. *Geochimica et Cosmochimica Acta*, **160**, 70–90, https://doi.org/10.1016/j.gca.2015.03.014

Falkenroth, M., Schneider, B. and Hoffmann, G. 2019. Beachrock as sea-level indicator – a case study at the coastline of Oman (Indian Ocean). *Quaternary Science Reviews*, **2016**, 81–98, https://doi.org/10.1016/j.quascirev.2019.01.003

Faqira, M., Rademakers, M. and Afifi, A. 2009. New insights into the Hercynian Orogeny, and their implications for the Paleozoic Hydrocarbon System in the Arabian Plate. *GeoArabia*, **14**, 199–228.

Fike, D.A., Grotzinger, J.P., Pratt, L.M. and Summons, R.E. 2006. Oxidation of the Ediacaran ocean. *Nature*, **444**, 744–747, https://doi.org/10.1038/nature05345

Filbrandt, J.B., Al-Dhahab, S. *et al.* 2006. Kinematic interpretation and structural evolution of North Oman, Block 6, since the Late Cretaceous and implications for timing of hydrocarbon migration into Cretaceous reservoirs. *GeoArabia*, **11**, 97–140.

Fink, R., Virgo, S., Arndt, M., Visser, W., Littke, R. and Urai, J.L. 2016. Solid bitumen in calcite veins from the Natih Formation in the Oman Mountains: multiple phases of petroleum migration in a changing stress field. *International Journal of Coal Geology*, **157**, 39–51, https://doi.org/10.1016/j.coal.2015.07.012

Forbes, G.A., Jansen, H.S.M. and Schreurs, J. 2010. Lexicon of Oman Subsurface Stratigraphy. Reference Guide to the Stratigraphy of Oman's Hydrocarbon Basins. *Gulf PetroLink, GeoArabia, Special Publication*, **5**, 373.

Fournier, M., Lepvrier, C., Razin, P. and Jolivet, L. 2006. Late Cretaceous to Paleogene Post-obduction extension and subsequent

Neogene compression in the Oman Mountains. *GeoArabia*, **11**, 17–40.

Fournier, M., Chamot-Rooke, N. *et al.* 2010. Somalia plate kinematics, evolution of the Aden–Owen–Carlsberg triple junction, and opening of the Gulf of Aden. *Journal of Geophysical Research*, **115**, B04102, https://doi.org/10.1029/2008JB006257

Frisch, W., Meschede, M. and Blakey, R. 2011. *Plate Tectonics: Continental Drift and Mountain Building*. Berlin, Heidelberg: Springer-Verlag, 212, https://doi.org/10.1007/978-3-540-76504-2_8

Gaina, C., van Hinsbergen, D.J.J. and Spakman, W. 2015. Tectonic interactions between India and Arabia since the Jurassic reconstructed from marine geophysics, ophiolite geology, and seismic tomography. *Tectonics*, **34**, 875–906, https://doi.org/10.1002/2014TC003780

Garber, J.M., Rioux, M. *et al.* 2016. Fast subduction, slow exhumation: dating continental subduction beneath the Oman ophiolite. Geological Society of America Annual Meeting, 25–28 September 2016, Denver, Colorado, USA, https://doi.org/10.1130/abs/2016AM-284338

Gardner, R.A.M. 1988. Aeolianites and marine deposits of the Wahiba sands: character and paleoenvironments. *Journal of Oman Studies, Special Report*, **3**, 75–94.

Gavillot, Y., Axen, G.J., Stöckli, D.F., Horton, B.K. and Fakhari, D. 2010. Timing of thrust activity in the High Zagros fold-thrust belt, Iran, from (U-Th)/He thermochronology. *Tectonics*, **29**, TC4025, https://doi.org/10.1029/2009TC002484

Genna, A., Nehlig, P., Le Goff, E., Guerrot, C. and Shanti, M. 2002. Proterozoic tectonism of the Arabian Shield. *Precambrian Research*, **117**, 21–40, https://doi.org/10.1016/S0301-9268(02)00061-X

Gerya, T.V. 2011. Intra-oceanic subduction zones. *In*: Brown, D. and Ryan, P.D. (eds) *Arc-Continent Collision, Frontiers in Earth Sciences*. Berlin, Heidelberg: Springer-Verlag, 23–51, https://doi.org/10.1007/978-3-540-88558-0_2

Glennie, K.W., Boeuf, M.G.A., Hughes Clarke, M.W., Moody-Stuart, M., Pilaar, W.F.H. and Reinhardt, B.M. 1973. Late Cretaceous nappes in Oman Mountains and their geological evolution. *American Association of Petroleum Geologists Bulletin*, **57**, 5–27.

Glennie, K.W., Boeuf, M.G.A., Hughes Clarke, M.W., Moody-Stuart, M., Pilaar, W. and Reinhardt, B.M. 1974. Geology of the Oman Mountains *Koninklijk Nederlands Geologisch en Mijnbouwkundig Genootschap Transactions*, **31**, 423.

Gnos, E., Immenhauser, A. and Peters, T. 1997. Late Cretaceous/early Tertiary convergence between the Indian and Arabian plates recorded in ophiolites and related sediments. *Tectonophysics*, **271**, 1–19, https://doi.org/10.1016/S0040-1951(96)00249-1

Goffé, B., Michard, A., Kienast, J.R. and le Mer, O. 1988. A case of obduction related high P, low T metamorphism in upper crustal nappes, Arabian continental margin, Oman: P-T paths and kinematic interpretation. *Tectonophysics*, **151**, 363–386, https://doi.org/10.1016/0040-1951(88)90253-3

Gomez-Perez, I., Moss, J. and Morton, A. 2019. Early tectonic evolution of Oman: insights from clastic provenance data and implications for the structural evolution of Arabia. *AAPG Event, 2nd Edition, Structural styles of the Middle East*, 9–11 December 2019, Muscat, Oman.

Gomez-Rivas, E., Bons, P.D. *et al.* 2014. The Jabal Akhdar Dome in the Oman Mountains: evolution of a dynamic fracture system. *American Journal of Science*, **314**, 1104–1139, https://doi.org/10.2475/07.2014.02

Goodenough, K.M., Styles, M.T. *et al.* 2010. Architecture of the Oman-UAE ophiolite: evidence for a multiple-phase magmatic history. *Arabian Journal of Geosciences*, **3**, 439–458, https://doi.org/10.1007/s12517-010-0177-3

Goodenough, K.M., Thomas, R.J., Styles, M.T., Schofield, D.I. and MacLeod, C.J. 2014. Records of ocean growth and destruction in the Oman-UAE ophiolite. *Elements*, **10**, 109–114, https://doi.org/10.2113/gselements.10.2.109

Gorin, G.E., Raacz, L.G. and Walter, M.R. 1982. Late Precambrian-Cambrian sediments of Huqf Group, Sultanate of Oman. *Bulletin of the American Association of Petroleum Geologists*, **66**, 2609–2627.

Gradstein, F.M., Ogg, J.G. and Hilgen, F.J. 2012. On the geologic time scale. *Newsletters on Stratigraphy*, **45**, 171–188, https://doi.org/10.1127/0078-0421/2012/0020

Gray, D.R. and Gregory, R.T. 2003. Ophiolite obduction and the Semail ophiolite: the behavior of the underlying margin. *Geological Society, London, Special Publications*, **218**, 449–466.

Gray, D.R. and Gregory, R.T. 2004. Comment on "Eoalpine (Cretaceous) evolution of the Oman Tethyan continental margin: insights from structural field study in Jabal Akhdar (Oman Mountains)" by J-P Breton *et al. GeoArabia*, **9**, 143–147.

Gray, D.T., Gregory, R.T. and Miller, J.M. 2000. A new structural profile along the Muscat-Ibra transect, Oman: implications for emplacement of the Semail ophiolite. *Geological Society of America, Special Paper*, **349**, 513–523.

Gray, D.T., Hand, M., Mawby, J., Armstrong, R.A., Miller, J.McL. and Gregory, R.T. 2004*a*. Sm-Nd and zircon U-Pb ages from garnet-bearing eclogites, NE Oman: constraints on high-P metamorphism. *Earth and Planetary Science Letters*, **222**, 407–422, https://doi.org/10.1016/j.epsl.2004.03.016.

Gray, D.R., Miller, J.M., Foster, D.A. and Gregory, R.T. 2004*b*. Transition from subduction- to exhumation-related fabrics in glaucophane-bearing eclogites, Oman: evidence from relative fabric chronology and $^{40}Ar/^{39}Ar$ ages. *Tectonophysics*, **389**, 35-64, https://doi.org/10.1016/j.tecto.2004.06.016

Gray, D.R., Gregory, R.T., Armstrong, R.A., Richards, I.J. and Miller, J.McL. 2005*a*. Age and stratigraphic relationship of structurally deepest level rocks, Oman Mountains: U/Pb SHRIMP evidence for Late Carboniferous Neotethys rifting. *The Journal of Geology*, **113**, 611–626, https://doi.org/10.1086/449325

Gregory, R.T., Gray, D.R. and Miller, J.M. 1998. Tectonics of the Arabian margin associated with the emplacement of the Oman margin along the Ibra transect: new evidence from northeast Saih Hatat. *Tectonics*, **17**, 657–670, https://doi.org/10.1029/98TC02206

Grélaud, C., Razin, P. and Homewood, P. 2010. Channelized systems in an inner carbonate platform setting: differentiation between incisions and tidal channels (Natih Formation, Late Cretaceous, Oman). *Geological Society, London, Special Publications*, **329**, 163–186, https://doi.org/10.1144/SP329.8

Grobe, A., Urai, J.L., Littke, R. and Lünsdorf, N.K. 2016. Hydrocarbon generation and migration under a large overthrust: the carbonate platform under the Semail Ophiolite, Jebel Akhdar, Oman. *International Journal of Coal Geology*, **168**, 3–19, https://doi.org/10.1016/j.coal.2016.02.007

Grobe, A., Virgo, S., von Hagke, C., Urai, J.L. and Littke, R. 2018. Multiphase structural evolution of a continental margin during obduction orogeny: insights from the Jebel Akhdar Dome, Oman Mountains. *Tectonics*, **37**, 888–913, https://doi.org/10.1002/2016TC004442

Grobe, A., von Hagke, C., Littke, R., Dunkl, I., Wübbeler, F., Muchez, P. and Urai, J.L. 2019. Tectono-thermal evolution of Oman's Mesozoic passive continental margin under the obducting Semail Ophiolite: a case study of Jebel Akhdar, Oman. *Solid Earth*, **10**, 149–175, https://doi.org/10.5194/se-10-149-2019

Grotzinger, J.P., Al-Siyabi, H.A., Al-Hashimi, R.A. and Cozzi, A. 2002. New model for tectonic evolution of Neoproterozoic-Cambrian Huqf Supergroup basins, Oman. *5th Middle East Geosciences Conference, GEO 2002*. GeoArabia, Abstract, **7**, 241.

Guilmette, C., Smit, M.A. *et al.* 2018. Forced subduction initiation recorded in the sole and crust of the Semail Ophiolite of Oman. *Nature Geoscience*, **11**, 688–695, https://doi.org/10.1038/s41561-018-0209-2

Guiraud, R., Bosworth, W., Thierry, J. and Delplanque, A. 2005. Phanerozoic geological evolution of Northern and Central Africa: an overview. *Journal of African Earth Sciences*, **43**, 83–143, https://doi.org/10.1016/j.jafrearsci.2005.07.017

Gurnis, M., Hall, C. and Lavier, L. 2004. Evolving force balance during incipient subduction *Geochemistry, Geophysics, Geosystems*, **5**, Q07001, https://doi.org/10.1029/2003GC000681

Haase, K.M., Freund, S., Koepke, J., Hauff, F. and Erdmann, M. 2015. Melts of sediments in the mantle wedge of the Oman ophiolite. *Geology*, **43**, 275–278, https://doi.org/10.1130/G36354.1

Haase, K.M., Freund, S., Beier, C., Koepke, J., Erdmann, M. and Hauff, F. 2016. Constraints on the magmatic evolution of the oceanic crust from plagiogranite intrusions in the Oman ophiolite. *Contributions to Mineralogy and Petrology*, **171**, https://doi.org/10.1007/s00410-016-1261-9

Hacker, B.R. and Gnos, E. 1997. The conundrum of Samail: explaining the metamorphic history. *Tectonophysics*, **279**, 215–226, https://doi.org/10.1016/S0040-1951(97)00114-5

Hacker, B.R., Mosenfelder, J.L. and Gnos, E. 1996. Rapid emplacement of the Oman ophiolite: thermal and geochronical constrains. *Tectonics*, **15**, 1230–1247, https://doi.org/10.1029/96TC01973

Hall, C.E., Gurnis, M., Sdrolias, M., Lavier, L.L. and Müller, D.R. 2003. Catastrophic initiation of subduction following forced convergence across fracture zones. *Earth and Planetary Science Letters*, **212**, 15–30, https://doi.org/10.1016/S0012-821X(03)00242-5

Hanna, S.S. 1990. The Alpine deformation of the central Oman Mountains. *Geological Society, London, Special Publications*, **49**, 341–359.

Hanna, S.S. 1995. *Field guide to the Geology of Oman. Volume 1 Western Hajar Mountains and Musandam.* The Historical Association of Oman, Ruwi (Historical Association of Oman), 178.

Hannss, C. 1991. *The Capital Area of Northern Oman, Teil 1, Basic Relief Development.* Reichert Verlag, Wiesbaden, 117–124.

Hansman, R.J., Ring, U., Thomson, S.N., den Brock, B. and Stübner, K. 2017. Late Eocene uplift of the Al Hajar Mountains, Oman, supported by stratigraphic and low-temperature thermochronology. *Tectonics*, **36**, 3081–3109, https://doi.org/10.1002/2017TC004672

Hansman, R.J., Albert, R., Gerdes, A. and Ring, U. 2018. Absolute ages of multiple generations of brittle structures by U-Pb dating of calcite. *Geology*, **46**, 207–210, https://doi.org/10.1130/G39822.1

Haq, B.U., Hardenbol, J. and Vail, P.R. 1987. The chronology of fluctuating sea level since the Triassic. *Science*, **235**, 1156–1167, https://doi.org/10.1126/science.235.4793.1156

Hersi, O.S. and Al-Harthi, A. 2010. Lithofacies attributes of a transgressive carbonate system: the Middle Eocene Seeb Formation, Al Khoudh area, Muscat, Oman. *Sultan Qaboos University Journal for Science*, **15**, 41–54, https://doi.org/10.24200/squjs.vol15iss0pp41-54

Heward, A.P. and Penney, R.A. 2014. Al Khlata glacial deposits in the Oman Mountains and their implications. *Geological Society, London, Special Publications*, **392**, 279–301, https://doi.org/10.1144/SP392.15

Hoffman, P.F., Kaufman, A.J., Halverson, G.P. and Schrag, D.P. 1998. A Neoproterozoic Snowball Earth. *Science*, **281**, 1342–1346, https://doi.org/10.1126/science.281.5381.1342

Hoffmann, G., Rupprechter, M. and Mayrhofer, C. 2013. Review of the long-term coastal evolution of North Oman-subsidence v. uplift. *Zeitschrift der Deutschen Gesellschaft für Geowissenschaften (German Journal of Geosciences)*, **164**, 237–252, https://doi.org/10.1127/1860-1804/2013/0002

Hoffmann, G., Schneider, B., Mechernich, S., Falkenroth, M., Dunai, T. and Preusser, F. 2019. Quaternary uplift along a passive continental margin (Oman, Indian Ocean). *Geomorphology*, **350**, 106870, https://doi.org/10.1016/j.geomorph.2019.106870

Holland, M., Urai, J.L., Muchez, P. and Willemse, J.M. 2009*b*. Evolution of fractures in a highly dynamic thermal, hydraulic, and mechanical system – (I) Field observations in Mesozoic Carbonates, Jabal Shams, Oman Mountains. *GeoArabia*, **14**, 57–110.

Homewood, P., Razin, P., Grelaud, C., Droste, H., Vahrenkamp, V., Mettraux, M. and Mattner, J. 2008. Outcrop sedimentology of the Natih Formation, northern Oman: a field guide to selected outcrops in the Adam Foothills and Al Jabal al Akhdar areas. *GeoArabia*, **13**, 39–120.

Hughes Clarke, M.W. 1988. Stratigraphy and rock unit nomenclature in the oil-producing area of interior Oman. *Journal of Petroleum Geology*, **11**, 5–60, https://doi.org/10.1111/j.1747-5457.1988.tb00800.x

Hughes Clarke, M.W. 2006. *Oman's Geological Heritage*. Oman: Petroleum Development, 247 p.

Immenhauser, A., Schreurs, G., Peters, T., Matter, A., Hauser, M. and Dumitrica, P. 1998. Stratigraphy, sedimentology and depositional environment of the Permian to uppermost Cretaceous Batain Group, eastern Oman. *Eclogae Geolgica Helvetica*, **91**, 217–235.

Immenhauser, A., Schreurs, G., Gnos, E., Oterdoom, W. and Hartmann, B. 2000. Late Paleozoic to Neogene geodynamic evolution of the northeastern Oman margin. *Geological Magazine*, **137**, 1–18, https://doi.org/10.1017/S0016756800003526

Immerz, P.W., Oterdoom, H. and El-Tonbary, M. 2000. The Huqf/Haima hydrocarbon system of Oman and the termianl phase of the Pan-African Orogeny: evaporite depositions in a compressive setting. *4th Middle East Geosciences Conference, GEO 2000*, GeoArabia, Abstract, **5**, 113–114.

Ishikawa, T., Nagaishi, K. and Umino, S. 2002. Boninitic volcanism in the Oman ophiolite: implications for thermal condition during transition from spreading ridge to arc. *Geology*, **30**, 899–902, https://doi.org/10.1130/0091-7613(2002)030<0899:BVITOO>2.0.CO;2

Jacobs, J., Bingen, B., Thomas, R.J., Bauer, W., Wingate, M.T.D. and Feitio, P. 2008. Early Paleozoic orogenic collapse and voluminous late-tectonic magmatism in Dronning Maud Land and Mozambique: insights into the partially delaminated orogenic root of the East African-Antarctic Orogen *Geological Society, London, Special Publications*, **308**, 69–90, https://doi.org/10.1144/SP308.3

Kellerhals, P. and Matter, A. 2003. Facies analysis of a glaciomarine sequence, the Neoproterozoic Mirbat Sandstone Formation, Sultanate of Oman. *Eclogae Geologicae Helvetiae*, **96**, 49–70.

Kent, P.E. 1979. The emergent Hormuz salt plugs of Southern Iran. *Journal of Petroleum Geology*, **2**, 117–144, https://doi.org/10.1111/j.1747-5457.1979.tb00698.x

Kilner, B., Mac Niocaill, C. and Brasier, M. 2005. Low-latitude glaciation in the Neoproterozoic of Oman. *Geology*, **33**, 413–416, https://doi.org/10.1130/G21227.1

Kirschvink, J.L. 1992. Late Proterozoic low-latitude global glaciation: the Snowball Earth. *In*: Schopf, J.W. and Klein, C. (eds) *The Proterozoic Biosphere.* Cambridge University Press, Cambridge, 51–52.

Konert, G., Afifi, A.M., Al-Hajri, S.A. and Droste, H.J. 2001. Paleozoic stratigraphy and hydrocarbon habitat of the Arabian Plate. *GeoArabia*, **6**, 407–442.

Koopman, A., van der Berg, M., Romine, K. and Teasdale, J. 2007. Proterozoic to Cambrian plate-tectonics and its control on the structural evolution of the Ara Salt-Basin in Oman. *Abstract AAPG European Region Conference*, Athens, Greece: AAPG Search and Discovery Article #90072.

Kozur, H.W., Wardlaw, B.R., Baud, A., Béchennec, F., Marcoux, J. and Richoz, S. 2001. Middle Permian conodonts from Oman. *Permophiles*, **38**, 10–12.

Kusky, T., Robinson, C. and El-Baz, F. 2005. Tertiary-Quaternary faulting and uplift in the northern Oman Hajar Mountains. *Journal of the Gological Society of London*, **162**, 871–888 https://doi.org/10.1144/0016-764904-122

Leather, J. 2001. *Sedimentology, chemostratigraphy and geochronology of the lower Huqf Supergroup, Oman.* Ph.D. thesis, Trinity College Dublin, Ireland, 227.

Leather, J., Allen, P.A., Braiser, M.D. and Cozzi, A. 2002. A Neoproterozoic Snowball Earth under scrutiny: evidence from the Fiq glaciation of Oman. *Geology*, **30**, 891–894, https://doi.org/10.1130/0091-7613(2002)0302.0.CO.2

Le Guerroué, E., Allen, P.A. and Cozzi, A. 2005. Two distinct glacial successions in the Neoproterozoic of Oman. *GeoArabia*, **10**, 17–34.

Le Guerroué, E., Allen, P.A. and Cozzi, A. 2006. Chemostratigraphic and sedimentological framework of the largest negative carbon isotopic excursion in Earth history: the Neoproterozoic Shuram Formation (Nafun Group, Oman). *Precambrian Research*, **144**, 68–92, https://doi.org/10.1016/j.precamres.2006.01.007

Le Métour, J., Villey, M. and de Gramont, X. 1986. *Geological map of Quryat, sheet NF 40-4D, scale 1:100,000, Explanatory Notes*. Directorate General of Minerals, Oman Ministry of Petroleum and Minerals.

Leroy, S., Gente, P. *et al.* 2004. From rifting to spreading in the eastern Gulf of Aden: a geophysical survey of a young oceanic basin from margin to margin. *Terra Nova*, **16**, 185–192, https://doi.org/10.1111/j.1365-3121.2004.00550.x

Lippard, S.J., Shelton, A.W. and Gass, I.G. 1986. The ophiolite of northern Oman. *Geological Society of London, Memoir*, **11**, 178.

Loosveld, R.J.H., Bell, A. and Terken, J.J.M. 1996. The tectonic evolution of interior Oman. *GeoArabia*, **1**, 28–51.

Lovelock, P.E.R., Potter, T.L., Walsworth-Bell, E.B. and Wiemer, W.M. 1981. Ordovician rocks in the Oman mountains: the Amdeh formation. *Geologie en Minjbouw*, **60**, 487–495.

MacLeod, C.J., Lissenberg, L. and Bibby, L.E. 2013. 'Moist MORB' axial magmatism in the Oman ophiolite: the evidence against a mid-ocean ridge origin. *Geology*, **41**, 459, https://doi.org/10.1130/G33904.1

Mann, A., Hanna, S.S. and Nolan, S.C. 1990. The post-Campanian tectonic evolution of the Central Oman Mountains: tertiary extension of the Eastern Arabian Margin. *Geological Society, London, Special Publications*, **49**, 285–305.

Marquer, D., Mercolli, I. and Peters, T. 1998. Early Cretaceous intra-oceanic rifting in the Proto-Indian Ocean recorded in the Masirah Ophiolite, Sultanate of Oman. *Tectonophysics*, **292**, 1–16, https://doi.org/10.1016/S0040-1951(98)00063-8

Mattern, F. and Scharf, A. 2018. Postobductional extension along and within the Frontal Range of the Eastern Oman Mountains. *Journal of Asian Earth Sciences*, **154**, 369–385, https://doi.org/10.1016/j.jseaes.2017.12.031

Mattern, F. and Scharf, A. 2019. Transition from the Hajir Formation to the Muaydin Formation: a facies change coinciding with extensional, syndepositional faulting (Ediacaran, Jabal Akhdar, Central Oman Mountains). *Journal of African Earth Sciences*, **152**, 237–244, https://doi.org/10.1016/j.jafrearsci.2019.02.016

Mattern, F., Schneider, W., Wang, P. and Li, C. 1998. Continental strike-slip rifts and their stratigraphic signature: application to the Bangong/Nujiang zone (Tibet) and the South Penninic zone (Alps). *Geologische Rundschau*, **87**, 206–224, https://doi.org/10.1007/s005310050203

Mattern, F., Moraetis, D. *et al.* 2018*a*. Coastal dynamics of uplifted and emerged Late Pleistocene near-shore coral patch reefs at Fins (eastern coastal Oman, Gulf of Oman). *Journal of African Earth Sciences*, **138**, 192–200, https://doi.org/10.1016/j.afrearsci.2017.11.018

Mattern, F., Scharf, A., Al-Sarmi, M., Pracejus, B., Al-Hinaai, A. and Al-Mamari, A. 2018*b*. Compaction history of Upper Cretaceous shale and related tectonic framework, Arabian Plate, Eastern Oman Mountains. *Arabian Journal of Geosciences*, **11**, 444, https://doi.org/10.1007/s12517-018-3781-2

Mattes, B.W. and Conway Morris, S. 1990. Carbonate/evaporate deposition in the Late Precambrian – Early Cambrian Ara Formation of Southern Oman. *Geological Society, London, Special Publications*, **49**, 617–636.

Maury, R.C., Béchennec, F., Cotton, J., Caroff, M., Cordey, F. and Marcoux, J. 2003. Middle Permian plume-related magmatism of the Hawasina Nappes and the Arabian Platform: implications on the evolution of the Neotethyan margin in Oman. *Tectonics*, **22**, 12-11-12-13, https://doi.org/10.1029/2002TC001483

McQuarrie, N. and van Hinsbergen, D.J.J. 2013. Retrodeforming the Arabia-Eurasia collision zone: age of collision v. magnitude of continetal subduction. *Geology*, **41**, 315–318, https://doi.org/10.1130/G33591.1

Miller, H.M., Matter, J.M. *et al.* 2016. Modern water/rock reactions in Oman hyperalkaline peridotites aquifers and implications for microbial habitability. *Geochimica et Cosmochimica Acta*, **179**, 217–241, https://doi.org/10.1016/j.gca.2016.01.033

Miller, J.M., Foster, D.A., Gray, D.R. and Gregory, R.T. 1999. Geological and geochronological constraints on the exhumation of a high-pressure metamorphic terrane, Oman. *Geological Society, London, Special Publications*, **154**, 241–260.

Miller, J.M., Gray, D.R. and Gregory, R.T. 2002. Geometry and significance of internal windows and regional isoclinal folds in northeast Saih Hatat, Sultanate of Oman. *Journal of Structural Geology*, **24**, 359–386, https://doi.org/10.1016/S0191-8141(01)00061-X

Miller, N., Johnson, P.R. and Stern, R.J. 2008. Marine v. non-marine environments for the Jibalah Group, NW Arabian Shield: a sedimentologic and geochemical survey and report of possible metazoan in the Dhaiqa Formation. *The Arabian Journal for Science and Engineering*, **33**, 55–77.

Moores, E.M. and Vine, F.J. 1971. Troodos massif, Cyprus and other ophiolites as oceanic crust: evaluation and implications. *Philosophical Transactions of the Royal Society of London A*, **268**, 443–466, https://doi.org/10.1098/rsta.1971.0006

Moraetis, D., Mattern, F., Scharf, A., Frijia, G., Kusky, T.M., Yuan, Y. and Hussain, I.L. 2018. Neogene to Quaternary uplift history along the passive margin of the northeastern Arabian Peninsula eastern Hajar Mountains, Oman. *Quaternary Research*, **90**, 418–434, https://doi.org/10.1017/qua.2018.51

Morris, A., Meyer, M., Anderson, M.W. and MacLeod, C.J. 2016. Clockwise rotation of the entire Oman ophiolite occurred in a suprasubduction zone setting. *Geology*, **44**, 1055–1058, https://doi.org/10.1130/G38380.1

Mount, V.S., Crawford, R.I.S. and Bergman, S.C. 1998. Regional structural style of the Central and Southern Oman Mountains: Jabal Akhdar, Saih Hatat and the Northern Ghaba Basin. *GeoArabia*, **3**, 475–490.

Musson, R.M.W. 2009. Subduction in the Western Makran: the historian's contribution. *Journal of the Geological Society of London*, **166**, 387–391, https://doi.org/10.1144/0016-76492009-119

Nicolas, A. 1989. *Structures of Ophiolites and Dynamics of Oceanic Lithosphere*. Kluwer, Netherlands, 367 p.

Nicolas, A. and Boudier, F. 2017. Emplacement of Semail-Emirates ophiolite at ridge-trench collision. *Terra Nova*, **29**, 127–134, https://doi.org/10.1111/ter.12256

Nolan, S.C., Skelton, P.W., Clissold, B.P. and Smewing, J.D. 1990. Maastrichtian to Early Tertiary stratigraphy and paleogeography of the Central and Northern Oman Mountains. *Geological Society, London, Special Publications*, **49**, 307–325.

Oterdoom, W.H., Worthing, M.A. and Partington, M. 1999. Petrological and tectonostratigraphical evidence for a Mid Ordovician rift pulse on the Arabian Peninsula. *GeoArabia*, **4**, 467–500.

Pandit, M.K., Sial, A.N., Jamrani, S.S. and Ferreira, V.P. 2001. Carbon isotopic profile across the Bilara Group rocks of the Trans-Aravalli Marwar Supergroup in western India: impilcations for Neoproterozoic-Cambrian transition. *Gondwana Research*, **4**, 387–394, https://doi.org/10.1016/S1342-937X(05)70338-5

Paukert, A.N., Matter, J.M., Kelemen, P., Shock, E.L. and Havig, J.R. 2012. Reaction path modeling of enhanced in situ CO_2 mineralization for carbon sequestration in the peridotites of the Samail Ophiolite, Sultanate of Oman. *Chemical Geology*, **330–331**, 86–100, https://doi.org/10.1016/j.chemgeo.2012.08.013

Pavlopoulos, K., Moraetis, D., Scharf, A., Foumelis, M. and Fouache, E. 2018. Pleistocene uplift evidences and geomorphological processes, along the Daghmar-Dibbab coastal area, (NE Oman). EGU 2018–5791, vol. 20, Vienna.

Pearce, J.A., Alabaster, T., Shelton, A.W. and Searle, M.P. 1981. The Oman ophiolite as a Cretaceous arc-basin complex: evidence and implications. *Philosophical Transactions of the Royal Society of London A*, **300**, 299–317, https://doi.org/10.1098/rsta.1981.0066

Perrin, M., Plenier, G., Dautria, J-M., Cocuaud, E. and Prévot, M. 2000. Rotation of the Semail ophiolite (Oman): additional paleomagnetic data from the volcanic sequence. *Marine Geophysical Researches*, **21**, 181–194, https://doi.org/10.1023/A:1026738313805

Peters, T. and Mercolli, I. 1997. Formation and evolution of the Masirah Ophiolite (Sultanate of Oman). *Ofioliti*, **22**, 15–34.

Peters, T., Immenhauser, A., Mercolli, I. and Meyer, J. 1995. *Geological map of Masirah North and Masirah South*. Explanatory notes, scale 1:50,000, sheet K768 North and K768 South. Directorate General of Minerals, Oman Ministry of Petroleum and Minerals, Muscat, Oman.

Pillevuit, A. 1993. Les Blocs Exotiques du Sultanat d'Oman: evolution paléogéographique d'une marge passive flexural. *Mémoires de Géologie (Lausanne)*, **17**, 1–249.

Pillevuit, A., Marcoux, J., Stampfli, G. and Baud, A. 1997. The Oman Exotics: a key to the understanding of the Neotethyan geodynamic evolution. *Geodinamica Acta, Paris*, **10**, 209–238, https://doi.org/10.1080/09853111.1997.11105303

Poupeau, G., Saddiqi, O., Michard, A., Goffé, B. and Oberhänsli, R. 1998. Late thermal evolution of the Oman Mountains subophiolitic windows: apatite fission-track thermochronology. *Geology*, **26**, 1139–1142, https://doi.org/10.1130/0091-7613(1998)026<1139:LTEOTO>2.3.CO;2

Pratt, B.R. and Smewing, J.D. 1990. Jurassic and Early Cretaceous platform margin configurations and evolution, central Oman Mountains. *Geological Society, London, Special Publications*, **49**, 69–88.

Prigent, C., Agard, P., Guillot, S., Godard, M. and Dubacq, B. 2018*a*. Mantle wedge (de)formation during subduction infancy: evidence from the base of the Semail Ophiolite Mantle. *Journal of Petrology*, **59**, 2016–2092, https://doi.org/10.1093/petrology/egy090

Prigent, C., Guillot, S., Agard, P. and Ildefonse, B. 2018*b*. Fluid-assistant deformation and strain localization in the cooling mantle wedge of a young subduction zone (Semail Ophiolite). *Journal of Geophysical Research: Solid Earth*, **123**, 7529–7524, https://doi.org/10.1029/2018JB015492

Prigent, C., Guillot, S., Agard, P., Lemarchand, D., Soret, M. and Ulrich, M. 2018*c*. Transfer of subduction fluids into the deforming mantle wedge during nascent subduction: evidence from trace elements and boron isotopes (Semail ophiolite, Oman). *Earth and Planetary Science Letters*, **484**, 213–228, https://doi.org/10.1016/j.epsl.2017.12.008

Quick, J.E. 1991. Late Proterozoic transpression on the Nabitah Fault system – implications for the assembly of the Arabian Shield. *Precambrian Research*, **53**, 119–147, https://doi.org/10.1016/0301-9268(91)90008-X

Rabu, D., Béchennec, F., Beurrier, M. and Hutin, G. 1986. *Geological map of Nakhl, sheet NF40-3E, scale: 1:100,000, with Explanatory Notes*. Directorate General of Minerals, Oman Ministry of Petroleum and Minerals.

Rabu, D., Le Métour, J., Béchennec, F., Beurrier, M., Villey, M. and Bourdillon-Jeudy De Grissac, C. 1990. Sedimentary aspects of the Eo-Alpine cycle on the northeast edge of the Arabian Platform (Oman Mountains). *Geological Society, London, Special Publications*, **49**, 49–68.

Rabu, D., Nehlig, P. *et al.* 1993. Stratigraphy and structure of the Oman Mountains. *Orleans, France, Document du Bureau de Récherches Géologiques et Minières*, **221**, 21–47.

Ravaut, P., Bayer, R., Hassani, R., Rousset, D. and Yahya'ey, A.A. 1997. Structure and evolution of the northern Oman margin: gravity and seismic constraints over the Zagros-Makran-Oman collision zone. *Tectonophysics*, **279**, 253–280, https://doi.org/10.1016/S0040-1951(97)00125-X

Regard, V., Bellier, O. *et al.* 2005. Cumulative right-lateral fault slip rate across the Zagros-Makran transfer zone: role of the Minab-Zendan fault system in accommodating Arabia-Eurasia convergence in southeast Iran. *Geophysical Journal International*, **162**, 177–203, https://doi.org/10.1111/j.1365-246X.2005.02558.x

Rickards, B.R., Booth, G.A., Paris, F. and Heward, A.P. 2010. Marine flooding events of the Early and Middle Ordovician of Oman and the United Arab Emirates and their graptolite, acritarch and chitinozoan associations. *GeoArabia*, **15**, 81–120.

Ricou, L.E. 1971. Le croissant ophiolitique péri-arabe, une ceinture de nappes mise en place au crétacé supérieur. *Revue de Geographie Physique et de Géologie Dynamique*, **13**, 327–350.

Ricou, L.E. 1994. Tethys reconstructed: plates, continental fragments and their boundaries since 260 Ma from Central America to South-eastern Asia. *Geodinamica Acta*, **7**, 169–218, https://doi.org/10.1080/09853111.1994.11105266

Rioux, M., Bowring, S.A., Kelemen, P.B., Gordon, S., Dudás, F. and Miller, R. 2012. Rapid crustal accretion and magma assimilation in the Oman-U.A.E. ophiolite: high precision U-Pb zircon geochronology of the gabbroic crust. *Journal of Geophysical Research*, **117**, B07201, https://doi.org/10.1029/2012JB009273

Rioux, M., Bowring, S., Kelemen, P., Gordon, S., Miller, R. and Dudás, F. 2013. Tectonic development of the Samail ophiolite: high-precision U-Pb zircon geochronology and Sm-Nd isotopic constraints on crustal growth and emplacement. *Journal of Geophysical Research: Solid Earth*, **118**, 2085–2101, https://doi.org/10.1002/jgrb.50139

Rioux, M., Garber, J., Bauer, A., Bowring, S., Searle, M.P., Kelemen, P. and Hacker, B. 2016. Synchronous formation of the metamorphic sole and igneous crust of the Semail ophiolite: new constraints on the tectonic evolution during ophiolite formation from high-precision U-Pb zircon geochronology. *Earth and Planetary Science Letters*, **451**, 185–195, https://doi.org/10.1016/j.epsl.2016.06.051

Robertson, A.H.F. 1987*a*. The transition from passive margin to Upper Cretaceous foreland basin related to ophiolite emplacement in the Oman Mountains. *Geological Society of America Bulletin*, **99**, 633–653, https://doi.org/10.1130/0016-7606(1987)99<633:TTFAPM>2.0.CO;2

Robertson, A.H.F. 1987*b*. Upper Cretaceous Muti Formation: transition of a Mesozoic carbonate platform to a foreland basin in the Oman Mountains. *Sedimentology*, **34**, 1123–1142, https://doi.org/10.1111/j.1365-3091.1987.tb00596.x

Robertson, A.H.F. and Searle, M.P. 1990. The northern Oman Tethyan continental margin: stratigraphy, structure, concepts and controversies. *Geological Society, London, Special Publications*, **49**, 3–25.

Rodgers, D.W. and Gunatilaka, A. 2002. Bajada formation by monsoonal erosion of a subaerial forebulge, Sultanate of Oman. *Sedimentary Geology*, **154**, 127–146, https://doi.org/10.1016/S0037-0738(02)00126-4

Roger, J., Platel, J.-P., Cavelier, C. and Bourdillon-de-Grisac, C. 1989. Données nouvelles sur la stratigraphie et l'histoire géologique du Dhofar (Sultanat d'Oman). *Bulletin de la Société Géologique de France*, **2**, 265–277, https://doi.org/10.2113/gssgfbull.V.2.265

Roger, J., Béchennec, F., Janjou, D., Le Métour, J., Wyns, R. and Beurrier, M. 1991. *Geological map of Ja'alan, sheet NF 40-08E, scale:100,000, with Explanatory Notes*. Directorate General of Minerals, Oman Ministry of Petroleum and Minerals.

Rougeulle, A. 2017. Medieval Qalhāt, historical vs archaeological data. *Arabian Humanities: Revue internationale d'archéologie et de sciences sociales sur la péninsule Arabique [International Journal of Archaeology and Social Sciences in the Arabian Peninsula]*, **9**, https://doi.org/10.4000/cy.3442

Rousseau, M., Dromart, G., Garcia, J-P., Atrops, F. and Guillocheau, F. 2005. Jurassic evolution of the Arabian carbonate platform edge in the central Oman Mountains. *Journal of the Geological Society, London*, **162**, 349–362, https://doi.org/10.1144/0016-764903-178

Rousseau, M., Dromart, G., Droste, H. and Homewood, P. 2006. Stratigraphic organization of the Jurassic sequence in Interior Oman, Arabian Peninsula. *GeoArabia*, **11**, 17–50.

Ruban, D., Al-Husseini, M.I. and Iwasaki, Y. 2007. Review of Middle East Paleozoic plate tectonics. *GeoArabia*, **12**, 35–56.

Saddiqi, O., Michard, A., Goffé, B., Poupeau, G. and Oberhänsli, R. 2006. Fission-track thermochronology of the Oman Mountains continental widows, and current problems of tectonic interpretation. *Bulletin de la Société Géologique de France*, **177**, 127–134, https://doi.org/10.2113/gssgfbull.177.3.127

Scharf, A., Mattern, F. and Al Sadi, S. 2016. Kinematics of post-obduction deformation of the Tertiary Ridge at Al-Khod Village (Muscat Area, Oman). *Sultan Qaboos University Journal for Science*, **21**, 26–40, https://doi.org/10.24200/squjs.vol21iss1pp26-40

Scharf, A., Mattern, F., Moraetis, D., Callegari, I. and Weidle, C. 2019*a*. Postobductional kinematic evolution and geomorphology of a major regional structure – the Semail Gap Fault Zone (Oman Mountains). *Tectonics*, **38**, 2756–2778, https://doi.org/10.1029/2019TC005588

Scharf, A., Mattern, F. and Pracejus, B. 2019*b*. Two new microscopic ductile kinematic indicators from the Oman Mountains. *Journal of Structural Geology*, **119**, 107–117, https://doi.org/10.1016/j.jsg.2018.12.006

Scharf, A., Mattern, F., Bolhar, R., Bailey, C.M. and Ring, U. 2019*c*. Early Cenozoic listwaenite formation at a major extensional fault zone of the Oman Mountains (Fanja area) – insights from structural analysis and U-Pb carbonate dating. *AAPG Event, Structural styles in the Middle East*, 9–11 December 2019, Muscat, Oman.

Scharf, A., Mattern, F., Bolhar, R., Bailey, C.M. and Ring, U. 2020. U-Pb dating of postobductional carbonate veins in listwaenite of the Oman Mountains near Fanja. *International Conference on Ophiolites and the Oceanic Lithosphere: Results of the Oman Drilling Project and Related Research*, 12–14 January 2020, Sultan Qaboos University, Muscat, Sultanate of Oman.

Scharf, A., Mattern, F., Al-Wardi, M., Frijia, G., Moraetis, D., Pracejus, B., Bauer, W. and Callegari, I. 2021*a*. Appendices to: The Geology and Tectonics of the Jabal Akhdar and Saih Hatat Domes, Oman Mountains. *Geological Society, London, Memoirs*, **54**, 113–116, https://doi.org/10.1144/M54.7

Scharf, A., Mattern, F., Al-Wardi, M., Frijia, G., Moraetis, D., Pracejus, B., Bauer, W. and Callegari, I. 2021*b*. Thrusts, extensional faults and fold patterns of the major units. *Geological Society, London, Memoirs*, **54**, 49–60, https://doi.org/10.1144/M54.3

Scharf, A., Mattern, F., Al-Wardi, M., Frijia, G., Moraetis, D., Pracejus, B., Bauer, W. and Callegari, I. 2021*c*. Tectonostratigraphy of the eastern part of the Oman Mountains. *Geological Society, London, Memoirs*, **54**, 11–47, https://doi.org/10.1144/M54.2

Scharf, A., Mattern, F., Al-Wardi, M., Frijia, G., Moraetis, D., Pracejus, B., Bauer, W. and Callegari, I. 2021*d*. Introduction and tectonic framework. *Geological Society, London, Memoirs*, **54**, 1–10, https://doi.org/10.1144/M54.1

Scharf, A., Mattern, F., Al-Wardi, M., Frijia, G., Moraetis, D., Pracejus, B., Bauer, W. and Callegari, I. 2021*e*. Large-scale structure of the study area. *Geological Society, London, Memoirs*, **54**, 61–66, https://doi.org/10.1144/M54.4

Schlaich, M., Aigner, T. and Bendias, D. 2021. Facies distribution, depositional environment, and reservoir geometries of clastic sandbodies in the Lower Jurassic Mafraq Member (Oman Mountains, Sultanate of Oman). *In*: *Lower Triassic to Middle Jurassic Sequence of the Arabian Plate*, EAGE.

Schreurs, G. and Immenhauser, A. 1999. West-northwest directed obduction of the Batain Group on the eastern Oman continental margin at the Cretaceous-Tertiary boundary. *Tectonics*, **18**, 148–160, https://doi.org/10.1029/1998TC900020

Schröder, S., Schreiber, B.C., Amthor, J.E. and Matter, A. 2003. A depositional model for the terminal Neoproterozoic-Early Cambrian Ara Group evaporites in south Oman. *Sedimentology*, **50**, 879–898, https://doi.org/10.1046/j.1365-3091.2003.00587.x

Searle, M.P. 1985. Sequence of thrusting and origin of culminations in the northern and central Oman Mountains. *Journal of Structural Geology*, **7**, 129–143, https://doi.org/10.1016/0191-8141(85)90127-0

Searle, M.P. 2007. Structural geometry, style and timing of deformation in the Hawasina Window, Al Jabal al Akhdar and Saih Hatat culminations, Oman Mountains. *GeoArabia*, **12**, 99–130.

Searle, M.P. and Alsop, G.I. 2007. Eye-to-eye with a mega-sheath fold: a case study from Wadi Mayh, northern Oman. *Geology*, **35**, 1043–1046, https://doi.org/10.1130/G23884A.1

Searle, M.P. and Cox, J.S. 1999. Tectonic setting, origin and obduction of the Oman ophiolite. *Geological Society of America Bulletin*, **111**, 104–122, https://doi.org/10.1130/0016-7606(1999)111<0104:TSOAOO>2.3.CO;2

Searle, M.P. and Cox, J.S. 2002. Subduction zone metamorphism during formation and emplacement of the Semail Ophiolite in the Oman Mountains. *Geological Magazine*, **139**, 241–255, https://doi.org/10.1017/S0016756802006532

Searle, M.P. and Graham, G.M. 1982. "Oman Exotics" – oceanic carbonate build-ups associated with the early stages of continental rifting. *Geology*, **10**, 43–49, https://doi.org/10.1130/0091-7613(1982)10<43:OECBAW>2.0.CO;2

Searle, M.P. and Malpas, J. 1980. Structure and metamorphism of rocks beneath the Semail ophiolite of Oman and their significance in ophiolite obduction. *Earth and Environmental Science Transactions of the Royal Society of Edinburgh*, **71**, 247–262, https://doi.org/10.1017/S0263593300013614

Searle, M.P. and Malpas, J. 1982. Petrochemistry and origin of sub-ophiolitic metamorphic and related rocks in the Oman Mountains. *Journal of the Geological Society, London*, **139**, 235–248, https://doi.org/10.1144/gsjgs.139.3.0235

Searle, M.P., Waters, D.J., Martin, H.N. and Rex, D.C. 1994. Structure and metamorphism of blueschist – eclogite facies rocks from the northeastern Oman Mountains. *Journal of Geological Society, London*, **151**, 555–576, https://doi.org/10.1144/gsjgs.151.3.0555

Searle, M.P., Warren, C.J., Waters, D.J. and Parrish, R.R. 2003. Subduction zone polarity in the Oman Mountains: implications for ophiolite emplacement. *Geological Society, London, Special Publications*, **218**, 467–480, https://doi.org/10.1144/GSL.SP.2003.218.01.24

Searle, M.P., Warren, C.J., Waters, D.J. and Parrish, R.R. 2004. Structural evolution, metamorphism and restoration of the Arabian continental margin, Saih Hatat region, Oman Mountains. *Journal of Structural Geology*, **26**, 451–473, https://doi.org/10.1016/j.jsg.2003.08.005

Searle, M.P., Waters, D.J., Garber, J.M., Rioux, M., Cherry, A.G. and Ambrose, T.K. 2015. Structure and metamorphism beneath the obducting Oman ophiolite: evidence from the Bani Hamid granulites, northern Oman mountains. *Geosphere*, **11**, 1812–1836.

Siddall, M., Rohling, E.J., Almogi-Labin, A., Hemleben, Ch., Meischner, D., Schmelzer, I. and Smeed, D.A. 2003. Sea-level fluctuations during the last glacial cycle. *Nature*, **423**, 853–858, https://doi.org/10.1038/nature01690

Skelton, P.W., Nolan, S.C. and Scott, R.W. 1990. The Maastrichtian transgression onto the northwestern flank of the Proto-Oman Mountains: sequences of rudist-bearing beach to open shelf facies. *Geological Society, London, Special Publications*, **49**, 521–547.

Smewing, J.D., Abbotts, I.L., Dunne, L.A. and Rex, D.C. 1991. Formation and emplacement ages of the Masirah ophiolite, Sultanate of Oman. *Geology*, **19**, 453–456, https://doi.org/10.1130/0091-7613(1991)019<0453:FAEAOT>2.3.CO;2

Soret, M., Agard, P., Dubacq, B., Plunder, A. and Yamato, P. 2017. Petrological evidence for stepwise accretion of metamorphic soles during subduction infancy (Semail ophiolite, Oman and UAE). *Journal of Metamorphic Geology*, **35**, 1–30, https://doi.org/10.1111/jmg.12267

Spencer, C.J., Cavosie, A.J. *et al.* and the Edinburgh Ion Microprobe Facility (EIMF). 2017. Evidence for melting mud in Earth's mantle from extreme oxygen isotope signature in zircon. *Geology*, **45**, 975–978, https://doi.org/10.1130/G39402.1

Stampfli, G.M. and Borel, G.D. 2002. A plate tectonic model for the Paleozoic and Mesozoic constrained by dynamic plat boundaries and restored synthetic oceanic isochrons. *Earth and Planetary Science Letters*, **196**, 17–33, https://doi.org/10.1016/S0012-821X(01)00588-X

Stampfli, G.M., Marcoux, J. and Baud, A. 1991. Tethyan margins in space and time. *Palaeogeography, Palaeoclimatology,*

Palaeoecology, **87**, 373–409, https://doi.org/10.1016/0031-0182(91)90142-E

Stern, R.J. 2004. Subduction initiation: spontaneous and induced. *Earth and Planetary Science Letters*, **226**, 275–292, https://doi.org/10.1016/S0012-821X(04)00498-4

Stern, R.J. and Bloomer, S.H. 1992. Subduction zone infancy: examples from the Eocene Izu-Bonin-Mariana and Jurassic California arcs. *Geological Society of America Bulletin*, **104**, 1621–1636, https://doi.org/10.1130/0016-7606(1992)104<1621:SZIEFT>2.3.CO;2

Stern, R.J. and Gerya, T. 2018. Subduction initiation in nature and models: a review. *Tectonophysics*, **746**, 173–198, https://doi.org/10.1016/j.tecto.2017.10.014

Stern, R.J. and Johnson, P. 2010. Continental lithosphere of the Arabian Plate: a geologic, petrologic, and geophysical synthesis. *Earth-Science Reviews*, **101**, 29–67, https://doi.org/10.1016/j.earscirev.2010.01.002

Stern, R.J., Avigad, D., Miller, N.R. and Beyth, M. 2006. Evidence for the Snowball Earth hypothesis in the Arabian-Nubian Shield and the East African Orogen. *Journal of African Earth Sciences*, **44**, 1–20, https://doi.org/10.1016/j.afrearsci.2005.10.003

Stern, R.J., Reagan, M., Ishizuka, O., Ohara, Y. and Whattam, S. 2012. To understand subduction initiation, study forearc crust: to understand forearc crust, study ophiolites. *Lithosphere*, **4**, 469–483, https://doi.org/10.1130/L183.1

Stewart, S.A. 2016. Structural geology of the Rub' Al-Khali Basin, Saudi Arabia. *Tectonics*, **35**, 2417–2438, https://doi.org/10.1002/2016TC004212

Stuart-Smith, P.G., Romine, K., Taedale, J.M. and Schreurs, J. 2007. Basement control on structure and evolution of the salt basins of Oman. *GeoArabia*, **12**, 193 (abstract of a presentation made at GEO 2006, Bahrain, 27–29 March).

Styles, M.T., Ellison, R.A. *et al.* 2006. *The Geology and Geophysics of the United Arab Emirates*, vol. **2**. Ministry of Energy, United Arab Emirates, Abu Dhabi.

Tavani, S., Corradetti, A., Sabbatino, M., Seers, T. and Mazzoli, S. 2021. Geological record of the transition from induced to self-sustained subduction in the Oman Mountains. *Journal of Geodynamics*, **133**, 101674, https://doi.org/10.1016/j.jog.2019.101674

Tilton, G.R., Hopson, C.A. and Wright, J.E. 1981. Uranium-lead isotopic ages of the Semail ophiolite, Oman, with applications to Tethyan oceanic ridge tectonics. *Journal of Geophysical Research*, **86**, 2763–2775, https://doi.org/10.1029/JB086iB04p02763

Torsvik, T.H. and Cocks, L.R.M. 2017. *Earth History and Paleogeography*. Cambridge University Press, Cambridge, 317.

Tucker, M.E. and Wright, V.P. 1990. *Carbonate Sedimentology*. Blackwell Science, Oxford, p. 482.

van Buchem, F.S.P., Pittet, B. *et al.* 2002*a*. High-resolution sequence stratigraphic architecture of the Barremian-Aptian carbonate systems in northern Oman and the United Arab Emirates Kharaib and Shu'aiba formations. *GeoArabia*, **7**, 461–500.

van Buchem, F.S.P., Razin, P., Homewood, P.W., Oterdoom, W.H. and Philip, J. 2002*b*. Stratigraphic organization of carbonate ramps and organic-rich intrashelf basins: Natih Formation (middle Cretaceous) of northern Oman. *American Association of Petroleum Geologists Bulletin*, **86**, 21–53.

van Hinsbergen, D.J.J., Peters, K. *et al.* 2015. Dynamics of intraoceanic subduction initiation: 2. Suprasubduction zone ophiolite formation and metamorphic sole exhumation in context of absolute plate motion. *Geochemistry, Geophysics, Geosystems*, **16**, 1771–1785, https://doi.org/10.1002/2015GC005745

van Hinsbergen, D.J.J., Maffione, M., Koornneef, L.M.T. and Guilmette, C. 2019. Kinematic and paleomagnetic restoration of the Semail ophiolite (Oman) reveals subduction initiation along an ancient Neotethyan fracture zone. *Earth and Planetary Science Letters*, **518**, 183–196, https://doi.org/10.1016/j.epsl.2019.04.038

Vernant, P., Nilforoushan, F. *et al.* 2004. Present-day crustal deformation and plate kinematics in the Middle East constrained by GPS measurements in Iran and northern Oman. *Geophysical Journal International*, **157**, 381–398, https://doi.org/10.1111/j.1365-246X.2004.02222.x

Villey, M., Le Métour, J. and de Gramont, X. 1986. *Geological map of Fanjah, sheet NF 40-3F, scale: 1:100,000, with Explanatory notes*. Directorate General of Minerals, Oman Ministry of Petroleum and Minerals.

Warburton, J., Burnhill, T.J., Graham, R.H. and Isaac, K.P. 1990. The evolution of the Oman Mountains foreland basin. *Geological Society, London, Special Publications*, **49**, 419–427.

Warren, C.J., Parrish, R.R., Searle, M.P. and Waters, D.J. 2003. Dating the subduction of the Arabian continental margin beneath the Semail Ophiolite. *Geology*, **31**, 889–892, https://doi.org/10.1130/G19666.1

Warren, C.J., Parrish, R.R., Waters, D.J. and Searle, M.P. 2005. Dating the geologic history of Oman's Semail ophiolite: insight from U-Pb geochronology. *Contributions to Mineralogy and Petrology*, **150**, 403–422, https://doi.org/10.1007/s00410-005-0028-5

Watchorn, F., Nichols, G.J. and Bosence, D.W.J. 1998. Rift-related sedimentation and stratigraphy, southern Yemen (Gulf of Aden). *In*: Purser, B.H. and Bosence, D.W.J. (eds) *Sedimentation and Tectonics of Rift Basins: Red Sea-Gulf of Aden*. Chapman & Hall, London, 165–191.

Whattam, S.A. and Stern, R.J. 2011. The 'subduction initiation rule': a key for linking ophiolites, intra-oceanic forearcs, and subduction initiation. *Contribution to Mineralogy and Petrology*, **162**, 1031–1045, https://doi.org/10.1007/s00410-011-0638-z

Whitehouse, M.J., Pease, V. and Al-Khirbash, S. 2016. Neoproterozoic crustal growth at the margin of the East Gondwana continent – age and isotopic constraints from the easternmost inliers of Oman. *International Geology Review*, **58**, 2046–2064, https://doi.org/10.1080/00206814.2016.1207207

Wilde, A., Simpson, L. and Hanna, S. 2002. Preliminary study of Cenozoic hydrothermal alteration and platinum deposition in the Oman Ophiolite. *In*: Jessell, M.J. (ed.) *General Contributions, Journal of the Virtual Explorer*, **6**, 7–13.

Wohlwend, S., Hart, M. and Weissert, H. 2016. Chemostratigraphy of the Upper Albian to mid-Turonian Natih Formation (Oman) – how authigenic carbonate changes a global pattern. *The Depositional Record*, **2**, 97–117, https://doi.org/10.1002/dep2.15

Worthing, Michael A. and Wilde, Andrew R. 2002. Basanites related to Late Eocene extension from NE Oman. *Journal of the Geological Society*, **159**, 469–483, https://doi.org/10.1144/0016-764901-088

Wyns, R., Béchennec, F., Le Métour, J. and Roger, J. 1992. *Geological map of Tiwi, sheet NF40-8B, scale 1:100,000, with Explanatory notes*. Directorate General of Minerals, Oman Ministry of Petroleum and Minerals.

Yamato, P., Agard, P., Goffé, B., De Andrade, V., Vidal, O. and Jolivet, L. 2007. New, high-precision P-T estimates for obduction, nappe stacking and exhumation processes. *Journal of Metamorphic Geology*, **25**, 657–682, https://doi.org/10.1111/j.1525-1314.2007.00722.x

Yuan, Y., Kusky, T.M. and Rajendran, S. 2016. Tertiary and Quaternary marine terraces and planation surfaces of Northern Oman: interaction of flexural bulge migration associated with the Arabian-Eurasian collision and eustatic sea level changes. *Journal of Earth Science*, **27**, 955–970, https://doi.org/10.1007/s12583-015-0656-2

Zharkov, M.A. 1984. *Paleozoic Salt-Bearing Formations of the World*. Springer, Berlin.

Chapter 6

Conclusions, differences between the Jabal Akhdar and Saih Hatat domes and unanswered questions

Andreas Scharf[1]*, Frank Mattern[1], Mohammed Al-Wardi[1], Gianluca Frijia[2], Daniel Moraetis[3], Bernhard Pracejus[1], Wilfried Bauer[4] and Ivan Callegari[4]

[1]Department of Earth Sciences, College of Science, Sultan Qaboos University, PO Box 36, PC 123, Al-Khod, Muscat, Sultanate of Oman

[2]Department of Physics and Earth Sciences, University of Ferrara, Via Saragat 1, 44122, Ferrara, Italy

[3]Department of Applied Physics and Astronomy, College of Sciences, University of Sharjah, PO Box 27272, Sharjah, United Arab Emirates

[4]Department of Applied Geosciences, German University of Technology GUtech, PO Box 1816, PC 130, Halban, Sultanate of Oman

*Correspondence: scharfa@squ.edu.om

Abstract: This chapter provides the conclusions/outlines of the tectonics, affecting the Southeastern Oman Mountains, including the Jabal Akhdar and Saih Hatat domes. The main tectonic events include amongst others (1) Neoproterozoic rifting, (2) two distinct early Paleozoic compressive events, (3) large-scale open 'Hercynian' folding and formation of a pronounced unconformity during the late Paleozoic, (4) rifting preceding the opening of the Neo-Tethys Ocean during the late Paleozoic, (5) late Cretaceous obduction of the Semail Ophiolite and the response of the Arabian lithosphere as well as (6) post-obductional tectonics. Also of major geological significance are the three major glaciations (Sturtian, Marinoan and Late Paleozoic Gondwana glaciation) which have been recorded in the rocks of northern Oman. Moreover, major lithological, structural and metamorphic differences exist between the Jabal Akhdar and Saih Hatat domes. It appears likely that a major fault, striking parallel to the eastern margin of the Jabal Akhdar Dome, probably originating during Neoproterozoic terrain accretion, acted as a divide between both domes until present. This fault was multiple times reactivated and could explain the differences between the two domes. A catalogue of unanswered questions is included in chronological order to express that many geological aspects need further investigation and future research projects.

Conclusions

The Jabal Akhdar and Saih Hatat domes expose a pile of autochthonous Neoproterozoic–Cretaceous rocks, consisting of mostly (meta)sedimentary rocks. These rocks were deposited on Cryogenian crystalline basement rocks that were formed during the collision of juvenile, mostly volcanogenic, terranes. The sedimentary sequences unconformably overlying the crystalline basement record intracontinental extension and rifting at *c.* 645 Ma, possibly related to the Nabitah Orogeny at 680–600 Ma. Subsidence and basin deepening during the Late Neoproterozoic was followed by the Cadomian, East African (Angudan) and Late Cretaceous Semail Orogeny (*sensu* Mattern *et al.* 2020). The rocks in northern Oman record a minimum of three major glaciations (Sturtian, Marinoan and Late Paleozoic Gondwana glaciation). Uplift and some faulting within the Arabian part of Gondwana during the Late Paleozoic ('Hercynian' Orogeny) could be related to the future breakaway of the Cimmerian super-terrane (Ruban *et al.* 2007; Torsvik and Cocks 2017). Late Paleozoic Pangea rifting and subsequent spreading formed the Hawasina Basin, floored by extended Arabian continental crust. During the same time and during the Mesozoic, mostly shallow-marine limestones (Hajar Supergroup) accumulated on the Arabian Platform of regular lithospheric thickness.

Along the Hawasina Basin in the Neo-Tethys Ocean, an intraoceanic subduction zone formed during the Albian. The initial location of the probably forced subduction with a WNW-ward dipping slab may have been along a NNE-striking transform fault and/or fracture zone. During the Cenomanian, the subduction zone became self-sustained and new oceanic lithosphere formed. In the course of obduction, the entire ophiolite rotated *c.* 90° clockwise from its formation at *c.* 96 Ma until its final emplacement onto the Arabian Shelf with coeval N-ward movement of Arabia at *c.* 70 Ma. The slab therefore rotated towards a dip direction to the NE. Above the slab at *c.* 96 Ma, new oceanic lithosphere formed within the Neo-Tethys Ocean. This hot, young and buoyant oceanic lithosphere was obducted onto the extended Arabian margin as the Semail Ophiolite. Obduction ensued during or slightly after its formation. The hot oceanic lithosphere overthrust the distal Hawasina Basin sedimentary rocks and formed a metamorphic sole within the WNW- to NE-dipping subduction zone. The Semail Ophiolite was obducted SW-ward above the Hawasina thrust sheets and reached the Arabian Platform during the Campanian.

Loading by the obducted allochthonous units formed the Aruma Foreland Basin within the Arabian Platform with an associated forebulge. The craton-ward part of the Aruma Foreland Basin was uplifted by a few hundred metres. The filled foreland basin was later overthrust by the allochthonous rock masses, which were finally emplaced after a minimum thrust range of 550 km during the Maastrichtian. The Arabian margin and platform went into the subduction zone in the course of ophiolite thrusting. The continental lithosphere of Arabia was too buoyant to be subducted deeply. The attached oceanic lithosphere was therefore torn off during the Early Campanian (slab break-off). As a result, the continental lithosphere was rapidly exhumed from at least 80 km depth at the northeastern Saih Hatat area.

Mega-sheath folds formed during a late stage of obduction and after exhumation of the Arabian margin. The northern Saih Hatat Dome emerged and was deeply eroded above sea-level during the Maastrichtian, resulting in deposition of pre-Permian clasts into the post-obductional Al-Khod Formation.

From: Scharf, A., Mattern, F., Al-Wardi, M., Frijia, G., Moraetis, D., Pracejus, B., Bauer, W. and Cellegari, I. 2021. *The Geology and Tectonics of the Jabal Akhdar and Saih Hatat Domes, Oman Mountains*. Geological Society, London, Memoirs, **54**, 105–111, https://doi.org/10.1144/M54.6

A relative tectonic quiescence during the Late Paleocene and Early Eocene led to deposition of mostly shallow-marine limestone in the northeastern and eastern study area. Late Eocene SSW-directed compression resulted in the main exhumation in the range 4–6 km of the Jabal Akhdar Dome. This compression may be related to a decreasing subduction velocity along the Makran Subduction Zone. Late Paleogene uplift may have been the result of gravitational collapse of the Jabal Akhdar and Saih Hatat domes and extension along the Frontal Range Fault and Semail Gap. Further Oligocene–Miocene extension may be related to rifting in the Red Sea and Gulf of Aden, as well as to orogenic collapse. Present-day convergence between Arabia and Eurasia, and the isostatic rebound of the Arabian Crust, resulted in and still results in some gentle active tectonics in the study area, evident from faults, incisions of wadi sediments, hot springs and uplifted marine terraces.

Differences between the Jabal Akhdar and Saih Hatat domes

The Jabal Akhdar and Saih Hatat domes are the two largest culminations in the Oman Mountains and provide insights into the relatively old geological development of the Arabian margin and platform. Large areas between and around them are blanketed either by the allochthonous nappes or post-obductional sedimentary rocks. Major differences between the two domes exist in terms of their lithologies and/or lithostratigraphies, structures and metamorphism, as summarized in Table 6.1. Figure 6.1 depicts and compares the simplified lithostratigraphic columns of these domes.

In the Jabal Akhdar Dome, five Neoproterozoic–earliest Cambrian formations exist, while in the Saih Hatat Dome, only two formations have been mapped (fig. A.1, Scharf *et al.* 2021*a*, Chapter 7, this Memoir). The age of these formations is poorly determined in parts of the Jabal Akhdar Dome, while the ages from the formations of the Saih Hatat Dome are unconstrained (section on 'Units below the Hercynian Unconformity' in Scharf *et al.* 2021*b*, Chapter 2, this Memoir). A correlation between the formations of the domes is therefore viewed differently from one author to another (e.g. is Hiyam Formation the equivalent of the Hajir Formation or the Kharus Formation?). The base of the lowest formations in both domes is not exposed, that is, the contact to older formations or to the crystalline basement. It is therefore not possible to determine the thickness of these lower formations. However, the Hatat Formation seems to be thicker than the Mistal Formation (Fig. 6.1).

Rocks of the Jabal Akhdar Dome were affected by an extensional event during the transition of Nafun Group I to Nafun Group II (transition of Hajir to Mu'aydin formations; Mattern and Scharf 2019; sections on 'Units below the Hercynian Unconformity' in Chapter 2 and 'Units below the Angudan Unconformity' in Scharf *et al.* 2021*c*, Chapter 5, this Memoir). No such extensional structures have been mapped in the Saih Hatat Dome. Furthermore, Callegari *et al.* (2020) have shown that the Cadomian Orogeny (F1; section on 'F1 folds' in Scharf *et al.* 2021*d*, Chapter 3, this Memoir) affected the Jabal Akhdar Dome, while no such deformation has been reported from the Saih Hatat Dome. Future research will reveal whether such structures are also present in the Saih Hatat Dome. Pre-Permian folds within the Hatat Formation of the Saih Hatat Dome were mapped by Villey *et al.* (1986). These steep-dipping and NNE-trending folds might have been formed during the Angudan Orogeny. Similarly oriented folds exist in the Jabal Akhdar Dome.

The reason(s) for the differences in lithostratigraphy and possibly thickness of the Neoproterozoic–earliest Cambrian formations in both domes is unclear, but it is assumed that a major fault, oriented parallel to and near the Semail Gap, may have been responsible for this (Mattern *et al.* 2018; Mattern and Scharf 2019; Scharf *et al.* 2019; section on 'Main extensional fault systems' in Chapter 3). This fault, which may have formed during Cryogenien terrain accretion (Romine *et al.* 2004), has been reactivated; areas on both sides of the fault have been uplifted or lowered at different times. Alternatively, the Semail Gap may have been the boundary between West and East Oman during the Angudan collision (Gomez-Perez *et al.* 2019).

Table 6.1. *Comparison of lithologies, structures and metamorphism between the Jabal Akhdar and Saih Hatat domes*

	Jabal Akhdar Dome (JAD)	Saih Hatat Dome (SHD)
Cenozoic	Significant rapid cooling and/or exhumation during the Late Eocene–Miocene	Minor cooling and/or exhumation during the Late Eocene–Miocene
	Minor rapid cooling and/or exhumation during the latest Cretaceous to Paleocene–Eocene	Significant rapid cooling and/or exhumation during the latest Cretaceous to Paleocene–Eocene
	No clasts from pre-Permian formations of the JAD have been identified in the Al-Khod Formation	Some clasts in the Al-Khod Formation are derived from pre-Permian formations of the SHD
Mesozoic	No high-pressure/low-temperature metamorphism	High-pressure/low-temperature metamorphism up to eclogite-facies conditions
	Top-to-the NE fabrics might be related to exhumation of a subduction zone	Intense deformation related to subduction/exhumation
	Wasia Group is overall present, except in the NE	Wasia Group is missing; the Kahmah Group is also missing in the NE
	Thin Mahil Formation	Thick Mahil Formation
Paleozoic	Relatively thin and monotonous Saiq Formation with few volcanic rocks	Thick Saiq Formation and diverse in lithology with plenty of volcanic rocks
	No Paleozoic and Pre-Permian formations in the JAD	Amdeh and Al Khalata formations are exposed in the SHD
	Cadomian (F1) and Angudan (F2) folds are mapped	No Cadomian (F1) folds are mapped
	No separation between the Angudan and Hercynian unconformities	Angudan and Hercynian unconformities are individually mapped
Neoproterozoic–earliest Cambrian	Synsedimentary extensional deformation at the transition of the Hajir and Mu'aydin formations	No such deformation has been recorded in the SHD
	Five Neoproterozoic–earliest Cambrian formations	Two Neoproterozoic–earliest Cambrian formations
	The lowest formation seems to be thinner than in the SHD	The lowest formation seems to be thicker than in the JAD

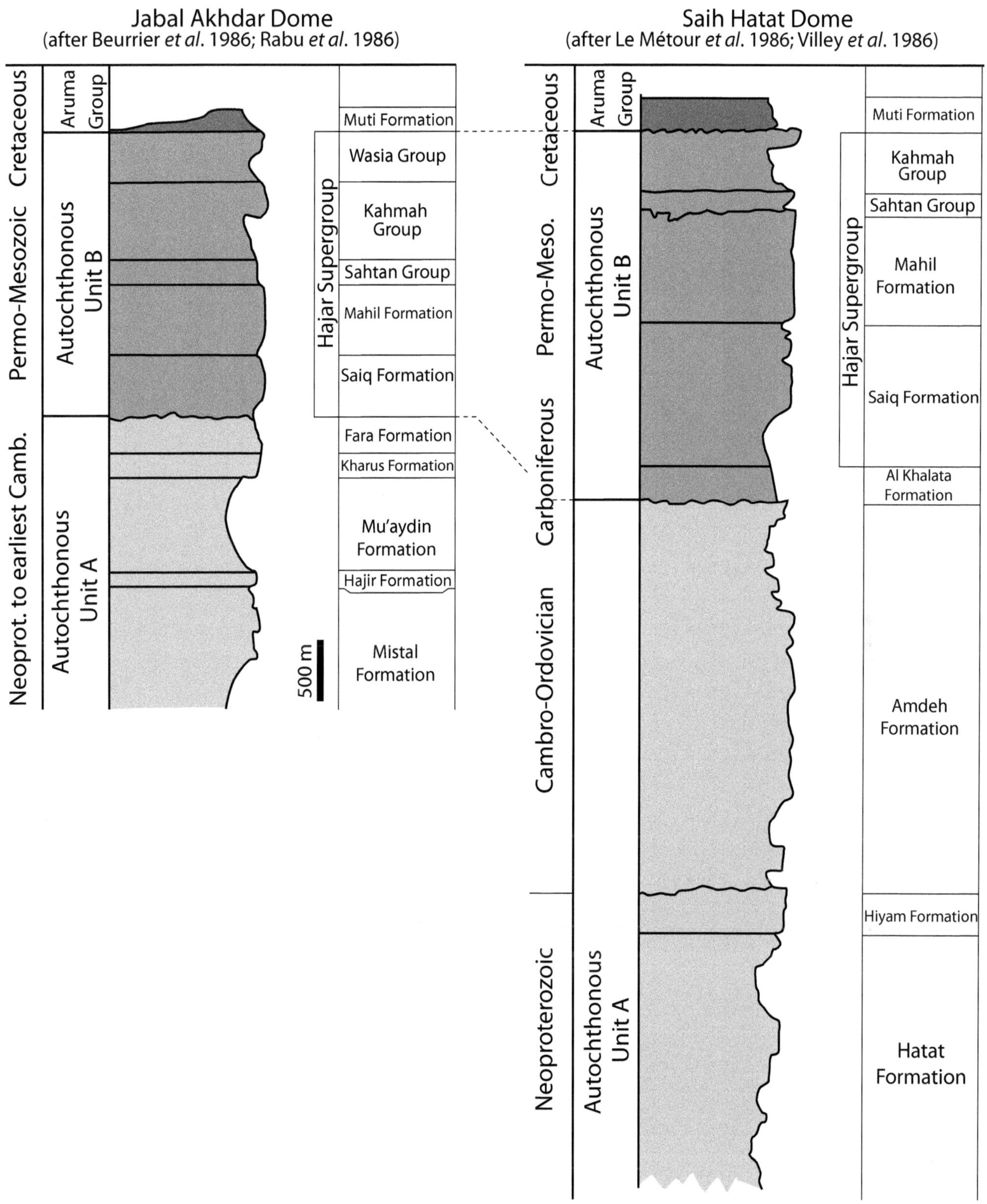

Fig. 6.1. Schematic comparison of the lithostratigraphic columns of the Jabal Akhdar and Saih Hatat domes. Note that the Al Khlata Formation has been identified by Heward and Penney (2014) and added to the column of the Saih Hatat Dome.

The Neoproterozoic–earliest Cambrian formations are topped in the Saih Hatat Dome by the Angudan Unconformity, and in the Jabal Akhdar Dome by the Hercynian Unconformity. In the Saih Hatat Dome, two additional Paleozoic formations (Amdeh and Al Khlata formations) below and above the Hercynian Unconformity are exposed. Similar to during the Neoproterozoic, the major fault in the Semail Gap area may have played an important role in the distribution of different formations within the Jabal Akhdar and Saih Hatat areas during the Paleozoic. During the Paleozoic, the Saih Hatat side must have been downfaulted with respect to the Jabal Akhdar side, providing the large amount of accommodation space for the thick siliciclastic of the Amdeh Formation (Mattern *et al.* 2018) and eventually the Al Khlata Formation (see section on 'Units above the Angudan Unconformity' in Chapter 5).

The lithology of the Hajar Supergroup is different between the two domes. In the Jabal Akhdar Dome, the Saiq Formation is overall monotonous and consists mostly of limestone and dolomite with a basal conglomerate and rare volcanic rocks. In the Saih Hatat Dome, the Saiq Formation is highly variable in its lithology. The basal section is characterized by overall massive amount of volcanic rocks related to the rifting of Gondwana (see Chauvet *et al.* 2009; section on 'Pangea rifting and formation of the Neo-Tethys Ocean and Hawasina Basin' in Chapter 5). Furthermore, the lithology in the Saih Hatat Dome is highly diverse, ranging from conglomerate via tuffaceous micaschist, dolomite, limestone, marl, basalt, andesite, tuffite and siltstone to mudstone (e.g. Chauvet *et al.* 2009). Moreover, the Permian Saiq Formation is much thicker in the Saih Hatat Dome (*c.* 1000 m) than in the Jabal Akhdar Dome (*c.* 450 m). A similar trend is also recorded in the Triassic Mahil Formation, where it is 700–800 m thick in the Saih Hatat Dome and only 600 m thick in the Jabal Akhdar Dome. The abundant amount of volcanic rocks, the large diversity in lithology of the Saiq Formation and the increase in thickness of the Saiq and Mahil formations towards the Saih Hatat Dome may be explained by the different distances of the two domes from the rift axis of the Pangea rifting, with the Saih Hatat area being closer to the rift axis (section on 'Pangea rifting and formation of the Neo-Tethys Ocean and Hawasina Basin' in Chapter 5). Thermal subsidence of the heated crust provided more accommodation space proximal to the rift zone (Saih Hatat area). This explains the increase in thickness of the post-rift lithologies towards the Saih Hatat Dome (upper part of the Saiq and the Mahil formations).

A further difference in the Hajar Supergroup of the domes is that the Wasia Group (Nahr Umr and Natih formations) is missing in the Saih Hatat Dome. This has been related to different erosion during upbulging (forebulge) of the Arabian lithosphere. Towards the Saih Hatat Dome, upbulging was more pronounced; this explains the erosion of the uppermost formations of the Arabian shelf deposits immediately in front of the advancing allochthonous nappes during the Late Cretaceous (see section on 'Stage III: obduction development, subduction of the Arabian margin and bulging of the Arabian lithosphere' in Chapter 5). The Wasia Group is also missing at the northeastern section of the Jabal Akhdar Dome, while at the southern margin of the Jabal Akhdar Dome only the top of the Natih Formation is partly eroded (fig. A.1).

In the course of the Late Cretaceous emplacement of the Semail Ophiolite, the Arabian margin was subducted and eventually exhumed. This resulted in the presence of the high-pressure/low-temperature rocks in the northeastern Saih Hatat Dome and in intense multiple folding (e.g. Saih Hatat fold nappe, upper–lower plate discontinuity and sheath folds; see sections on 'Upper–lower plate discontinuity or Hulw–As-Sifah Window' and 'F4 folds' in Chapter 3, and 'Intraoceanic thrusting, formation and obduction of the Semail Ophiolite' in Chapter 5). High-pressure/low-temperature metamorphism is recorded in Amdeh Formation and Hajar Supergroup of the northeastern Saih Hatat Dome. The rocks of the Jabal Akhdar Dome do not preserve such pressure-dominated minerals. According to Breton *et al.* (2004), top-to-the-NE shearing in the Hajar Supergroup from the Jabal Akhdar Dome occurred during exhumation of a Campanian subduction zone. This shearing is more pronounced in the northeastern Jabal Akhdar Dome than in the southern Jabal Akhdar Dome. The Jabal Akhdar Dome is therefore part of a distal subduction slab, while the Saih Hatat Dome was more proximal (see section on 'Stage IV: exhumation of high-pressure/low-temperature rocks' in Chapter 5). Grobe *et al.* (2018) reported that the top-to-the-NE shearing of the Hajar Supergroup in the Jabal Akhdar Dome formed coevally with thinning of the ophiolite. This shearing may have been related to a lowering of the convergence rates during the Early Campanian (Grobe *et al.* 2018).

During the ophiolite emplacement, the Arabian continental crust was rapidly exhumed at the Saih Hatat Dome, resulting in deposition of the post-obductional Al-Khod Formation with pre-Permian clasts, derived from the Saih Hatat Dome (section 5.3.3.4). This rapid exhumation during the latest Cretaceous has been also recorded using low-temperature geochronology (see section on 'Stage IV: exhumation of high-pressure/low-temperature rocks' in Chapter 5). The Jabal Akhdar Dome only underwent a little exhumation during the same time period (e.g. Hansman *et al.* 2017). The Saih Hatat Dome was therefore a major topographic feature as early as during the Late Cretaceous, while the Jabal Akhdar Dome had a relatively low topographic relief. This changed during the Late Eocene; only minor exhumation has been recorded in Saih Hatat Dome (1 km), while it was significant at the Jabal Akhdar Dome (4–6 km; Hansman *et al.* 2017; see section on 'Uplift and cooling history' in Chapter 5). The Saih Hatat Dome therefore experienced major cooling and/or exhumation during the Late Cretaceous and only moderate cooling and/or exhumation during the Late Eocene–Miocene; the Jabal Akhdar Dome experienced the opposite. It is unclear why the Late Eocene–Miocene cooling and/or exhumation was much more pronounced in the Jabal Akhdar Dome compared with the Saih Hatat Dome. A possible reason may be the differently activated deep-rooted structures. The present-day morphology is different between the two domes. The Saih Hatat Dome has a highly eroded core (Hatat Bowl), while the Jabal Akhdar Dome is overall less eroded. Furthermore, the mountains in the Jabal Akhdar Dome (up to *c.* 3000 m) are higher than in the Saih Hatat Dome (up to *c.* 2000 m). Erosion probably had more time to act on the rocks in the Saih Hatat Dome (since the latest Cretaceous), while rocks of the Jabal Akhdar Dome have been on the surface much less (since the late Eocene?).

Unanswered questions

While the overall architecture and development of the Oman Mountains is well-known from the Neoproterozoic until today, many questions remain. We provide a list of unsolved problems in chronological order, hopefully to trigger interest in the international geoscientific community to address these in future research projects.

Neoproterozoic issues

(1) The stratigraphic correlation or differences in the Neoproterozoic formations between the Jabal Akhdar and Saih Hatat domes is still unexplained. Detailed studies of the Neoproterozoic formations of both areas could lead to plausible correlations. At the same time, sedimentological and age analyses of the respective formations may explain why some are only present in one of the two domes (see sections on 'Units below the Hercynian Unconfirmity' in Chapter 2 and 'Differences between the Jabal Akhdar and Saih Hatat domes' in this chapter).

(2) According to Gomez-Perez *et al.* (2019), East and West Oman collided during the Angudan event. East Oman was probably part of India, while West Oman was part of Arabia prior to collision. Where is the suture zone between these terranes in northern Oman? Could this boundary be along the Semail Gap and explain the different Neoproterozoic formations in both the Jabal Akhdar and Saih Hatat domes (see sections on 'Pre-'Hercynian' deformation' in Chapter 5 and 'Differences between the Jabal Akhdar and Saih Hatat domes' in this chapter)?

(3) Do the possible diamictites in the Hatat Formation of the Saih Hatat Dome belong to the Sturtian or the Marinoan glaciation (see section on 'Units below the Hercynian Unconfirmity' in Chapter 2)?
(4) Are there any Cadomian structures in the Saih Hatat Dome (see sections on 'F1 folds' in Chapter 3 and 'Units below the Hercynian Unconfirmity' in Chapter 2)?

Pre-'Hercynian' issues

(5) Is there some Ara Salt below the Saih Hatat Dome? Was the Ghaba Salt Basin extending further north, east of the Semail Gap as proposed by Rickards *et al.* (2010) (see section on 'Units above the Angudan Unconformity' in Chapter 5)?

Pre-ophiolite obduction issues

(6) Could some of the basal conglomerates of the Saiq Formation of the Jabal Akhdar Dome possibly belong to the Al Khlata or Gharif formations (see section on 'Akhdar Group' in Chapter 2)?
(7) What is the age of Neo-Tethys rifting? Is it Late Carboniferous, Early Permian or Mid Permian? Why is the Batain Basin older than the Hawasina Basin, as concluded by Immenhauser *et al.* (2000) (see section on 'Pangea rifting and formation of the Neo-Tethys Ocean and Hawasina Basin' in Chapter 5)?

Ophiolite obduction issues

(8) Is the Hawasina Basin underlain by fully developed oceanic crust (e.g. Pillevuit *et al.* 1997) or thinned continental crust (e.g. Maury *et al.* 2003) (see section on 'Pangea rifting and formation of the Neo-Tethys Ocean and Hawasina Basin' in Chapter 5)?
(9) What is the age of subduction initiation? Is it *c.* 104 Ma or *c.* 96 Ma as concluded by Guilmette *et al.* (2018) or Rioux *et al.* (2016), respectively (see section on 'Stage I: subduction initiation' in Chapter 5)?
(10) What was the strike of the subduction zone during its initiation (NNE or ESE; see section on 'Stage I: subduction initiation' in Chapter 5)?
(11) Was subduction initiation spontaneous or induced (see section on 'Stage I: subduction initiation' in Chapter 5)?
(12) What was the spreading rate at the spreading centre during formation of new oceanic lithosphere, how long did spreading last, and what was the velocity and amount of intraoceanic thrusting and obduction? These questions take into account the position and orientation of the Late Cretaceous spreading centre within the Neo-Tethys Ocean and therefore the distance to Arabia. If clockwise rotation of the Semail Ophiolite during obduction was involved (van Hinsbergen *et al.* 2019; fig. 5.4; see section on 'Stage I: subduction initiation' in Chapter 5), the NNE section of the ophiolite was further away from Arabia than the SSW part. The amount and velocity of obduction was therefore heterogeneous across the Semail Ophiolite (i.e. larger amount and velocity in the Southeastern Oman Mountains and slower amount and velocity in the Northern Oman Mountains).
(13) The Semail Ophiolite today measures *c.* 8 km in thickness in the NE (i.e. between the Jabal Akhdar and Saih Hatat domes). The thickness is substantially reduced towards the SW. The Semail Ophiolite may have been as thick as 15–20 km or even 30–40 km during intraoceanic thrusting (e.g. Nicolas 1989; Hacker and Gnos 1997). If the thickness of the ophiolite was 30–40 km in the beginning, how did the ophiolite thin by *c.* 300% to its present-day thickness? This question was raised and discussed by Hacker and Gnos (1997), but is still not resolved. Recent studies showed that the overburden of the entire allochthonous units in the JAD did not amount to more than 10–12 km (Aldega *et al.* 2017; Grobe *et al.* 2019; section on 'Stage IV: exhumation of high-pressure/low-temperature rocks' in Chapter 5). Thinning of the Semail Ophiolite must therefore have taken place between its Cenomanian formation and the final emplacement on Arabia at *c.* 70 Ma.
(14) How far can the ophiolite thrust/obduction model of Nicolas and Boudier (2017) and Boudier and Nicolas (2018), in the context of a collision of an active spreading centre with a continent-oriented subduction, be supported by field data (see section on 'Stage II: self-sustained subduction, supra-subduction zone setting, and formation of oceanic lithosphere and metamorphic sole' in Chapter 5)?
(15) Did the Arabian continental lithosphere represent the lower plate that was subducted to the north (as advocated by most authors), or was the continent the upper plate under which subduction ensued (e.g. Gregory *et al.* 1998; Nicolas and Boudier 2017; Boudier and Nicolas 2018; see section on 'Stage II: self-sustained subduction, supra-subduction zone setting, and formation of oceanic lithosphere and metamorphic sole' in Chapter 5)?
(16) Did the forebulge during obduction migrate (Robertson 1987), or was it stationary with a narrowing basin during the arrival of the Hawasina nappes (Warburton *et al.* 1990)?
(17) Why is the Juweiza Formation (ophiolite clast-bearing sedimentary rocks of the foreland basin) absent from the JAD and SHD areas, but present in the Northern Oman Mountains (see section on 'Stage III: obduction development, subduction of the Arabian margin and bulging of the Arabian lithosphere' in Chapter 5)?
(18) What is the origin of the Batinah Mélange (see section on 'The Batinah Mélange' in Chapter 3)?

Post-ophiolite obduction issues

(19) What was the role of the (inherited?) deep structures during doming of the Jabal Akhdar and Saih Hatat domes? Is Cenozoic doming of the Jabal Akhdar Dome related to thin-skinned or thick-skinned tectonics (see section on 'Main extensional fault systems' in Chapter 3)?
(20) When exactly were the large post-obductional extensional faults active (e.g. Frontal Range Fault, Semail Gap; see section on 'Main extensional fault systems' in Chapter 3)?
(21) How far to the SE can the main extensional fault (e.g. Wadi Mansah Fault Zone) be traced along the southwestern margin of the Saih Hatat Dome (see section on 'Main extensional fault systems' in Chapter 3)?
(22) Why is listwaenite restricted to the northeastern Jabal Akhdar and northwestern Saih Hatat domes? Why is listwaenite absent from the other margins of the domes (see sections on 'Post-obduction units' in Chapter 2.3 and 'Main extensional fault systems' in Chapter 3)?
(23) What is the kinematic history of the Semail Gap, that is, when, how many times and in which way was it reactivated? What role did the Semail Gap play in the course of the exhumation of the Jabal Akhdar Dome (see section on 'Main extensional fault systems' in Chapter 3)?
(24) What is the heat source of the thermal springs (up to 70°C) along the margins of the Jabal Akhdar and Saih Hatat domes (exothermal serpentinization and/or heating by the geothermal gradient and/or heating by

Cenozoic basalt; see section on 'Quaternary deformation' in Chapter 5)?
(25) What is the cause for E–W-directed compression, beyond the Rusayl Embayment, during the Oligocene–Miocene as reported by Fournier *et al.* (2006; see section on 'Post-obductional deformation' in Chapter 5)?
(26) What is the present-day uplift rate of the Oman Mountains, and how much of this uplift can be attributed to elastic rebound, the Arabia–Eurasia collision or other aspects (see section on 'Quaternary deformation' Chapter 5)?
(27) What is the age of wadi incision within the study area? Is this incision related to uplift of the Oman Mountains or to decreasing sea-level, or both (see section on 'Quaternary deformation' Chapter 5)?

Author contributions **AS:** Conceptualization (lead), Writing original draft (lead), Visualization (lead), Supervision (lead), Investigation (lead); **FM:** Conceptualization (equal), Writing – review and editing (equal), Visualization (equal), Investigation (equal); **MAW:** Writing – review and editing (equal); **GF:** Writing – review and editing (equal); **DM:** Writing – review and editing (equal); **BP:** Writing – review and editing (equal); **WB:** Writing – review and editing (equal); **IC:** Writing – review and editing (equal).

Funding This research received no specific grant from any funding agency in the public, commercial, or not-for-profit sectors.

Data availability Data sharing is not applicable to this article as no datasets were generated or analysed during the current study.

References

Aldega, L., Carminati, E., Scharf, A., Mattern, F. and Al-Wardi, M. 2017. Estimating original thickness and extent of the Semail Ophiolite in the eastern Oman Mountains by paleothermal indicators. *Marine and Petroleum Geology*, **84**, 18–33, https://doi.org/10.1016/j.marpet.geo.2017.03.024

Beurrier, M., Béchennec, F., Rabu, D. and Hutin, G. 1986. Geological map of Rustaq, sheet NF 40-03D, scale 1:100 000, with explanatory notes. *Directorate General of Minerals, Oman Ministry of Petroleum and Minerals.*

Boudier, F. and Nicolas, A. 2018. Synchronous seafloor spreading and subduction at the paleo-convergent margin of Semail and Arabia. *Tectonics*, **37**, 2961–2982, https://doi.org/10.1029/2018TC005099

Breton, J.-P., Béchennec, F., Le Métour, J., Moen-Maurel, L. and Razin, P. 2004. Eoalpine (Cretaceous) evolution of the Oman Tethyan continental margin: insights from a structural field study in Jabal Akhdar (Oman Mountains). *GeoArabia*, **9**, 41–58.

Callegari, I., Scharf, A. *et al.* 2020. Gondwana accretion tectonics and implications for the geodynamic evolution of eastern Arabia: first structural evidence of the existence of the Cadomian Orogeny in Oman (Jabal Akhdar Dome, Central Oman Mountains). *Journal of Asian Earth Sciences*, **187**, 104070, https://doi.org/10.1016/j.jseaes.2019.104070

Chauvet, F., Dumont, T. and Basile, C. 2009. Structures and timing of Permian rifting in the central Oman Mountains (Saih Hatat). *Tectonophysics*, **475**, 563–574, https://doi.org/10.1016/j.tecto.2009.07.008

Fournier, M., Lepvrier, C., Razin, P. and Jolivet, L. 2006. Late Cretaceous to Paleogene Post-obduction extension and subsequent Neogene compression in the Oman Mountains. *GeoArabia*, **11**, 17–40.

Gomez-Perez, I., Moss, J. and Morton, A. 2019. Early Tectonic Evolution of Oman: Insights from Clastic Provenance Data and Implications for the Structural Evolution of Arabia. AAPG Event, 2nd Edition, Structural styles of the Middle East, 9–11 December 2019, Muscat, Oman.

Gregory, R.T., Gray, D.R. and Miller, J.M. 1998. Tectonics of the Arabian margin associated with the emplacement of the Oman margin along the Ibra transect: new evidence from northeast Saih Hatat. *Tectonics*, **17**, 657–670, https://doi.org/10.1029/98TC02206

Grobe, A., Virgo, S., von Hagke, C., Urai, J.L. and Littke, R. 2018. Multiphase structural evolution of a continental margin during obduction orogeny: insights from the Jebel Akhdar Dome, Oman Mountains. *Tectonics*, **37**, 888–913, https://doi.org/10.1002/2016TC004442

Grobe, A., von Hagke, C., Littke, R., Dunkl, I., Wübbeler, F., Muchez, P. and Urai, J.L. 2019. Tectono-thermal evolution of Oman's Mesozoic passive continental margin under the obducting Semail Ophiolite: a case study of Jebel Akhdar, Oman. *Solid Earth*, **10**, 149–175, https://doi.org/10.5194/se-10-149-2019

Guilmette, C., Smit, M.A. *et al.* 2018. Forced subduction initiation recorded in the sole and crust of the Semail Ophiolite of Oman. *Nature Geoscience*, **11**, 688–695, https://doi.org/10.1038/s41561-018-0209-2

Hacker, B.R. and Gnos, E. 1997. The conundrum of Samail: explaining the metamorphic history. *Tectonophysics*, **279**, 215–226, https://doi.org/10.1016/S0040-1951(97)00114-5

Hansman, R.J., Ring, U., Thomson, S.N., den Brock, B. and Stübner, K. 2017. Late Eocene uplift of the Al Hajar Mountains, Oman, supported by stratigraphic and low-temperature thermochronology. *Tectonics*, **36**, 3081–3109, https://doi.org/10.1002/2017TC004672

Heward, A.P. and Penney, R.A. 2014. Al Khlata glacial deposits in the Oman Mountains and their implications. *Geological Society, London, Special Publications*, **392**, 279–301, https://doi.org/10.1144/SP392.15

Immenhauser, A., Schreurs, G., Gnos, E., Oterdoom, W. and Hartmann, B. 2000. Late Paleozoic to Neogene geodynamic evolution of the northeastern Oman margin. *Geological Magazine*, **137**, 1–18, https://doi.org/10.1017/S0016756800003526

Le Métour, J., Villey, M. and de Gramont, X. 1986. Geological map of Quryat, sheet NF 40-4D, scale 1:100 000 with explanatory notes. *Directorate General of Minerals, Oman Ministry of Petroleum and Minerals.*

Mattern, F. and Scharf, A. 2019. Transition from the Hajir Formation to the Muaydin Formation: a facies change coinciding with extensional, syndepositional faulting (Ediacaran, Jabal Akhdar, Central Oman Mountains). *Journal of African Earth Sciences*, **152**, 237–244, https://doi.org/10.1016/j.jafrearsci.2019.02.016

Mattern, F., Pracejus, B. and Al Balushi, L. 2018. Heavy mineral beach placers of the Ordovician Amdeh Formation (Member 4, Wadi Qahza, Saih Hatat, eastern Oman Mountains): where is the main source area? *Journal of African Earth Sciences*, **147**, 633–646, https://doi.org/10.1016/j.afrearsci.2018.06.031

Mattern, F., Scharf, A. *et al.* 2020. Deformation of the Cambro-Ordovician Amdeh Formation (Members 1 and 2): characteristics, origins, and stratigraphic significance (Wadi Amdeh, Saih Hatat Dome, Oman Mountains). *Geosciences*, **10**, https://doi.org/10.3390/geosciences10020048

Maury, R.C., Béchennec, F., Cotton, J., Caroff, M., Cordey, F. and Marcoux, J. 2003. Middle Permian plume-related magmatism of the Hawasina Nappes and the Arabian Platform: implications on the evolution of the Neotethyan margin in Oman. *Tectonics*, **22**, 12-1–12-13, https://doi.org/10.1029/2002TC001483

Nicolas, A. 1989. *Structures of Ophiolites and Dynamics of Oceanic Lithosphere*. Kluwer, Netherlands, 367.

Nicolas, A. and Boudier, F. 2017. Emplacement of Semail-Emirates ophiolite at ridge-trench collision. *Terra Nova*, **29**, 127–134, https://doi.org/10.1111/ter.12256

Pillevuit, A., Marcoux, J., Stampfli, G. and Baud, A. 1997. The Oman Exotics: a key to the understanding of the Neotethyan geodynamic evolution. *Geodinamica Acta*, **10**, 209–238, https://doi.org/10.1080/09853111.1997.11105303

Rabu, D., Béchennec, F., Beurrier, M. and Hutin, G. 1986. Geological map of Nakhl, sheet NF40-3E, scale: 1:100 000, with explanatory notes. *Directorate General of Minerals, Oman Ministry of Petroleum and Minerals.*

Rickards, B.R., Booth, G.A., Paris, F. and Heward, A.P. 2010. Marine flooding events of the Early and Middle Ordovician of Oman and the United Arab Emirates and their graptolite, acritarch and chitinozoan associations. *GeoArabia*, **15**, 81–120.

Rioux, M., Garber, J., Bauer, A., Bowring, S., Searle, M.P., Kelemen, P. and Hacker, B. 2016. Synchronous formation of the metamorphic sole and igneous crust of the Semail ophiolite: new constraints on the tectonic evolution during ophiolite formation from high-precision U-Pb zircon geochronology. *Earth and Planetary Science Letters*, **451**, 185–195, https://doi.org/10.1016/j.epsl.2016.06.051

Robertson, A.H.F. 1987. The transition from passive margin to Upper Cretaceous foreland basin related to ophiolite emplacement in the Oman Mountains. *Geological Society of America Bulletin*, **99**, 633–653, https://doi.org/10.1130/0016-7606(1987)99<633:TTFAPM>2.0.CO;2

Romine, *et al.* 2004. *North Oman Haima-Huqf Tectonostratigraphic Study*, SRK Confidential Report to PDO, July 2004.

Ruban, D., Al-Husseini, M.I. and Iwasaki, Y. 2007. Review of Middle East Paleozoic plate tectonics. *GeoArabia*, **12**, 35–56.

Scharf, A., Mattern, F., Moraetis, D., Callegari, I. and Weidle, C. 2019. Postobductional kinematic evolution and geomorphology of a major regional structure – the Semail Gap Fault Zone (Oman Mountains). *Tectonics*, **38**, 2756–2778, https://doi.org/10.1029/2019TC005588

Scharf, A., Mattern, F., Al-Wardi, M., Frijia, G., Moraetis, D., Pracejus, B., Bauer, W. and Callegari, I. 2021*a*. Appendices to: The Geology and Tectonics of the Jabal Akhdar and Saih Hatat Domes. *Geological Society, London, Memoirs*, **54**, 109–113, https://doi.org/10.1144/M54.7

Scharf, A., Mattern, F., Al-Wardi, M., Frijia, G., Moraetis, D., Pracejus, B., Bauer, W. and Callegari, I. 2021*b*. Tectonostratigraphy of the eastern part of the Oman Mountains. *Geological Society, London, Memoirs*, **54**, 11–43, https://doi.org/10.1144/M54.2

Scharf, A., Mattern, F., Al-Wardi, M., Frijia, G., Moraetis, D., Pracejus, B., Bauer, W. and Callegari, I. 2021*c*. Tectonic evolution of the Oman Mountains. *Geological Society, London, Memoirs*, **54**, 63–99, https://doi.org/10.1144/M54.5

Scharf, A., Mattern, F., Al-Wardi, M., Frijia, G., Moraetis, D., Pracejus, B., Bauer, W. and Callegari, I. 2021*d*. Thrusts, extensional faults and fold patterns of the major units. *Geological Society, London, Memoirs*, **54**, 45–56, https://doi.org/10.1144/M54.3

Torsvik, T.H. and Cocks, L.R.M. 2017. *Earth History and Paleogeography*. Cambridge University Press, Cambridge, 317.

van Hinsbergen, D.J.J., Maffione, M., Koornneef, L.M.T. and Guilmette, C. 2019. Kinematic and paleomagnetic restoration of the Semail ophiolite (Oman) reveals subduction initiation along an ancient Neotethyan fracture zone. *Earth and Planetary Science Letters*, **518**, 183–196, https://doi.org/10.1016/j.epsl.2019.04.038

Villey, M., de Gramont, X. and Le Métour, J. 1986. *Geological map of Seeb, sheet NF 40-3C, scale: 1:100,000, with Explanatory notes*. Directorate General of Minerals, Oman Ministry of Petroleum and Minerals.

Warburton, J., Burnhill, T.J., Graham, R.H. and Isaac, K.P. 1990. The evolution of the Oman Mountains foreland basin. *Geological Society, London, Special Publications*, **49**, 419–427.

Appendices to: The Geology and Tectonics of the Jabal Akhdar and Saih Hatat Domes, Oman Mountains

Andreas Scharf[1]*, Frank Mattern[1], Mohammed Al-Wardi[1], Gianluca Frijia[2], Daniel Moraetis[1,3], Bernhard Pracejus[1], Wilfried Bauer[4] and Ivan Callegari[4]

[1]Department of Earth Sciences, College of Science, Sultan Qaboos University, P.O. Box: 36, PC 123, Al-Khod, Muscat, Sultanate of Oman

[2]Department of Physics and Earth Sciences, University of Ferrara, Via Saragat 1, 44122, Ferrara, Italy

[3]Department of Applied Physics and Astronomy, College of Sciences, University of Sharjah, P.O. Box: 27272, Sharjah, United Arab Emirates

[4]Department of Applied Geosciences, German University of Technology GUtech, P.O. Box: 1816, PC 130, Halban, Sultanate of Oman

*Correspondence: scharfa@squ.edu.om

Appendix A

See foldout. A more detailed topographic map can be found in the online publication.

Appendix B

Figure A.3 provides a correlation/comparison of Permian ages between the 'International Stratigraphic Chart' and the 'Tethyan Time Scale'.

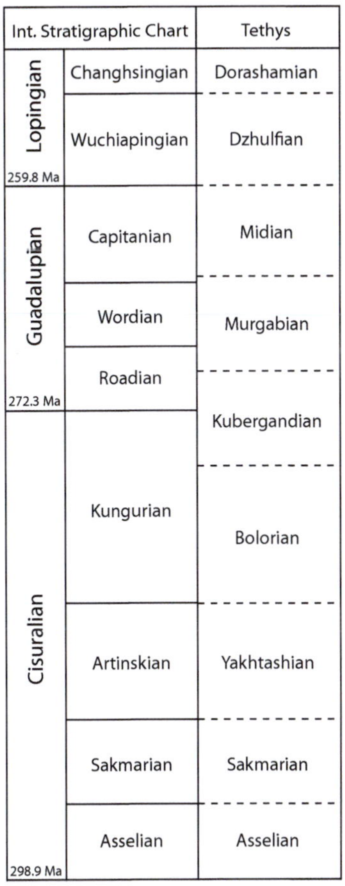

Fig. A.3. Comparison of the Permian Period between the 'International Stratigraphic Chart' and 'Tethyan Time Scale'. Modified from Angiolini *et al.* (2016).

Appendix C

Prof. Christopher Bailey (College of William & Mary, USA) creates and provides a number of helpful and interesting blogs about the fascinating geology of Oman:

(1) https://wmblogs.wm.edu/cmbail/inside-ghubrah-bowl-oman-dropstones-double-duckbills-pencil-structures-oh/
(2) https://wmblogs.wm.edu/cmbail/omans-geological-triple-point/
(3) https://wmblogs.wm.edu/cmbail/the-road-to-the-moho/
(4) https://wmblogs.wm.edu/cmbail/wadi-bani-ghafir-at-sidaq-gorge-water-versus-rock-in-oman/
(5) https://wmblogs.wm.edu/cmbail/omans-mega-sheath-folds/
(6) https://wmblogs.wm.edu/cmbail/dispatches-from-oman-juxtaposition/
(7) https://wmblogs.wm.edu/cmbail/dispatches-from-oman-fodder-for-the-tectonic-cannon/
(8) https://wmblogs.wm.edu/cmbail/dispatches-from-oman-wadi-jizzi-standing-at-the-bottom-of-the-tethys-ocean/
(9) https://wmblogs.wm.edu/cmbail/dispatches-from-oman-ophiolite-to-aflaj/
(10) https://wmblogs.wm.edu/cmbail/finding-sharqiya-sands-oman/

Appendix D (inconsistencies in maps)

Inconsistencies are found between the BRGM 1:100 000 and 1:250 000 maps within the Hawasina formations. In all cases we have followed the 1:100 000 maps rather the 1:250 000 map.

(1) The geological map of Nazwa (1:250 000; Béchennec *et al.* 1992*b*) depicts the Hawasina unit immediately E of Nizwa as Umar Group, while the geological map of Birkat Al-Mawz (1:100 000; Hutin *et al.* 1986) depicts the same area as Al-Aridh Group.
(2) Hawasina rocks west of Rustaq are differently mapped in the 1:250 000 and 1:100 000 map scale (Béchennec *et al.* 1992*a*; Beurrier *et al.* 1986, respectively). In the 1:250 000 map some rocks are presented as Matabat Fm. (Hamrat Duru Group) while the same area in the 1:100 000 map is Umar Group.

From: Scharf, A., Mattern, F., Al-Wardi, M., Frijia, G., Moraetis, D., Pracejus, B., Bauer, W. and Cellegari, I. 2021. *The Geology and Tectonics of the Jabal Akhdar and Saih Hatat Domes, Oman Mountains*. Geological Society, London, Memoirs, **54**, 113–115, https://doi.org/10.1144/M54.7

(3) Hawasina rocks in the Baid area (southwestern margin of the SHD) have been mapped differently. In the 1:100 000 map of Fanja (Villey *et al.* 1986*c*) some rocks are mapped as part of the Umar Group, while in the 1:250 000 map of Seeb (Béchennec *et al.* 1992*a*) mapped the same outcrops as Hamrat Duru Group.

Appendix E (Thermal and hyperalkaline springs)

Table A.1. *Thermal springs of the study area*

#	Spring	Province	Wilayat (district)	Latitude	Longitude	T (°C)	EC (µS/cm)
1	Al Khadhra	S Batinah	Rustaq	23.359035	57.311128	40	892
2	Al Kasfah	S Batinah	Rustaq	23.394033	57.411030	45	894
3	Saighi	S Batinah	Rustaq	23.387660	57.429604	39	812
4	Al Ghareedh	S Batinah	Nakhl	23.375225	57.826822	37	738
5	Sakhnan	S Batinah	Nakhl	23.373920	57.834505	36	709
6	Al Thawarah	S Batinah	Nakhl	23.377910	57.836682	37	687
7	Subaykhah	S Batinah	W. Maawil	23.504770	57.919691	51	1375
8	Saleel	S Batinah	W. Maawil	23.512153	57.935619	41	1237
9	Hammam Sharqi	Dakhliyah	Bidbid	23.454106	58.102635	41	759
10	Hammam Ansab or Al-Ali	Muscat	Bawsher	23.469453	58.323011	66	1420
11	Heed	Muscat	Bawsher	23.528464	58.382122	41	1084
12	Hammam Ghala	Muscat	Bawsher	23.529642	58.383682	42	1086
13	Falaj Bawsher	Muscat	Bawsher	23.535521	58.398906	36	1000
14	Al Ghar	Muscat	Bawsher	23.536471	58.398760	37	996
15	Sad	Muscat	Bawsher	23.541938	58.407320	37	960
16	Al Hobe	Dakhliyah	Semail	23.320060	57.947286	37	500
17	Hannah	Dakhliyah	Semail	23.292894	57.917994	38	524
18	Ghaiz Al Dhahia or Al-Mahl	Dakhliyah	Semail	23.288760	57.919080	45	652
19	Manabik	Dakhliyah	Semail	23.280847	57.916044	45	664
20	Qury	Dakhliyah	Semail	23.218039	57.870992	35	473
21	Hammam Sayjah	Dakhliyah	Semail	23.214513	57.871624	39	539
22	Hammam Afiah	Dakhliyah	Semail	23.104017	57.823819	n.a.	n.a.
23	Wadi Daiqa Gorge	Muscat	Quriyat	23.070630	58.774270	n.a.	n.a.

Locations (#) are shown in figure 4.1 (Scharf *et al.* 2021, Chapter 4, this Memoir) Note, the springs #2 (Al Kasfah) and #10 (Hammam Ansab) are the same as springs #33 and #37 in Table A.2. Wadi Daiqa Gorge (#23) after Alan Heward pers. comm. (an older name for the gorge, stemming from the journey of Lees and Cox is the 'Devils Gap'. That was the name attributed to it by sailors as it is a readily recognizable cleft on the skyline when looked at from a distance, Heward, pers. comm.). EC, electrical conductivity; n.a., not available.

Table A.2. *Hyperalkaline springs in the study area after Paukert* et al. *(2012, highlighted in bold print) and Chavagnac* et al. *(2013a, b)*

#	Site of Chavagnac *et al.* (2013*a*, *b*) and Paukert *et al.* (2012)	Spring's name	Latitude	Longitude	Altitude (m)	T (°C)	pH	EC (µS/cm)
24	2	Nidab	23°12′57″	58°08′25″	473	33.2	11.5	n.a.
25	3	Yellowstone du pauvre	22°49′10″	57°48′48″	479	20.8	10.1	n.a.
26	5	Little Grand Canon	22°50′49″	58°03′22″	533	28.3	11.8	n.a.
27	6					32.2	11.7	
	OM10_02L	Falaij	22°50′16″	58°03′24″	n.a.	30.8	11.5	2230
	OM10_03R	Falaij	22°50′33″	58°03′29″	n.a.	27.9	11.6	1967
28	8	Izki les 2 puits	22°55′40″	57°46′20″	530	23.4	7.8	n.a.
29	10	Le partage du midi	22°52′26″	57°31′12″	472	22.4	11.7	n.a.
30	11	Lac bleu de Bahla	22°59′28″	57°17′34″	562	26.3	10.8	n.a.
31	12	Ain Al-Waddah	22°59′28″	57°17′34″	562	33.3	11.7	n.a.
32	20	Le salaire de la peur	23°37′39″	57°06′55″	322	23.9	9.1	n.a.
33	21	Rustaq	23°23′36″	57°24′41″	373	45.2	8.2	n.a.
34	22	La poule au pot	23°34′19″	58°01′38″	94	35.8	11.3	n.a.
35	23	4 x 4 practice	23°33′49″	58°06′44″	140	26.3	10.4	n.a.
36	24	Gabbro lité	23°33′30″	58°06′25″	80	21.0	11.4	n.a.
37	25	L'ane blanc	23°28′12″	58°19′26″	N.A.	65.5	7.9	n.a.
38	26	La grande Ligurie	23°37′12″	57°06′48″	337	31.3	11.3	n.a.
39	30	Rencontre de la colombe	22°54′22″	58°25′35″	660	27.5	11.7	n.a.
	OM10_04S	Qafifah	22°54′16″	58°25′29″	n.a.	22.4	11.8	1685
	OM10_04U	Qafifah	22°54′17″	58°25′27″	n.a.	24.9	11.7	1778
	OM09_W05L	Qafifah	22°54′17″	58°25′27″	n.a.	23.8	11.6	1854
40	31	les lauriers roses	22°53′44″	58°23′41″	689	26.4	11.6	n.a.
41	32	Graviere	23°19′21″	58°13′42″	367	31.0	11.5	n.a.
	OM10_01A	Misbit	23°17′27″	58°13′55″	n.a.	31.6	11.2	1640
42	**OM09_W06O**	Dima	22°59′02″	58°35′40″	n.a.	30.3	11.5	1905

Locations (#) are shown in figure 4.1 (Scharf *et al.* 2021, Chapter 4, this Memoir). Note, the springs #33 (Rustaq) and #37 (L'ane blanc) of Chavagnac *et al.* (2013*a*, *b*) are the same as springs #2 and #10 in Table A.1. Note the springs #26, 27 (Little Grand Canon), #39 (Rencontre de la colombe) and #41 (Graviere) of Chavagnac *et al.* (2013*a*, *b*) are the same sites of Paukert *et al.* (2012; Falaij, Qafifah and Misbit, respectively). EC, electrical conductivity; n.a., not available.

References

Al-Husseini, M.I. 2014. Ediacaran-Cambrian Middle East geologic time scale 2014, proposed correlation of Oman's Abu Mahara Supergroup and Saudi Arabia's Jibalah Group. *GeoArabia*, **19**, 107–132.

Angiolini, L., Campagna, M. *et al.* 2016. Brachiopods from the Cisuralian-Guadalupian of Darvaz, Tajikistan and implications for Permian stratigraphic correlations. *Palaeoworld*, **25**, 539–568, https://doi.org/10.1016/j.palwor.2016.05.006

Béchennec, F., Le Métour, J., Rabu, D., Bourdillon-de-Grissac, C., de Wever, P., Beurrier, M. and Villey, M. 1990. The Hawasina Nappes; Stratigraphy, palaeogeography and structural evolution of a fragment of the South-Tethyan passive continental margin. *In*: Robertson, A.H.F., Searle, M.P. and Ries, A.C. (eds) *The Geology and Tectonics of the Oman Region*. Geological Society, London, Special Publications, **49**, 213–223, https://doi.org/10.1144/GSL.SP.1992.049.01.14

Béchennec, F., Roger, J., Le Métour, J. and Wyns, R. 1992*a*. *Geological map of Seeb, sheet NF 40-03, scale 1:250,000, with Explanatory Notes*. Directorate General of Minerals, Oman Ministry of Petroleum and Minerals.

Béchennec, F., Wyns, R., Roger, J., Le Métour, J. and Chevrel, S. 1992*b*. *Geological map of Nazwa, sheet NF 40-07, scale 1:250,000, with Explanatory Notes*. Directorate General of Minerals, Oman Ministry of Petroleum and Minerals.

Beurrier, M., Béchennec, F., Hutin, G. and Rabu, D. 1986. *Geological map of As Suwayq, sheet NF 40-03A, scale 1:100,000, with Explanatory notes*. Directorate General of Minerals, Oman Ministry of Petroleum and Minerals.

Chavagnac, V., Ceuleneer, G., Monnin, C., Lansac, B., Hoareau, G. and Boulart, C. 2013*a*. Mineralogical assemblages forming at hyperalkaline warm springs hosted on ultramafic rocks: a case study of Oman and Ligurian ophiolites. *Geochemistry, Geophysics, Geosystems*, **14**, 2474–2495, https://doi.org/10.1002/ggge.20146

Chavagnac, V., Monnin, C., Ceuleneer, G., Boulart, C. and Hoareau, G. 2013*b*. Characterization of hyperalkaline fluids produced by low-temperature serpentinization of mantle peridotites in the Oman and Ligurian ophiolites. *Geochemistry, Geophysics, Geosystems*, **14**, 2496–2522, https://doi.org/10.1002/ggge.20147

Droste, H. 2014. Petroleum geology of the Sultanate of Oman. *In*: Marlow, L., Kendall, C. and Yose, L. (eds) *Petroleum Systems of the Tethyan Region*. American Association of Petroleum Geologists, Memoirs, **106**, 713–755, https://doi.org/10.1036/12431870M1063039

Gradstein, F.M., Ogg, J.G. and Hilgen, F.J. 2012. On the geologic time scale. *Newsletters on Stratigraphy*, **45**, 171–188, https://doi.org/10.1127/0078-0421/2012/0020

Hughes Clarke, M.W. 2006. *Oman's Geological Heritage*. Petroleum Development Oman, London.

Hutin, G., Béchennec, F., Beurrier, M. and Rabu, D. 1986. *Geological map of Birkat Al Mawz, sheet NF 40-07B, scale 1:100,000, with Explanatory Notes*. Directorate General of Minerals, Oman Ministry of Petroleum and Minerals.

Paukert, A.N., Matter, J.M., Kelemen, P., Shock, E.L. and Havig, J.R. 2012. Reaction path modeling of enhanced in situ CO_2 mineralization for carbon sequestration in the peridotites of the Samail Ophiolite, Sultanate of Oman. *Chemical Geology*, **330–331**, 86–100, https://doi.org/10.1016/j.chemgeo.2012.08.013

Scharf, A., Mattern, F., Al-Wardi, M., Frijia, G., Moraetis, D., Bauer, W. and Callegari, I. 2021. Large-scale structure of the study area. *Geological Society, London, Memoirs*, **54**, https://doi.org/10.1144/M54.4

Villey, M., Le Métour, J. and de Gramont, X. 1986*c*. *Geological map of Fanjah, sheet NF 40-3F, scale: 1:100,000, with Explanatory notes*. Directorate General of Minerals, Oman Ministry of Petroleum and Minerals.

Index

Page numbers in *italics* refer to Figures and the *foldout*. page numbers in **bold** refer to Tables.